IONIC
LIQUIDS

IONIC LIQUIDS

Edited by
Douglas Inman
Imperial College of Science and Technology
London, United Kingdom

and
David G. Lovering
R.M.C.S.
Shrivenham, Swindon, United Kingdom

PLENUM PRESS • NEW YORK AND LONDON

Library of Congress Cataloging in Publication Data

Main entry under title:

Ionic liquids.

Includes index.
1. Fused salts — Addresses, essays, lectures. 2. Electrolyte solutions — Addresses, essays, lectures. I. Inman, Douglas. II. Lovering, David G.
QD189.I58   546'.34   80-16402
ISBN 0-306-40412-5

QD
189
I58

©1981 Plenum Press, New York
A Division of Plenum Publishing Corporation
227 West 17th Street, New York, N.Y. 10011

Printed in the United States of America

Foreword

As Chairmen of the Electrochemistry and Molten Salts Discussion Groups of the Chemical Society, it gave us great pleasure to welcome the conference Highly Concentrated Aqueous Solutions and Molten Salts, which our Groups cosponsored, at St. John's College, Oxford in July 1978.

During the meeting the editors of the present volume, and those giving lectures, came to the conclusion that the verbal presentations deserved to be expanded and to be more widely disseminated in a permanent form. Thus the articles which appear in this volume were commissioned and prepared.

A greater exchange of information between aqueous chemists and those concerned with molten salts is to be welcomed and to this end the present volume aims to focus attention on the borderline areas between the two in an attempt to facilitate a wider awareness of the concepts and methods appropriate to the respective specialities. Similarly, and particularly in the electrochemical field, a greater exchange of information between the academic and industrial practitioners of the subject is desirable.

The problems involved are not trivial but when the interactions in these largely (but not wholly) ionic liquids are better understood, this will surely be to the benefit of all concerned with solution chemistry.

Douglas Inman, Imperial College
Chairman, Electrochemistry Group

David Kerridge, University of Southampton
Chairman, Molten Salts Discussion Group

Preface

A number of recent events led to the appearance of this text at this particular time.

During 1978, the EUCHEM Conference on Molten Salts at Lysekil, Sweden, was closely followed by a meeting on Highly Concentrated Aqueous Solutions and Molten Salts in Oxford, jointly sponsored by the Electrochemistry Group and Molten Salts Discussion Group of the Chemical Society. The Second International Symposium on Molten Salts was held in Pittsburgh, U.S.A., later in the year as part of the Electrochemical Society's 154th meeting. The underlying reasons for this expanding interest and support of ionic liquids will become apparent.

Most of the chemical and biological processes of the natural world proceed in aqueous solutions. This factor clearly provided the scenario for the development of chemical sciences until the middle of this century, although it is worth recalling that Faraday verified his laws of electrolysis using molten lead chloride! To date, the structure of the ubiquitous liquid we know as water is still imperfectly understood in spite of a lifetime of study by, for example, Bernal. When a solute is introduced (especially when the resulting solutions are concentrated) many complicated processes and equilibria may arise, and these are considered in some of the ensuing chapters. The ordering that occurs in electrolyte solutions down to uni-molar levels, observed in the Russian X-ray studies of the 1930s, may now be predicted by computer modeling; Stillinger at Bell Laboratories was one of the first to obtain the services of sufficiently large computers to carry out the necessary calculations. A number of chapters consider our latest understanding of the structure of electrolyte solutions and melts based on both experimental and simulation studies.

The extraction of aluminum from molten cryolite has become an established process over nearly a century of operation. However, the impact of new molten salt technologies really dawned only when uranium extraction became of military importance some forty years ago. In retrospect, the lack of enthusiasm for such simple ionic liquids as molten alkali halides for example, is astonishing, even if materials and handling aspects do present unfamiliar difficulties. Nevertheless, applications in homogeneous nuclear reactors, fuel cells, refractory metal deposition, heat treatment, and metalliding are now being matched by theoretical advances. The quasilattice model for molten salts developed by Blander has recently been extended by Braunstein and Gal to highly concentrated aqueous solutions by incorporating water molecules into the coordination shell of the cation, and latterly the anion, sublattices. Čeleda, among others, has evolved an alternative quasilattice approach. Notwithstanding these advances, developments in aqueous solution and molten salt chemistry/electrochemistry have proceeded almost independently. We have recorded our concern about this situation in the past, but now at last we have a chance to do something positive by including chapters written by both molten salt and aqueous solution specialists. In many cases these authors are exploring the eschewed Mesopotamia of highly concentrated solutions through to dilute solutions of water in molten salts, discovering new links and unified treatments of solution theory, while pointing out the practical consequences of their deliberations. The experimental work in this area is extremely difficult, requiring controlled atmospheres of water vapor; meanwhile one is acutely aware that slightly damp and dirty melts may be behaving in less than predictable fashion.

By the very nature of this volume, its subject matter is rather diverse. Nevertheless, we hope that the themes of theory and application in *Ionic Liquids* will commend themselves to the reader.

Furthermore, we would like to point out that we have encouraged authors to give a selective and personal view rather than an encyclopedic treatment of their chosen topic. In this way they are free to emphasize those aspects of the field which they themselves consider to be of most pressing relevance. We trust that this will best stimulate new interest and awareness across the whole spectrum of solution concentrations.

What of the future? Molten salts in energy generation and storage appear to be attracting increasing attention. Applications extend across new storage batteries, fuel cells, heat transfer and storage, coal gasification, nuclear reactors, organic scrap pyrolysis, and materials recycling. Doubtless these, and many developments involving ionic liquids yet to come, owe much to the gradual realization that fossil fuel resources really are finite. Efficiency, recycling, and ecological soundness are at last being given the emphasis they deserve. The ever-tightening screws of scarcity and price promise to focus the minds, even of politicians and administrators, on the rewards of research and investment in these new technologies.

David G. Lovering
Douglas Inman

Contents

Research on Molten Salts: Introduction

A. R. Ubbelohde

In this introductory chapter to the other contributions made on molten salts and related themes, it is pleasing to record the early researches of Harold Hartley and two or three of his colleagues, matching other famous researches on dilute aqueous solutions of strong electrolytes. As is well illustrated by the assumptions made in the Debye–Hückel theory, electrolytic solutions then being studied were so dilute that mathematical complications could be kept under elegant control. In this earlier work on the physical chemistry of ionic liquids one could think of water as a solvent with dielectric constant of about 80, and although forces between the ions were long range, the high dilution and the high dielectric constant greatly simplified many of the problems concerning electrostatic interactions between the ions.

By contrast, in the group of researches presented herein the situation is very different. Long-range forces between the ions are acting in condensed assemblies; practically every problem is a many-body problem. Theoretical attacks on such systems might at first sight seem forbidding through their complexity. However, it is useful to record certain guiding principles which often help to simplify them.

i. When dealing with molten salts (or highly concentrated electrolyte solutions), the general electrostriction pulls the molecules together much more tightly than in neutral melts or solutions. In dealing with some properties, effects due to electrostrictions can be thought of as if an external pressure of several thousand atmospheres were acting on the liquid.

ii. In these melts ions are brought into contact, so that the repulsion contact radius becomes an important parameter when comparing the beha-

A. R. Ubbelohde • Department of Chemical Engineering, Imperial College, London, England.

vior of different ions with the same electrostatic charge. However, this parameter is not quite as simple as in the crystalline state, for reasons outlined in iii–v below.

iii. Because of the high density of packing of ions, and the long-range action of electrostatic forces, the net field acting on any particular ion is arrived at by many-body compensation. Compensation is most far reaching in ionic crystals when these are free of defects, since the lattice symmetry then imposes very considerable cancellation of electrostatic forces due to neighbors acting on any particular ion. This part-neutralization of the field around each ion can be seen from the well-known contrast between the contact distance $r_\pm$ in an ion pair in a gas (where there is zero compensation) with that in a perfect crystal lattice where compensation can be high owing to the symmetry. The shrinkage Δ of contact radii on comparing $r_\pm$(gas) with $r_\pm$(crystal) is:

	NaCl	CsCl	NaBr	CsBr
Δ (a.u.)	0.30	0.50	0.34	0.57

Because of their positional disordering, there is rather less electrostatic compensation in melts or concentrated solutions than in the crystals, and the above are upper limits.

iv. The net field acting in the neighborhood of any ion may be far from symmetrical in space. In many-body assemblies interesting consequences are found even in the crystalline state when this has low lattice symmetry. Intense local interactions can then favor solid–solid transformations because of profound perturbations of the vibration spectrum due to local assymetries. In melts (or strong solutions of salts), such intense local perturbations are likely to be much more common than in the crystals. It may be helpful to think of "ion pairs" or "ion clusters" as a consequence of such intense local perturbations, though in fact the many-body forces from the surrounding melt result in less-clear-cut characterization of "ion pairs" than may be feasible in dilute solutions of electrolytes.

v. Crystals of salts might be expected to show *similitude of behavior* on the basis that in a group of similar ionic crystals the charges are always the same and only the sizes of the ions vary. Tests of similitude in ionic melts are often made by comparing selected melting parameters throughout the group (cf. Reference 1). In fact, however, similitude between ionic melts proves to be only roughly applicable. On proceeding to ions of higher atomic number in the same column of the periodic system, although the charges carried by the ions may be similar, the polarizability of their electron clouds increases. This is important in discussing the molten state. Many of the less-obvious features of comparative behavior in concentrated solutions and in molten salts depend even more on polarizability than on the "spherical contact radius" of the ion. It is sometimes feasible to establish trends in ionic behavior with increasing atomic number, but constant charge, by applying polarizability correction factors in the series. In any particular case one must, however, decide whether a "contact radius" is still meaningful, in view of the shrinkage effects in very unsymmetrical electrostatic fields. This comment about monatomic ions applies *a fortiori* to polyatomic ions referred to below.

vi. Until fairly recently, little detailed physical chemistry was concerned with ions carrying a charge greater than 2, in melts or solutions. Opportunities for crossing this frontier in knowledge have, however, developed in two directions:

a. With monatomic ions, transitional metals frequently prefer the "charge-reduced form" in melts or solutions, as with vanadyl, zirconyl, or uranyl ions. However, cations of the rare earths do not show such a tendency to the same extent. Exceptionally high contact surface potentials may be looked for when these ions are present, with strong perturbing consequences.

b. Polyatomic ions are known in great variety in crystalline salts. Most of these compounds are inorganic, but some important groups of organic salts have also received attention from physical chemists, including electrochemists. One vital question is to know what happens to the chemical stability of various polyatomic ions when their crystals are melted. Bond rupture is always facilitated on forming the liquid state, for reasons that are often well known. Nevertheless, atoms in certain inorganic polyatomic ions are so strongly bound that no appreciable decomposition occurs for tens and even hundreds of degrees above the melting points of the crystals, especially when such melting points are low. For example, melts of polyatomic inorganic ions have been extensively studied for salts such as nitrates and nitrites.* Other low-melting inorganic salts are also known in which ion stability is adequate to permit refined measurements. For ions such as sulfate SO_4^{2-}, carbonate CO_3^{2-}, and phosphate PO_3^{-2} the bond stability in the melts is still quite adequate, but the rather high melting points entrain experimental difficulties so that relatively much less is known about the physical chemistry of such melts.

Another family of inorganic polyatomic anions which offer good prospects of yielding chemically stable ionic melts are the borofluorides with the anion $(BF_4)^-$. Studies on their possible dissociation in the melts are not yet very searching.

With salts of organic ions, the possibilities range over a much wider field than for inorganic salts. However, little or no systematic investigation has been made concerning the stability of organic ions once the salts have been melted. Well-investigated melts of salts with large polyatomic cations are not plentiful. A much wider range of salts have large anions. Possibly the negative ionic charge helps to stabilize the electron acceptor system of the individual bonds in a polyatomic ion by resonance effects. For salts with organic anions more significant has been the discovery (cf. Reference 2) that many instances of chemical instability on melting are due not to *intrinsic* disruption of the chemical bonds in the polyatomic ion, but to *extrinsic* attack by impurities. For example, a large family of alkane carboxylates $[M^+(COOR)^-$ with R ranging from CH_3 to $C_{12}H_{14}]$ give melts that are perfectly stable over a considerable range of temperatures provided molecular oxygen and

* See Reference 1, p. 200ff.

polyvalent cations are carefully excluded. It is amusing to reflect that these salts are just those we were taught as undergraduates in first-year chemistry to heat in order to make ketones. But in fact, when suitable precautions are taken, they do not decompose in this way, and can provide a whole range of novel anions that are stable. Interesting variations of configuration can be introduced by skillful organic synthesis, thus permitting a great variety of physicochemical measurements on novel ionic melts or concentrated solutions of electrolytes.

Brief reference must be made to the mechanisms of melting of salts, since this helps at times in dealing with statistical problems of systems with many-body interactions. It also helps to focus some less-familiar problems with molten salts.

It is important to start with the appropriate model for the system under discussion. For ionic melts, and often for concentrated ionic solutions, the most flexible procedure is to start with the crystal. Starting with an ionic crystal, the simplest model of a liquid is then produced by introducing the statistical equilibrium concentration of the *positional defects* of the ions. Both lattice vacancies—the so-called Schottky defects—and vacancies combined with ions at interstitial sites in the crystals—sometimes known as Frenkel defects—have been proposed as principal sources of positional disordering in crystals with ions of the inert gas type such as Group I and Group II cations, and Group VII and Group VIII anions. When these defects are introduced the lattice has to be expanded somewhat, to reduce the repulsion potentials created. A melt formed in this way is simply a highly disordered crystal lattice, and such models of melts are called quasicrystalline. A model of this kind was used, for example, in the Lennard-Jones–Devonshire theory of melting. More sophisticated quasicrystalline melts are feasible by introducing a large number of dislocations into the crystal lattice. Multidislocation melts have rather different transport properties from the simpler quasicrystalline melts. Another important point is that the lattice vacancies in any of these models of positional randomization can take up individual *small* molecules such as H_2O by a kind of sorption, without calling for much work against electrostatic forces, though only up to limit concentrations of the same order as the vacancies in any one-component melt. This kind of model may explain why melts of the larger ions may show considerably higher retention of water molecules. However, if concentrations higher than correspond roughly with filling of all the vacancies have to be achieved, the vapor pressure, i.e., the fugacity of the solvent, must rise very steeply. Detailed studies of such solubility thresholds could be rewarding.

While quasicrystalline molten salts may present the simplest models, many more complex situations can arise. Discussion of these is approached most simply by superposing other means of disordering the crystal lattice, additional to positional disorder. In order of progressive complication, the next case to be considered is that of polyatomic ions which are stable even in the melt but which are not spherical in their repulsion envelopes. As already stated, well-known salts of this kind giving melts at fairly low temperatures include the ions NO_3^-, NO_2^-, OH^-, CN^-, CO_3^{2-}, SO_4^{2-}, and many acid salts. Because these ions are no longer spherical, they can pack closely into irregular "birds-nest" clusters in the ionic melts. As a result, diverse vacancies in the molten state are both smaller and considerably larger than for simpler ionic crystals. The more complicated models can no longer be referred to any crystal lattice, because they contain regions with the superimposed clustering. They are often described as *anticrystalline*.

Obviously, interactions between ions in anticrystalline melts and solvents such as H_2O can be considerably more complicated than for simple quasicrystalline melts. Even more unusual model structures are needed to describe the melts of salts such as the alkane carboxylates M^+OOCR^- (Reference 2); unusual solvent effects may be associated with these structures, though this has not yet been widely investigated.

Thinking along these lines is still rather unfamiliar. Two special examples may be cited in order to indicate how one must remain alert to unfamiliar situations, which arise because the interactions determining the behavior of any one ion are many bodied.

Example 1. "Freedom to rotate" relative to neighboring molecules in the assembly depends in a sensitive way upon nonspherical repulsion envelopes of the ions, as well as solvent molecules in two-component systems. As a consequence, mixture laws even for two components only may show unexpectedly steep changes with changes of concentration, with freedom to rotate only at one side of the melting point diagram. Again, supercooled melts of pure alkali nitrates often crystallize spontaneously following general rules for supercooling of melts,[3] whereas mixtures such as 0.6 mol KNO_3 with 0.4 mol $Ca(NO_3)_2$ in which rotation is probably inhibited by the enhanced electrostriction fail to crystallize and the mixtures supercool into glasses.

Example 2. Formation of complex ions in melts or concentrated solutions often follows quite different chemical equations from those applicable for dilute solutions. This is of course due to the effects from more than one neighboring molecule on the ligands within any one complex. Examples which are in contrast with those in dilute aqueous solutions may be found throughout inorganic chemistry, e.g., in melts of NaF:

$$(AlF_6)^{3-} \rightleftharpoons (AlF_4)^- + 2F^-$$

$$(P_2O_7)^{4-} \rightleftharpoons (PO_4)^{3-} + (PO_3)^-$$

$$(B_4O_7)^{2-} \rightleftharpoons 3(BO_2)^- + (BO)^+$$

In this connection it may be added that enhanced interaction between molecules in two-component molten mixtures may present freak instances of strong interactions due to resonance effects, as well as strong electrostatic interactions of the classical type. In particular, group I hydroxides probably retain H_2O molecules quite strongly because of chain resonance interactions $\sim H_2O \sim (OH)^- \sim H_2O \sim$. The interesting question is to determine whether (and at what dilutions) different physical properties give evidence of any steep change from many-body to pairwise interactions. At least three groups of properties can be considered, i.e., thermodynamic, transport, and optical, and the range of precise methods for measurement now available is quite extensive.

In fact, what the present book is doing is to advance into the novel territory where classical physical chemistry, statistical thermodynamics of many-body systems, and a knowledge of peculiarities of recondite inorganic chemistry including chain effects have common interests. The prospects seem exciting for scientists who enjoy hybrid disciplines.

References

1. A. R. Ubbelohde, *The Molten State of Matter*, Wiley, New York (1978), p. 189.
2. A. R. Ubbelohde, *Rev. Int. Hautes Temp. Refract.* **13**, 5, (1976).
3. E. Rhodes, W. E. Smith, and A. R. Ubbelohde, *Trans. Faraday. Soc.* **63**, 1943, (1967).

The Structure of Concentrated Aqueous Solutions

J. E. Enderby

1. Introduction

A knowledge of the detailed nature of the ion–water coordination in the vicinity of the ion is crucial to a full understanding of concentrated aqueous solutions. The neutron first-order difference method described by Soper *et al.*[1] allows detailed ion–water conformations to be obtained directly and without recourse to modeling techniques. It avoids all the difficulties associated with *total* neutron or X-ray scattering data[2] and we are now in a position to answer in a definitive way specific questions relating to the number of water molecules in the first coordination shell and the orientation of the molecule with respect to the ion–oxygen axis.

The background to the method has been fully described in an earlier paper[1] and need not be repeated here. Two heavy-water (D_2O) solutions of the salt MX_n, identical in all respects except for the isotopic state of M (or X), are prepared. The neutron differential scattering cross-section of each of them is then measured by standard techniques[3] and one is then able to extract from the first-order difference, a *weighted radial distribution function* given by

$$G_M(r) = A_1 g_{MO}(r) + B_1 g_{MD}(r) + C_1 g_{MCl}(r) + D_1 g_{MM}(r) + E_1$$

where $E_1 = -(A_1 + B_1 + C_1 + D_1)$ and M is the cation whose isotopic state has been changed.

The individual radial distribution functions [$g_{MO}(r)$, etc.] are labeled in an obvious way and the coefficients $A_1, \ldots, D_1$ are determined by the

J. E. Enderby • Department of Physics, H. H. Wills Laboratory, University of Bristol, Bristol, England.

isotopic state of the samples and the concentrations. In all cases A_1 and B_1 are much greater than C_1 and D_1 so that for practical purposes $G_M(r)$ gives the distribution of the water molecules around the M ion. In a similar way, by changing the anion isotope, a weighted radial distribution function given by

$$G_X(r) = A_2 g_{XO}(r) + B_2 g_{XD}(r) + C_2 g_{XM}(r) + D_2 g_{XX}(r) + E_2$$

where $E_2 = -(A_2 + B_2 + C_2 + D_2)$ is obtained. Once again C_2 and D_2 are much smaller than A_2 and B_2.

2. Cationic Hydration

Two detailed studies involving Ni^{2+} (Reference 4) and Ca^{2+} (Reference 5) are now complete. The sample parameters are given in Table 1 and a summary of the results so far obtained is listed in Table 2.

Let us consider the general form of $G_{Ni}(r)$ shown in Figure 1. It is particularly significant that at $r \sim 3.0$ Å, $G_{Ni}(r) \sim -E_1$; thus g_{NiO} and g_{NiD} are both small at this value of r, which reflects the stability of the first hydration shell. The two peaks located at 2.07 and 2.67 Å we identify with Ni–O and Ni–D correlations, respectively, on the grounds that the ratio of the areas beneath them when weighted by r^2 is almost exactly $A:B$. An integral over $4\pi r^2 G_{Ni}(r)$ for $1.8 < r < 3.0$ Å yields the number of water molecules in the first coordination shell which, for convenience, we refer to

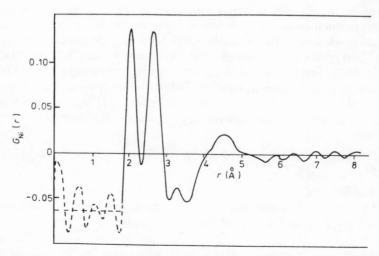

FIGURE 1. $G_{Ni}(r)$ for a 4.41 m solution of $NiCl_2$ in D_2O.

TABLE 1. Scattering Length and Sample Parameters (Cation Hydration)

Electrolyte solution	Isotopes	Abundance, (%)	Scattering lengths, 10^{-12} cm	c^a	Molality	A ($\times 10^{-2}$), b	B ($\times 10^{-2}$), b	C ($\times 10^{-3}$), b	D ($\times 10^{-3}$), b
NiCl$_2$ · D$_2$O	Ni^{natural}	—	1.03						
	^{63}Ni	94.9	−0.79						
				0.0270	4.41	1.74	4.00	5.05	0.32
				0.0192	3.05	1.26	2.90	2.52	0.15
				0.0093	1.46	0.64	1.46	0.61	0.04
				0.0056	0.85	0.385	0.885	0.22	0.013
				0.0028	0.42	0.194	0.446	0.054	0.00338
				0.00057	0.086	0.040	0.092	0.0023	0.00015
CaCl$_2$ · D$_2$O	Ca^{natural}	—	0.466						
	^{44}Ca	95.4	0.18						
				0.0275	4.49	0.30	0.68	0.09	0.02

$^a c$ is the atomic fraction of Ni^{2+} or Ca^{2+}.

TABLE 2. Hydration of Cations Obtained from Neutron Diffraction

Ion	Solute	Molality	Ion–oxygen distance (Å)	Ion–deuterium distance (Å)	θ^a ($\overrightarrow{OD}$ = 0.94 Å)	θ^a ($\overrightarrow{OD}$ = 1.00 Å)	Hydration number
Ni^{2+}	$NiCl_2$	4.41	2.07 ± 0.02	2.67 ± 0.02	$34° \pm 8°$	$42° \pm 8°$	5.8 ± 0.2
		3.05	2.07 ± 0.02	2.67 ± 0.02	$34° \pm 8°$	$42° \pm 8°$	5.8 ± 0.2
		1.46	2.07 ± 0.02	2.67 ± 0.02	$34° \pm 8°$	$42° \pm 8°$	5.8 ± 0.3
		0.85	2.09 ± 0.02	2.76 ± 0.02	$10° \pm 10°$	$27° \pm 10°$	6.6 ± 0.5
		0.46	2.10 ± 0.02	2.80 ± 0.02	$0° \pm 10°$	$17° \pm 10°$	6.8 ± 0.8
		0.086	2.07 ± 0.03	2.80 ± 0.03	$0° \pm 10°$	$0° \pm 20°$	6.8 ± 0.8
Ca^{2+}	$CaCl_2$	4.49	2.40 ± 0.03	2.93 ± 0.05	$45° \pm 15°$	$51° \pm 15°$	5.5 ± 0.2

[a] θ is the angle between the plane of the water molecule and the MO axis calculated for two values of $\overrightarrow{OD}$. DÔD is assumed in both cases to be 105.5°.

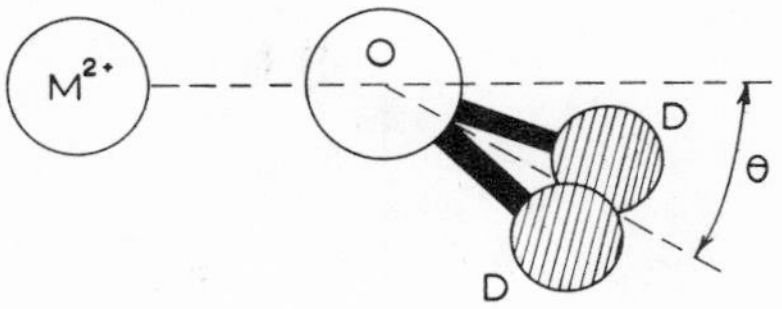

FIGURE 2. The cation-water conformation found for two divalent ions (Ni^{2+} and Ca^{2+}). The distances and angles are listed in Table 2.

as the *hydration number*. The first comment to make on the data summarized in Table 2 is that there is, within experimental error, no change in the hydration number as the concentration of $NiCl_2$ is reduced. However, since the technique allows both the Ni–O and the Ni–D separation to be determined, we are able to investigate the concentration dependence of the conformation of $Ni-D_2O$ provided we know the geometry of the water molecule. For all plausible values of the bond length and angle, the data show unambiguously that for concentrations in excess of 1 *m* the angle of tilt, θ, between the Ni–O axis and the plane of the water molecule is substantial (Figure 2). At the highest concentrations θ varies from 34° to 42° for bond lengths in the range 0.94–1.00 Å and is insensitive to the choice of bond angle. As the concentration of $NiCl_2$ is reduced the $N\overset{\frown}{i}O$ distance remains fixed and although the errors in $N\overset{\frown}{i}D$ become appreciable, the available evidence strongly suggests that this distance increases. We believe that we are seeing the first clear *structural* evidence for the distortions of the hydration spheres as the packing fraction of the hydrated ions is increased and that these observations, when combined with energy calculations, will allow a realistic estimate to be made of the repulsive part of the interionic potential.

Let us now consider the situation for Ca^{2+} (Table 2). For the most plausible value of the $O\overset{\frown}{D}$ bond length (1.00 Å) the angle of tilt found here is almost exactly that predicted by the tetrahedral model (i.e., an oxygen lone pair pointing at the Ca^{2+} ion). It will be clearly of interest to see whether θ is concentration dependent as in the Ni^{2+} case and such an experiment is in progress. It is worth noting that the $C\overset{\frown}{a}O$ distance found in this study is exactly that predicted by the molecular orbital method applied to a single water molecule attached to a calcium ion.[6]

3. Anionic Hydration

A range of solutions have been studied in which the $^{35}Cl \rightarrow {}^{37}Cl$ substitution was made.[7] The sample parameters and principal results are given in Tables 3 and 4 and a typical weighted distribution function $G_{Cl}(r)$ is shown in Figure 3. There are four conclusions from these data which we wish to emphasize:

TABLE 3. Scattering Length and Sample Parameters (Anion Hydration)

Electrolyte solution	Isotopes	Abundance (%)	Scattering lengths, 10^{-12} cm	c	Molality	A ($\times 10^{-2}$), b	B ($\times 10^{-2}$), b	C ($\times 10^{-3}$), b	D ($\times 10^{-3}$), b
LiCl	^{35}Cl	99.35	1.17	0.059	9.95	1.66	3.83	−1.34	4.38
	^{37}Cl	90.4	0.35						
	^{35}Cl	99.35	1.17	0.023	3.57	0.695	1.60	−0.2	0.6
	^{37}Cl	90.41	0.35						
NaCl	^{35}Cl	99.35	1.17	0.0331	5.32	0.99	2.27	0.65	1.37
	^{37}Cl	90.4	0.35						
RbCl	^{35}Cl	99.35	1.17	0.0279	4.36	0.82	1.88	1.00	0.90
	^{37}Cl	90.4	0.35						
CaCl$_2$	^{35}Cl	99.35	1.17	0.0275	4.49	1.60	3.68	1.16	3.77
	^{37}Cl	90.4	0.35						
NiCl$_2$	^{35}Cl	99.35	1.17	0.027	4.35	1.57	3.63	2.40	3.60
	^{37}Cl	90.4	0.35						

TABLE 4. Results for Cl⁻ Hydration

Solute	Molality	Cl–D(1), Å	Cl–O, Å	Cl–D(2), Å	ψ, deg	Coordination number
LiCl	3.57	2.25 ± 0.02	3.34 ± 0.05	—	0	5.9 ± 0.2
LiCl	9.95	2.22 ± 0.02	3.29 ± 0.04	3.50–3.68	0	5.3 ± 0.2
NaCl	5.32	2.26 ± 0.04	3.20 ± 0.05	—	0–20	5.5 ± 0.4
RbCl	4.36	2.26 ± 0.04	3.20 ± 0.05	—	0–20	5.8 ± 0.3
CaCl₂	4.49	2.25 ± 0.02	3.25 ± 0.04	3.55–3.65	0–7	5.8 ± 0.2
NiCl₂	4.35	2.29 ± 0.02	3.20 ± 0.04	3.40–3.50	22–32	5.7 ± 0.2

i. For the nontransition counterions, the linear configuration Cl⁻–D–O–D is favored (Figure 3) and the ion–water conformation is essentially the same for solutions (i.e., it does not depend on the nature of the counterion). There is no evidence for a tendency to the dipole configuration ($\psi \sim 52°$) in any of the solutions so far studied.

ii. Apart from a highly concentrated solution of LiCl, the hydration numbers found are close to, but always a little less than, six. In view of the data given in Table 2, this implies that a substantial degree of water sharing between the ions occurs.

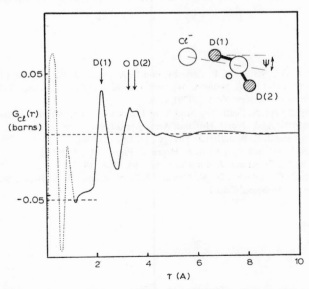

FIGURE 3. $G_{Cl}(r)$ for a 4.49 m solution of CaCl₂ is D₂O. This is the typical form of $G_{Cl}(r)$ for a range of solutions. The conformation of Cl–D₂O consistent with this $G_{Cl}(r)$ is shown on the top right.

iii. In order to index correctly the $D(1)$, O, and $D(2)$ features in $G_{Cl}(r)$, the $O\dot{D}$ bond length must be at least 1.00 Å. There is clear evidence that small ions like Li^+ tend to increase this distance by several per cent.

iv. The nonlinear configuration found for Cl^- in $NiCl_2$ probably reflects a rather special counterion effect. This will be discussed in greater detail in another paper.

4. Concluding Remarks

We have concentrated in this paper on the first-order difference method which yields direct information on the ion–water aspect of the solutions problem. As discussed by Enderby and Neilson,[2] the next step is to use second-order differences which yield directly the ion–ion correlation functions. Although technically difficult, some work along these lines has been carried out and a review of the current situation can be found in Enderby and Neilson.[2]

ACKNOWLEDGMENTS

I wish to thank my colleagues Dr. Neilson, Dr. Cummings, and Mr. Newsome for access to unpublished data and for many helpful discussions.

References

1. A. K. Soper, G. W. Neilson, J. E. Enderby, and R. A. Howe, *J. Phys. C.* **10**, 1793 (1977).
2. J. E. Enderby and G. W. Neilson, *Water: A Comprehensive Treatise*, Vol. 6, Ed. F. Franks, Plenum Press, New York (1979), p. 1.
3. J. E. Enderby, "Neutron Scattering Studies of Liquids," in *The Physics of Simple Liquids*, H. N. V. Temperley, ed., North-Holland, Amsterdam (1968), p. 613.
4. G. W. Neilson and J. E. Enderby, *J. Phys. C.* **11**, L625 (1978).
5. S. Cummings, J. E. Enderby, and R. A. Howe, *J. Phys. C* **13**, 1 (1980).
6. P. Kollman and I. D. Kuntz, *J. Am. Chem. Soc.* **94**, 9236 (1972).
7. S. Cummings, J. E. Enderby, G. W. Neilson, J. R. Newsome, R. A. Howe, W. S. Howells and A. K. Soper, to be published.

3

Nonvibrational Nondiffusional Modes of Motion in Hydrated Calcium Nitrate Melts

Ian M. Hodge and C. Austen Angell

1. Background to Relaxation Studies

Interest in highly concentrated aqueous solutions, in which all or most of the water is directly bound to ions, is relatively recent and has focused to a large extent on hydration coordination numbers and the extent to which cation-bound hydration water molecules are replaced by anions. For example, infrared and Raman spectroscopy of nitrate vibrations in nitrate melts[1,2] have provided evidence that the nitrate ion competes effectively with water for the first coordination shell of Ca^{2+} in $Ca(NO_3)_2$[3] solutions and their mixtures with KNO_3.[4] From the relative intensities of vibrational absorptions due to cation-perturbed and -unperturbed anion normal modes, equilibrium constants can be estimated for water–anion exchange.

Dielectric information about the dynamics of cation-bound water and anions is much less definitive, principally because the motions at and above room temperature occur in an awkward frequency range between translational diffusional frequencies ($\sim 10^{-11}$ sec in high-fluidity melts) and far-infrared lattice vibration frequencies. At these frequencies several problems of a fundamental nature also occur; for example, the background frequency-dependent loss due to Debye–Falkenhagen conductivity relaxation cannot be unambiguously separated from the dielectric relaxation in

Ian M. Hodge and C. Austen Angell • Department of Chemistry, Purdue University, West Lafayette, Indiana 47907. Present address for Dr. Hodge: B. F. Goodrich Research and Development Center, 9921 Brecksville Road, Brecksville, Ohio 44141.

water. The situation is not much better for ultrasonic relaxation, because at room temperature and megahertz frequencies resolution of different modes is poor.

Despite these problems, a number of microwave dielectric measurements have been made on a number of solutions.[5] For dilute solutions, it is assumed that rotation of cation-bound water is dielectrically inactive in order to account for the complex composition dependence of the observed dielectric relaxation time.[5] However, Giese[6] has suggested H_2O–anion exchange as a source of dielectric activity for cation-bound water in alkali halide solutions and shown that this is capable of accounting for the magnitude and sign of the change in relaxation time with increasing solute concentration. Relaxation rates for cation-bound water in hydrate melts at and above room temperature have also been determined from nmr linewidths[7-10] and T_1 and T_2 measurements,[11] and directly from ultrasonic relaxation measurements.[12-14] At room temperature the relaxation times fall in the range 10^{-5}–10^{-8} sec and have activation energies ranging from 8 to 12 kcal mol^{-1}.[7-10] Exchange of water with an anion in the primary cation coordination shell has often been invoked to account for these data, but no definite evidence for this, or any alternative, mechanism has yet emerged.

Better information can be obtained if the time scales for both diffusion and shorter time scale motions are lengthened simply by lowering the temperature to an appropriate value. It is still not generally appreciated that in many aqueous solutions the time scale of diffusion can be decreased by 13 orders of magnitude by temperature changes of 100°C or so. Indeed, many aqueous solutions when quenched at readily accessible rates can form glasses,[15] in which diffusional motions are on the time scale of seconds or minutes and longer, and which are quite rigid by normal mechanical criteria. There are two big advantages in cooling aqueous solutions to the glassy state for dielectric measurements. First, the time scales of the motions of interest are moved from the experimentally difficult microwave region to the manageable domain of audio and subaudio frequencies. Second, those non-vibrational, activated modes of motion which are characterized by different activation energies become progressively more uncoupled as the temperature decreases, so that the resolution of physically distinct processes improves immensely, sometimes to such an extent that separated absorptions are observed. The work reported here establishes that localized motion of cation-bound water can be observed well separated from the long-range diffusional and conductivity modes, thus enabling a more detailed study of the localized modes. Similar cooling techniques, using the addition of acetone to suppress crystallization, have been used to separate the pmr relaxations of cation-bound and "free" water.[16]

A number of restrictions on the types of systems suitable for this cooling technique exist. Foremost among these is the requirement that the solution be glass forming at normal quench rates (e.g., plunging of the sample into liquid nitrogen), and that the glass transition temperature T_g (at which the diffusional–conductivity time scale approaches seconds) be sufficiently high to enable observation of localized motions in the poorly conducting glassy, rather than highly conducting supercooled liquid, state. Problems associated with annealing effects on the conductivity background are also encountered. However, for those systems in which these factors are not a serious problem, good decoupling of a number of dielectrically active modes can be achieved and information on the dynamics of cation-bound water and other species can be obtained. We report here data acquired for aqueous solutions of $Ca(NO_3)_2$ and KNO_3–$Ca(NO_3)_2$ mixtures, some of which have received attention previously.[17]

It is a central feature of these measurements that resolution of the relaxations of interest can only be clearly resolved at very low probe frequencies. This is achieved by keeping the measuring frequency constant at 1 Hz whilst changing the temperature, and thus the relaxation times, until the temperature-dependent average time scale of each of the motions matches the measuring frequency. When this occurs, a maximum in the energy absorption from the applied field is observed. Different modes are generally characterized by different activation energies and different inherent relaxation times, so that absorption maxima for different motions occur at different temperatures for constant frequency. This technique is commonly used to detect glassy relaxations in organic high polymers.[18]

To establish a connection with what is probably a more familiar phenomenon, it should be pointed out that the spectra just described have many features in common with resonance spectroscopy at far-infrared and higher frequencies. The dielectric loss function, $\varepsilon''(\omega)$, is in fact directly proportional to the optical absorption coefficient $\alpha(\omega)$:

$$\varepsilon''(\omega) = \frac{n(\omega)c\alpha(\omega)}{\omega} \tag{1}$$

where ω is the angular frequency, $n(\omega)$ is the frequency-dependent real component of the refractive index, and c is the velocity of light *in vacuo*. The chief difference between optical and relaxation spectroscopies is that in the latter the frequency of maximum energy dissipation is a strong function of temperature, since it reflects an activated kinetic process.

2. Procedure for Obtaining Relaxation Spectra

Solutions of $Ca(NO_3)_2 \cdot RH_2O$ and $6KNO_3 \cdot 4[Ca(NO_3)_2 \cdot RH_2O]$ were prepared from reagent-grade salts and distilled deionized water. It was assumed that the commerical tetrahydrate of $Ca(NO_3)_2$ had an R value of exactly 4, whereas it is known that, in practice, it varies from 4.02 to 4.08. The R values given here may, therefore, be too high by up to 0.08, this being tolerated since we are concerned here with general trends over a range of R values of 0–14. Solutions characterized by a water-to-calcium ratio of approximately 3, 4, 5, 6, 8, 10, 12, and 14 for the $Ca(NO_3)_2$ system, and 0, ~ 0, 1, and 3 for the mixed-nitrate solutions, were prepared. The mixture denoted by ~ 0 is one containing an unknown small amount of water $(R < 0.2$ probably).[17] The mixed-nitrate solutions were investigated in order to extend the water-to-calcium ratio down to zero and retain glass-forming capability. There is spectroscopic evidence[4] to support the claim that K^+ does not compete at all with Ca^{2+} for H_2O, so that addition of KNO_3 would be expected to have a minor influence on the dynamics of Ca^{2+}-bound water. Data on the single-salt and mixed-salt systems which are each characterized by an R value of 3 confirm this expectation.

A gold-plated brass cell of cell constant ~ 4 pF (described elsewhere[17]) was used, and the glasses obtained by plunging the filled cell into liquid nitrogen. A Berberian–Cole low-frequency bridge,[19] modified with low-impedance circuits to reduce transient recovery times, was used at 1 Hz in the three-terminal mode. Occasional cracking of the glasses resulted in arbitrary changes in the apparent cell constant[17, 20] and a corresponding large scatter in ε'', but the loss tangent $\tan \delta$ ($\equiv \varepsilon''/\varepsilon'$, where ε' is the relative permittivity of the glass) is independent of cell constant and gave reproducible data at low temperatures. At higher temperatures near the glass transition temperature T_g, the magnitude of $\tan \delta$ was a function of quenching rate through T_g and annealing time at temperatures immediately below T_g. Studies of this effect[21] indicate that this is due to different background conductivities being frozen in or annealed out, since only the magnitude of $\tan \delta$ and never the peak positions was affected. This variation in conductivity background was minimized by subtracting it out to give spectra whose positions were reproducible and whose magnitude varied by $\sim 20\%$ or so at worst.

Spectra are reported as a function of $1/T$ in order to preserve the orientation of isothermal frequency spectra, i.e., inherently faster processes lie to the right.

Loss tangent spectra of the $Ca(NO_3)_2$ glasses are shown in Figure 1 as a function of $1/T$, and those of the mixtures are shown in Figure 2a with the $3R$ $Ca(NO_3)_2$ spectrum also included to illustrate its similarity to the mixed-nitrate spectrum at the same R. The spectra obtained after the conductivity backgrounds have been subtracted out are shown in Figure 2b for the mixed-nitrate glasses.

It is clear that for $R < 8$ there are two relaxations in addition to that due to conductivity (α) in the $Ca(NO_3)_2$ system, one denoted β occurring

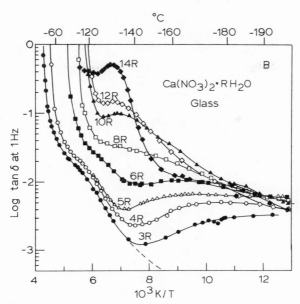

FIGURE 1. Loss tangent spectra as a function of $1/T$ for the indicated aqueous calcium nitrate glasses.

close to the conductivity edge and moving to lower temperatures (i.e., becoming inherently faster) with increasing water content. The second relaxation, termed γ, occurs at lower temperatures and moves in the opposite direction towards higher temperatures (i.e., becoming inherently slower) with increasing water content. At $\sim R \approx 8$ the two merge, and at higher R values a single secondary relaxation denoted δ is observed to move slowly to higher temperatures (become slower) with increasing water content up to the glass-forming limit[15] of 14R. These features, and others to be discussed later, are displayed in Figure 3 as a plot of temperature of maximum loss at 1 Hz vs. R.

Similar features are encountered in the mixed-nitrate system. The positions as a function of R of the higher temperature β relaxation are also plotted in Figure 3, and are seen to continue to lower R values, i.e., the same general trend observed in the $Ca(NO_3)_2$ system. Note, however, that for $R = 1$ and ~ 0 the temperature of maximum loss becomes independent of R. The low-temperature γ relaxation observed in the $Ca(NO_3)_2$ system is also observed in the 3R mixed-nitrate system, although it is displaced to lower temperatures (Figure 3). An extrapolation of the $Ca(NO_3)_2$ γ relaxation to $R = 0$ (see Figure 3) gives a temperature (45 K) close to that reported by Hayler and Goldstein[22] (50 K at 10 kHz, extrapolated to 42 K at 1 Hz). This suggests a common origin for the relaxation in each system.

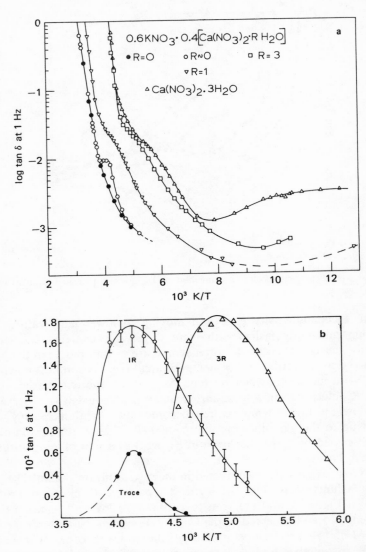

FIGURE 2. (a) Loss tangent spectra as a function of $1/T$ for the indicated aqueous glasses of $6KNO_3 4[Ca(NO_3)_2 RH_2O]$. The spectrum for $Ca(NO_3)_2 \cdot 3H_2O$ is also shown. (b) Residual loss tangent spectra obtained after subtracting the conductivity contribution to the spectra shown in (a).

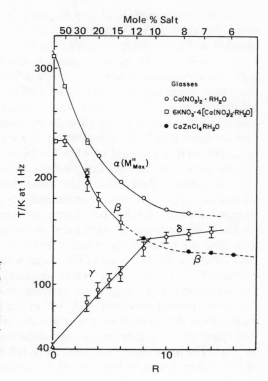

FIGURE 3. The R dependences of peak positions (temperature at 1 Hz) for relaxations observed in the two nitrate systems (Figures 1 and 2) and for the β relaxation observed in the $CaZnCl_4 \cdot RH_2O$ system.[21] The α transition points correspond to peak maxima in the loss modulus, corresponding closely to tan $\delta = 1$.

3. Interpretation of Relaxation Data

The present glassy state dielectric data bear some striking similarities to the room temperature (and above) ultrasonic relaxation data for $Ca(NO_3)_2$ solutions reported by Darbari and Petrucci.[12] These authors observed a relaxation at ~ 20 MHz for $Ca(NO_3)_2 \cdot 4H_2O$ which moved to higher frequencies with the addition of water, i.e., the same shift with water content observed for the β relaxation in the present work. Furthermore, their ultrasonic data for $4R$ and $4.3R$ $Ca(NO_3)_2$ suggest that another relaxation process occurs at higher frequencies. This would be consistent with the inherently faster γ relaxation observed in the present work being ultrasonically active and the resolution of the two relaxations at room temperature and megahertz frequencies being much poorer than in the glassy state at 1 Hz. At higher water contents ($4.7R$ and $5R$) this trend was not visible within the experimental precision of two significant figures. It seems likely that the apparent disappearance of the faster ultrasonic relaxation at higher water contents is due to its merging with the principal relaxation, since the present work indicates that the two relaxations come closer together at higher water contents. The fact that both the relaxations can be better resolved at lower frequencies in the glassy state is also consistent with the conclusion reached by the workers using ultrasonic techniques, that these processes are ruled by an enthalpy of activation rather than an activation volume.[12] Darbari and Petrucci noted that the relaxation frequency they observed lay close to the frequency calculated from Eigen's rate constant data[23] for substitution of water around Ca^{2+} in dilute solution, although in the light of what has been said it is not clear which of the two ultrasonic relaxations Eigen's data correspond to.

The present data exhibit the same dielectric trends observed in the microwave region at room temperature by Pottel and co-workers.[5] The increase in inherent relaxation frequency with water content observed here for the β relaxation is in agreement with the decrease in dielectric relaxation time with increasing water content observed for very concentrated LiCl solutions in which all of the water would be expected to be ionically bound $(R \lesssim 6)$. At higher water contents this trend is reversed and the relaxation time begins to increase with water content,[5] reflecting the same reversal observed at $R \approx 8$ in the present data. The fact that in each system the reversal occurs at a water content which exceeds by $\sim 2R$ that required for saturation of the cation hydration shell ($4R$ for Li^+, $6R$ for Ca^{2+}) is presumably not fortuitous, [cf. Marcus' comments (Chapter 7 of this volume) concerning hydration numbers and ion contact].

It is of course of considerable interest that cation-bound water is dielectrically active in the glassy state, and the question which naturally arises is how this activity is attained. One explanation, given in Section 1, is exchange of H_2O and NO_3^- in the primary coordination shell of Ca^{2+}, similar to that advanced by Giese[6] to account for the near-room-temperature microwave dielectric data or alkali halide solutions, and by others to account for their ultrasonic and nmr relaxation data. Although it might seem improbable that such an exchange could occur in the glassy state, coordination changes around Co^{2+} have been observed optically in the glassy state.[24]

Such an exchange mechanism would, of course, be sensitive to the counteranion and is therefore amenable to verification by observing the effects of changing the anion. Unfortunately, most aqueous solutions of calcium salts have glass temperatures which are significantly lower than those of $Ca(NO_3)_2$ solutions,[15] with the result that the relaxation observed in the nitrate system would be masked by the conductivity loss. However, solutions of salts with the stoichiometry $CaZnCl_4 \cdot RH_2O$ do have glass temperatures which are sufficiently high to permit observation of secondary relaxations similar to those found in the nitrate system.[21] It is unfortunate that $CaZnCl_4$ solutions are not glass forming at $R < 8$, so that overlap with the $Ca(NO_3)_2$ system is not attainable. Nevertheless, the positions of the β relaxations for high R $CaZnCl_4$ glasses (which are resolvable from the δ relaxation up to $16R$ in this system) continue smoothly the trend observed at lower R values for the $Ca(NO_3)_2$ system (see Figure 3), suggesting that the influence of the anion is quite small. In keeping with ultrasonic data which demonstrate the stability of $ZnCl_4^{2-}$ at quite high water contents,[13, 14] any influence of Zn^{2+} hydration on the $CaZnCl_4$ data is apparently quite negligible: the positions of the Zn^{2+}-bound water relaxation observed in $ZnCl_2 \cdot RH_2O$ glasses[21] are far removed from any relaxations observed in the $CaZnCl_4$ system. These observations on the $CaZnCl_4$ system suggest that water–anion exchange is perhaps not the motion which is responsible for the dielectric activity of cation-bound water, although this possibility cannot be ruled out. It is more likely that this exchange is associated with the viscosity–diffusivity–conductivity motions which give rise to the α relaxation.

An alternative source of dielectric activity for the β relaxation is rotation of H_2O about the oxygen–cation axis. In the past such rotations have always been taken to be dielectrically inactive because of an assumed coincidence of the rotation axis and the water dipole moment vector. However, recent neutron diffraction data obtained by Enderby and co-workers for calcium chloride and nickel chloride solutions[25] indicate unambiguously that in these systems such a coincidence does not exist, and that the angle between the two vectors varies continuously, from $\sim 40°$ for 4.4 m $NiCl_2$ (12.6R) and $\sim 50°$ for 4.5 m $CaCl_2$ (12.3R), to $\sim 0°$ for 0.086 m $NiCl_2$ (450R). Rotation about the oxygen–cation axis in the more concentrated of these solutions would therefore have a pronounced dielectric activity which should decrease with increasing water content. Since it would seem likely that similar cation–water geometries would also occur in the nitrate systems studied here, rotational motion of cation-bound water would appear to be a viable mechanism for dielectric relaxation.

Unfortunately it is not possible to test the predicted decrease in dielectric activity per water molecule of the β relaxation in $Ca(NO_3)_2$ with increasing water content because overlap with both the conductivity edge (α relaxation) and the low-frequency γ relaxation increases significantly as R increases. However, for the mixed-nitrate glasses ($R \leq 3$) the well-defined conductivity background of the anhydrous mixture can be shifted and subtracted to yield reasonably symmetric peaks which do not overlap significantly with the γ relaxation (Figure 2b). The peak heights and widths for the $1R$ and $3R$ glasses are seen to be comparable, indicating that the dielectric activity per water molecule does indeed decrease with increasing R, in agreement with the deductions made from Enderby's structural data and the assumption of rotation about the cation–oxygen axis.

To make this concept more definite we shall now demonstrate that changes in the angle (θ) between the rotational axis and dipole moment vector can, plausibly, give rise to changes in dielectric strength of the order observed. Let it be assumed that the dipole moment being relaxed is given by the component of the water dipole moment (μ_w) which lies at right angles to the rotation axis, magnitude $\mu_w \sin \theta$. Let it also be assumed that the peak heights of tan δ (Figure 2b) directly reflect the dispersion in permittivity, $\varepsilon_0 - \varepsilon_\infty$, and that the latter is proportional to $R[\mu_w \sin \theta]^2/kT$. The peak heights and half-widths for the 1R and 3R relaxations are more or less the same, so that

$$\frac{3}{T_{3R}} \sin^2 \theta_{3R} = \frac{1}{T_{1R}} \sin^2 \theta_{1R}, \quad \text{or} \quad \sin^2 \theta_{3R} = 0.286 \sin^2 \theta_{1R}$$

Examples of calculated $(\theta_{1R}, \theta_{3R})$ pairs are $(90°, 32°)$, $(60°, 28°)$, and $(30°, 15°)$. The values of θ_{3R} are somewhat smaller than those deduced by Enderby[25] at higher R values for $NiCl_2$ and $CaCl_2$ solutions, although this could well be accounted for by the presence of nitrate ions in the calcium coordination shell. However, the point to be emphasized is that the calculated changes in θ are physically reasonable and that the observed changes in dielectric strength per water at these low R values (and presumably higher R values as well), can be accounted for in terms of a rotational mechanism.

Support for a rotational source of dielectric activity also comes from a recent analysis by Braunstein and co-workers[26] of proton nmr spin-lattice relaxation times and transport properties in calcium nitrate solutions covering the same concentration range studied here. These workers estimated the rotational intramolecular correlation times of water by subtracting the contribution of the intermolecular translational correlation times estimated from diffusivity and viscosity data from their measured values of the spin-lattice relaxation time; they found that the translational times were always greater than the rotational times. This suggests that when the translational times approach values corresponding to the glass transition temperature (1–10 sec) the rotational times are sufficiently shorter that they approach ~ 1 sec (corresponding to a relaxation frequency of 1 Hz) at temperatures significantly below the glass temperature. This is consistent with a rotational source of dielectric activity for the β relaxation and anion–water exchange (translation) within the cation coordination shell as the source of the inherently slower α relaxation.

In contrast to the behavior of the β relaxation, the γ relaxation becomes slower with the addition of water (Figure 3). As has been pointed out above the similar positions and behavior in both the calcium nitrate and mixed-nitrate glasses suggest a common origin for both. It seems probable that it reflects some type of motion of the nitrate ion which is slowed down by hydrogen bonding interactions with water. Since it is inherently quite fast, occurring below liquid nitrogen temperatures in the anhydrous mixed-nitrate even at 10 kHz,[22] some type of rotational motion is possible, perhaps involving a change from unidentate to bidentate coordination. This motion would also be expected to be perturbed by the presence of extra K^+ cations, as found (Figure 3). It is also possible that the relaxation is largely due to an anharmonic overdamped vibration involving the nitrate ion. In either case, the relaxation may be thought of as involving some type of motion of $Ca^{2+}-NO_3^-$ ion pairs.

The intensity of the δ relaxation at high water contents ($R \gtrsim 8$) increases greatly with R (Figure 1). This suggests that it is associated with water which is not directly bound to Ca^{2+} (but which must be directly bound to the nitrate at these R values). Its high intensity relative to the β and γ relaxations indicates a large change in dipole movement for the motion. It is conceivable that this nitrate-bound water may resemble bulk water in its relaxation behavior, if a nitrate oxygen anchors a water proton to more or less the same extent as does an oxygen on a second water molecule and if a rotational motion is primarily involved. This naive concept is at least consistent with the relatively weak R dependence of this relaxation compared with the β and γ relaxations. A rotational source of dielectric activity for the δ relaxation is also suggested by the recent analysis of Braunstein.[26] This indicated that the pmr translational relaxation times decrease much more rapidly than the rotational times with increasing water concentration, so that the translation time which determines the conductivity approaches the rotational time with increasing dilution. Since the conductivity tail and δ relaxation observed in the present work do indeed approach more closely with increasing water content (Figure 3), a rotational source of activity for the δ relaxation is again suggested.

The presence of both β and δ relaxations may be responsible for the increase in dielectric relaxation width with increasing salt content observed by Gotlob and reported by Pottel[5] for room temperature microwave measurements, although an increasing contribution of a frequency-dependent conductivity process (such as that due to the Debye–Falkenhagen effect) cannot be ruled out as a major source of the broadening.

The freezing out of diffusional degrees of freedom by the formation of glasses, and the lowering of relaxation frequencies to the subaudio range, enables a fine resolution of separate processes to be achieved which is not feasible at room temperature.

In conclusion, resolvable, localized relaxation processes associated with cation-coordinated water (α and β relaxations), ion pairs (γ relaxation), and anion-bound water (δ relaxation) may be observed using the experimental technique described. It could well prove to be useful for nonaqueous solutions as well, and is clearly amenable to experiments in which water in the coordination shell is replaced by other solvent species such as methanol or DMF. The technique offers the promise of making significant contributions to the understanding of relaxation behavior of water in concentrated aqueous solutions.

References

1. R. E. Verral, in *Water: A Comprehensive Treatise*, Vol. 3, Ed. F. Franks, Plenum Press, New York (1973), Chap. 5, p. 211.
2. T. H. Lilley, in *Water: A Comprehensive Treatise*, Vol. 3, Ed. F. Franks, Plenum Press, New York (1973), Chap. 6, p. 265.
3. R. E. Hester and R. A. Plane, *J. Chem. Phys.* **40**, 411 (1960).
4. C. T. Moynihan and A. Fratiello, *J. Am. Chem. Soc.* **89**, 5546 (1967).
5. R. Pottel, in *Water: A Comprehensive Treatise*, Vol. 3, Ed. F. Franks, Plenum Press, New York (1973), Chap. 8, p. 401 and references given therein.
6. K. Giese, *Ber. Bunsenges. Phys. Chem.* **76**, 495 (1972).
7. R. G. Wawro and T. J. Swift, *J. Am. Chem. Soc.* **90**, 2792 (1968).
8. N. A. Matwiyoff and H. Taube, *J. Am. Chem. Soc.* **90**, 2796 (1968).
9. A. Fratiello, V. Kubo, S. Peak, B. Sanchez, and R. E. Schuster, *Inorg. Chem.* **10**, 2552 (1971).
10. R. D. Green and N. Sheppard, *J. Chem. Soc. Faraday Trans. 2*, **68**, 821 (1972).
11. H. G. Hertz, in *Water: A Comprehensive Treatise*, Vol. 3, Ed. F. Franks, Plenum Press, New York (1973), Chap. 7, p. 301 and references given therein.
12. G. S. Darbari and S. Petrucci, *J. Phys. Chem.* **73**, 921 (1969).
13. G. S. Darbari, M. R. Richelson, and S. Petrucci, *J. Chem. Phys.* **53**, 859 (1970).
14. K. Tamura, *J. Phys. Chem.* **81**, 820 (1977).
15. C. A. Angell and E. J. Sare, *J. Chem. Phys.* **52**, 1058 (1970).
16. A. Fratiello, V. Kubo, R. E. Lee, and R. E. Schuster, *J. Phys. Chem.* **74**, 3726 (1970) and references contained therein.
17. I. M. Hodge and C. A. Angell, *J. Phys. Chem.* **82**, 1761 (1978).
18. N. G. McCrum, B. E. Read, and G. Williams, *Anelastic and Dielectric Effects in Polymeric Solids*, Wiley, New York (1967).
19. J. G. Berberian and R. H. Cole, *Rev. Sci. Instrum.* **40**, 811 (1969).
20. G. P. Johari and M. Goldstein, *J. Chem. Phys.* 2372 (1970).
21. I. M. Hodge, unpublished results.
22. L. Hayler and M. Goldstein, *J. Chem. Phys.* **66**, 4736 (1977).
23. M. Eigen and G. Moass, *Z. Phys. Chem.* (*Leipzig*) **49**, 163 (1966).
24. C. A. Angell, A. Barkatt, C. T. Moynihan, and H. Sasabe, *Proceedings of the International Conference on Molten Salts*, Ed. J. P. Pemsler, Electrochemical Society, New York (1976), p. 195. C. A. Angell and A. Barkatt, *J. Phys. Chem.* **82**, 1972 (1978).
25. J. E. Enderby and G. W. Neilson, in *Water. A Comprehensive Treatise*, Vol. 6, Ed. F. Franks, Plenum Press, New York (1978) and references contained therein.
26. C. Girard, J. Braunstein, A. L. Bacarella, B. M. Benjamin, and L. L. Brown, *J. Chem. Phys.* **67**, 1555 (1977).

4

The Computer Simulation of Ionic Liquids

David Adams and Graham Hills

1. Introduction

Every respectable branch of science is based on a carefully woven mixture of theory and experiment, of hypothesis and observation. Great steps are made when new ideas suggest new experiments and new observations require new concepts. That this happy marriage does not exist for liquids or solutions is plainly evident in the petering out of attempts to improve on any of the classically correct theories of Born, Debye, Hückel, Bjerrum, and Onsager. All such theories invoke concepts of charged hard spheres embedded in a structureless dielectric interacting through a simple Coulomb potential. All attempts to recognize the "graininess" of the problem, i.e., the molecular composition and structure of the system, rapidly degenerate into hand-waving assertions of hydration numbers, field-dependent dielectric constants, solvent-separated ion pairs, etc.

An acceptance that the present understanding of solutions is entirely empirical is clouded by the close association of measurements with thermodynamics. Undergraduates and even others might be forgiven for thinking that the thermodynamics of ionic solutions was a highly theoretical study. Certainly anyone attempting to evaluate, say, the partial molar relative ionic entropy of sodium chloride in dilute solution will need his thermodynamic wits about him. But such weighty considerations are barely relevant to the understanding of an ion in solution, of a concentrated ionic solution, or of an ionic liquid. The interpretation of thermodynamic parameters in terms of solvent structure is seldom fruitful.

David Adams and Graham Hills • Department of Chemistry, The University, Southampton, SO9 5NH, United Kingdom.

FIGURE 1. Woodcock's representation of liquid KCl, the dark spheres representing chloride ions and white spheres representing potassium ions.[5]

To a chemist, the conceptual model must be that of Figure 1 in which highly mobile centers of mass (molecules) interact with one another via complex intermolecular forces. The averaging of these microscopic molecular motions and interactions over space and time results in the macroscopic observations described accurately by the thermodynamic parameters. However, whilst that is easy to state, it is far from easy to accomplish, and therein lies the stumbling block to the development of satisfactory theories of ionic solutions, of ionic or any other liquids.

That stumbling block is the difficulty of successfully applying the correct and elegant procedures of statistical mechanics designed for just these purposes. The appropriate integrations over time and space turn out to be extremely difficult and beyond normal algebraic means of simplification. With a few exceptions, such as the near-ideal gas or the near-perfect solid, the methods of classical statistical mechanics fail. Even compressed gases and simple liquids prove too difficult to describe, and it is only since the advent of large and efficient digital computers that it is possible to attempt the integrations and summations required to evaluate the properties of more complex fluids such as real liquids and real solutions. What follows is a review of progress to date in this direction. Much of the important work has been done in the last ten years and the impetus of such work is accelerating as bigger and better computational facilities become available.

2. Computer Simulation Techniques

The value of computer simulation techniques in the field of ionic liquids lies in their ability to evaluate at any instant of time the motions of individual molecules, their interactions with their near or distant neighbors, and the averages of these quantities over time or space. Their limitation lies in the finite capacity of even large computers, which presently limits the size of the representative systems to several hundred or, at most, a few thousand molecules. Even this is perhaps a less severe limitation than that stemming from our ignorance of actual interaction energies. Under all conditions we are presently forced to make a sweeping assumption that such interactions are relatively simple and pairwise additive. Permanent and induced multipoles over and above simple permanent monopoles and dipoles are seldom considered. In the case of water and aqueous solutions the quadrupole is all important and has had to be included.

Even with this second limitation, considerable progress can nevertheless be made. Comparison of computer-simulation results with real experimental data can be used to find effective pair potentials by "cut-and-try" methods. Also, the exact computer solutions using a particular pair potential can be compared with analytic approximations based on the same potential. In these circumstances, the computer simulation becomes the experimental test of the analytic theory. Where these theories are already complex, as for example in the hypernetted chain and Percus–Yevick theories,[1] computational methods may also be required to effect mathematical solutions of aspects of the theories. In brief, therefore, the involvement of computers in the theory of liquids and solutions is here to stay.

Two principal techniques have been used in the computer simulation of liquids and solutions. They are (1) the Monte Carlo (MC) method and (2) the method of molecular dynamics (MD).

3. The Monte Carlo Method

The Monte Carlo method numerically evaluates integrals by using a random number procedure. Single and double integrals are quite easily solved numerically by standard methods such as Simpson's rule or Gaussian quadrature. However, multidimensional integrals present severe problems since the number of points at which the integrand has to be found is the number of points in one dimension *raised to the power* of the dimensionality, and even the crudest formulas become intractable when there are many variables in the integrals. One solution is to sample the integrand, not at regularly spaced values of its arguments, but at many randomly chosen values. The answer obtained will not be particularly accurate but it can be improved indefinitely simply by using more and more random selections. Any answer found from a finite number of samplings will be subject to random errors but will not be systematically in error. This basic method can obviously be improved by identifying those values of the variables where the integrand is large and concentrating more attention on such important regions; not surprisingly this is known as importance sampling.

The multidimensional integrals produced by statistical mechanics are particularly challenging. For example, the average energy of a system of N point particles is

$$\langle E \rangle = \tfrac{3}{2}NkT + \frac{\int \cdots \int U(\mathbf{r}^N)\exp[-U(\mathbf{r}^N)/kT]\,d\mathbf{r}^N}{\int \cdots \int \exp[-U(\mathbf{r}^N)/kT]\,d\mathbf{r}^N} \tag{1}$$

where the integrals are over the $3N$ coordinates of the particles, $U(\mathbf{r})$ is the potential energy, k is Boltzmann's constant, and T the absolute temperature. Note that N will normally be a large number. In practice, values from 32 to several thousand have been used, so that the integrals are at least 96-fold and standard methods are out of the question. A second difficulty is that U is a steeply varying function of all the particle coordinates, and as any two particles are brought close together the potential energy will rise steeply to very high (if not infinite) values. The integrals themselves involve $\exp[-U/kT]$, which makes them badly behaved in the extreme. Clearly, a rather sophisticated method of importance sampling is required.

The special method devised is due to Metropolis, Rosenbluth, Rosenbluth, Teller, and Teller in a classic paper published in 1953[2] and hardly improved on since. Indeed, their method is so nearly universal that it is known simply as the Monte Carlo method. The method generates a sequence of configurations, the probability of a given configuration being proportional to $\exp[-U(\mathbf{r}^N)/kT]$ and thus automatically concentrates on important configurations. This is achieved by a random walk procedure. The next configuration is selected by first choosing one particle at random and giving it a small random shift in position. The difference in energy between this and the previous configuration, δU, can be quickly computed. The new configuration is retained if either δU is negative or $\exp(-\delta U/kT)$ is larger than a random number in the range 0–1. If it is not, then the previous configuration is readopted. This method produces a chain of configurations, and ensemble average properties are computed as averages along the chain. Usually the bulk of the computing time is taken up by the calculation of the very many changes in energy, δU, and all averages of interest are computed at the same time. The process can be shown to lead to the exact ensemble averages in the limit of an infinitely long chain. In practice, chain lengths are almost always in the range 10^5-10^7 though as ever-faster computers become available chain lengths will almost certainly increase.

It is worth stressing at this point that the sequence of configurations along a chain has no connection whatsoever with the time evolution of the system. The chain is merely a device for solving the multiple integrals of equilibrium statistical mechanics. It must also be understood that the method has to be applied for each specific value of temperature and density of interest; it does not provide, say, average energy as a *function* of temperature or density. Another (irritating) feature of the method is that the averages obtained only *tend* to the true values as the chain length increases, and moreover their reliability only improves as the square root of the chain length. It is at least possible that a chain only 10,000 steps in length might by chance give a result closer to the true value than that from a chain of a million steps. Estimates of the likely error in a result can be made by dividing the chain into several subchains each with its own subaverage and examining the statistical variations between the subaverages. This can give a useful guide but even here some caution must be exercised because the subaverage values will not always be independent of each other.

In most cases the properties of a system of N particles or molecules, with N in the range, say, of 32–4000, is not of great interest. What is generally required are properties in the thermodynamic limit of $N \to \infty$ while N/V is fixed. Results much closer to the thermodynamic limit, for a given number of particles, are obtained by the method of periodic boundary conditions, introduced by Metropolis *et al.*[2] The small system, typically a less than 5-nm-sided cube, is made to behave as though it were part of an infinite system by surrounding it by periodically repeated images of itself. Thus all surfaces are eliminated and the bulk fluid is approximated by a periodically repeating sample. This is illustrated in Figure 2. It is good practice, in a series of calculations, to repeat a few calculations with a different number of particles in the periodically repeating volume or "cell" to check that the results obtained are not still number dependent. When the interaction potential between the particles falls rapidly with increasing separation then a significant number dependence is only to be expected close to a phase transition or critical point. Nevertheless, the limited size of the volume simulated does restrict the range of problems that can be tackled. It is possible, as we shall see, to simulate simple phase boundaries,

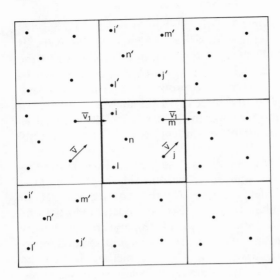

FIGURE 2. Computer simulation box and boundary conditions. (For clarity, only five particles are shown in each square.)

but phase transitions and critical points are not normally within the scope of the method. To be more specific, the restriction is not on the volume but on the number of particles or molecules, which is limited by the size and speed of the computer. For the forseeable future the upper limit in three dimensions seems likely to remain at about 10,000 particles, and for reasons of economy most calculations will use under 1000.

Thus the simulation of *dilute* ionic solutions as mixtures of ions and solvent molecules is not easily realizable. For any reasonable number of ions the corresponding number of solvent molecules would be too great. Dilute electrolyte solutions as such will therefore continue to be simulated as ions in a dielectric continuum, although the local interactions and the near-ion fields can still be evaluated on a molecular basis. This is the basis of many theories of electrolyte solutions from the Debye–Hückel theory onwards, and Monte Carlo calculations have been used to test these theories. These methods have been the subject of a recent review[3] and need not be repeated here. However, the work of Turq, Lantelme, and Friedman[4] deserves special mention. They have simulated the dynamics of dilute ionic solutions by treating the ions as Brownian particles. The solvent molecules form a dielectric continuum and affect the dynamics of the ions by giving rise to both a viscous drag on the motions of the ions and a randomly fluctuating force on each ion. This is the Langevin model of Brownian motion with the additional aspect that the ions interact with each other as point charges in a dielectric medium. The properties that can be studied by this method are rather limited, but it is an interesting way of studying the mobility of ions in dilute solution.

Returning to conventional computer simulation methods, we have already stressed that the Monte Carlo method is only capable of providing approximate results. For a given potential model, the computed averages will be free of systematic error but will be subject to random errors. The importance of these random errors will depend on the quantity of interest. The average potential energy, obtained from equation (1), is normally the most accurately obtained quantity. The ensemble averages of various components of the total potential energy can also be found accurately. In the case of ionic liquids such as molten salts it is therefore possible to evaluate not only the total energy but also the individual contributions to it from the Coulombic and the various short-range interactions.

The specific heat can be obtained from the fluctuations in the total energy,

$$C_v \equiv \frac{\partial E}{\partial T} = \frac{3}{2}Nk + \frac{\langle U^2 \rangle - \langle U \rangle^2}{kT} \tag{2}$$

but the accuracy here is dramatically less. Indeed, unless the Monte Carlo calculation is of exceptional length then the results obtained for C_v are unlikely to be known to be better than $\pm 10\%$ or even 20% and will not generally be worth quoting. This applies to all fluctuation expressions. There are two reasons for this: first, the expression involves the small difference between large numbers each with a random error, and second, quantities such as $\langle U^2 \rangle$ are never produced with quite as great an accuracy as $\langle U \rangle$. The expression for hydrostatic pressure,

$$P = \frac{NkT}{V} - \frac{1}{3V}\left\langle \sum_{i<j} r \frac{\partial \phi}{\partial r}\bigg|_{r_{ij}} \right\rangle \tag{3}$$

again involves a small difference between large numbers. The two terms on the right-hand side of equation (3), the pressure of a perfect gas of the same density and temperature and the intermolecular virial, are generally both large. Also the ensemble average virial is usually calculated with less accuracy than the energy. The net result of this is that accurate results for the pressure of a simulated fluid are just not attainable. For a molten salt, for example, the pressure will be uncertain by several hundred atmospheres or bars. This is of course a great problem when the low-pressure region is of interest and the common question, "Can you obtain the boiling point?" is easily answered "No!" Fortunately, pressure is such a steep function of density that computer simulation results are still very useful for testing theoretical expressions, for over a wide density range the variation in pressure will be much larger than a few hundred bar. Some Monte Carlo isotherms for KCl, obtained by Woodcock and Singer,[5] are shown in Figure 3. They show that the molar volume at zero pressure can be located fairly accurately and in this case there is good agreement with experiment.

It is also possible to reverse the problem and use the Monte Carlo method with the constant-pressure ensemble and find the ensemble average

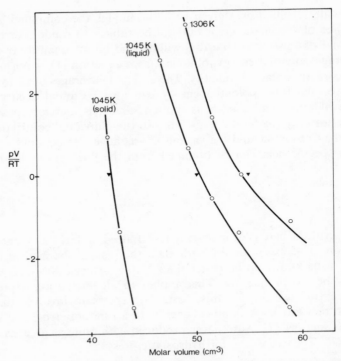

FIGURE 3. Isotherms of potassium chloride: ▼, experimental data; ○, Monte Carlo results. Negative pressures are observed in computer experiments because the small size of the model prevents vaporization.

volume at an applied pressure.[6] For molten salts, errors in the specific volumes are typically ± 0.2 cm^3 mol^{-1}.

One feature of the Monte Carlo method which some people find puzzling is that if the density is sufficiently low negative pressures are obtained. This is the consequence of the inability of the very small volume of fluid in the Monte Carlo cell to divide into two phases, vapor and liquid. Instead it remains in a metastable liquid state. This can happen with the constant-pressure ensemble where the application of a low pressure ought to result in the fluid volume becoming extremely large. Normally, however, the simulated fluid remains firmly in the liquid phase.

Computer simulation must therefore be used with extreme caution in determining the relative stability of different phases. If a simulation starts in the solid phase and melts, then evidently the liquid is the more stable phase. This occurs when the solid is *mechanically* unstable and literally shakes itself apart. If the solid is only thermodynamically unstable it may fail to melt. Occasionally the unstable solid fails to melt completely, a strange unphysical structure being produced, and this seems to be one of the rare occasions when the periodic boundary conditions have a major effect on the simulation. The spontaneous freezing of a supercooled liquid is an extremely rare event. Whatever the relative stability of the two phases the freezing of a pure liquid without a nucleus to crystallize around is kinetically unfavorable and unlikely ever to be observed in the time scale of a computer experiment.

The vapor–liquid phase transition can be located approximately using the constant-pressure ensemble but the method is not very reliable. The only means of reliably determining the coexistence curve for two phases is to determine the conditions when the two phases have the same temperature, pressure, and Gibbs free energy, rather than trying to observe the phase transition directly in a computer experiment. Unfortunately, the ordinary Monte Carlo method does not give information on the entropy, free energy, or chemical potential. Special methods are available and they have been well covered in a recent review of the Monte Carlo technique.[7] The simplest method is to use Monte Carlo to sample the grand canonical ensemble and find the ensemble average density as a function of the applied chemical potential. The method has been used to locate the Lennard-Jones fluid liquid–vapor coexistence curve[8] but has not been used so far for molten salts or mixtures of any sort. This method is not suitable when one of the phases is solid and the only available method then is to create an artificial thermodynamic route from solid to dilute vapor and to make several Monte Carlo calculations along it so that the free energy can be obtained by a standard thermodynamic equation.[9]

The reason why the ordinary Monte Carlo method fails to give free energy is that it only evaluates averages such as that of equation (1); it does not compute the partition function itself. As we have stressed, the method only samples configurations with a large Boltzmann weighting, it does not sample the entire ensemble. Entropy and free energy can be written as ensemble averages such as $\langle \exp(+U/kT) \rangle$, but although these expressions are exact they cannot be properly evaluated by Monte Carlo for $\exp(+U/kT)$ is only large in precisely those configurations that the method will never sample at all in calculations of finite length.[10]

There is one further feature of Monte Carlo calculations that needs to be explained. As we have seen, the method involves importance sampling. If the averages calculated along the chain are to be of any value it is necessary that the chain should *begin* with an important configuration. This requirement is met by a preliminary calculation to "age" or "equilibrate" the simulation. It may be shown that whatever the starting configuration the method will always eventually lead to configurations with a high Boltzmann weighting. For simple liquids the "equilibration" is generally rapid and the preliminary calculation will be $\sim 20\%$ or less of the total length of the calculation. However, it is becoming clear that for more complicated liquids, where the molecules are not spherical, not only are longer chains of configurations necessary but also much longer "equilibration" is essential; the preliminary calculation may even exceed in length the production part.

Monte Carlo calculations imply an act of faith. It is necessary to assume that the method samples all important regions of configuration space, otherwise the averages obtained will not be true ensemble averages. We have already mentioned one case when this is not so, i.e., the method fails to sample the thermodynamically stable phase when the calculation is started in a different phase. In that case, the failure can be turned to advantage and the metastable phase studied, but there are more awkward instances. It was found several years ago that Monte Carlo calculations on mixtures of hard spheres were not satisfactory. The simulation was found to "stick" in certain regions of the configuration space of the mixture and the excess properties of mixing could not be determined.[10] Calculations on pure fluids of nonspherical molecules can tend to "log jam" in some arrangements and so much longer calculations are necessary in order to obtain reliable results for the ensemble averages. For this reason Monte Carlo is tending to give way to the more powerful technique of molecular dynamics, even when only equilibrium properties are required.

4. Molecular Dynamics

The method of molecular dynamics, MD, is more easily understood than that of Monte Carlo, MC. It is simply the numerical solution of Newton's equations of motion. As before, periodic boundary conditions are used so that, like MC, a portion of the bulk fluid is simulated and much the same numbers of particles are used. MC computes ensemble averages whereas MD computes time averages. Of course, time averages should be the same as ensemble averages and it is gratifying to note that strictly comparable calculations by the two methods give nearly identical results. Newton's law, that the rate of change of momentum on a particle is equal to the net force acting on it, gives $3 \times N$ simultaneous second-order differential equations that have to be solved to determine how the positions of the N particles vary with time. There are several numerical solutions for such equations, all relying on the replacement of the differential equations by difference equations so that the solutions are in the form of the particle positions and particle velocities at discrete intervals of time. So, confusingly like Monte Carlo, an MD calculation will be said to be so many steps in length. Unlike Monte Carlo, however, the steps are *time steps* and the method does follow the evolution of the system in time. From one time step to the next, every particle in the simulation moves in the way ordained by the forces acting on it. An MD computer program has to calculate the forces acting on every particle at least once every time step. So while considerable effort on the part of the programmer is necessary to solve the equations of motion, particularly when the simulation includes molecules with fixed bond lengths and angles, it is the calculation of the forces which requires most of the computer time. A more technical account of the MD study of ionic systems has been given by Sangster and Dixon.[11]

Like Monte Carlo, the statistical reliability of the averages improves only as the square root of the number of steps. Virtually all MD calculations are between 1000 and 100,000 time steps in length, and although calculations are tending to become longer, 10^5 time steps is still a rare calculation of great length. Like MC it is essential with MD first to "equilibrate" the calculation. MD simulates an isolated system and the total energy remains constant. The temperature is calculated from the average kinetic energy, i.e.,

$$\tfrac{3}{2}NkT = \left\langle \sum_{i=1}^{N} \tfrac{1}{2}m_i\, \mathbf{v}_i^2 \right\rangle \tag{4}$$

The simulator is usually more interested in a particular temperature than a particular total energy, so that part of the "equilibration" is a cut-and-try process of periodically altering the particle velocities so as to obtain an average velocity close to that of interest. This velocity rescaling cannot be done in the production stage when averages are being calculated, though some simulators do make slight adjustments to the velocities in the course of a calculation to correct for drifts in the total energy.

MD can be used to calculate both equilibrium and time-dependent properties. Indeed, such a large amount of useful information can be obtained that it is standard practice to store the configuration at each time step on magnetic tape for later analysis. MC is limited by the size of the periodic cell. MD is similarly limited and is also limited to the simulation of very short time spans. The longest calculations have been limited to $\sim 10^{-9}$ sec and such calculations can rarely be performed. For example, to simulate argon, time steps of 10^{-14} sec are adequate. For simple fluids, such as argon, all processes within the liquid are short lived and 10^{-11} sec is a sufficient total time span in which to study them. Unfortunately, for more complicated fluids the simulation needs to be much longer to obtain sufficiently good statistics. Some properties, including energy, pressure, radial distribution function, diffusion coefficient, and velocity autocorrelation function can be quickly computed. For example the total potential energy is half the sum over the potential energies of the individual particles or molecules; when the average potential energy is calculated it is effectively by a double summation over all time steps and all particles. However, many other properties, such as specific heat, shear viscosity, and thermal conductivity cannot be written as simple sums over all particles; they are indivisible properties of the simulated fluid as a whole, and very much longer simulations are required to determine them with accuracy. Even with the fastest modern computers such calculations are almost prohibitively long.

One property which both techniques of computer simulation can easily study is the structure of the fluid. It might be questioned whether a liquid possesses a structure as such since its molecules are undergoing continual changes in position. Nevertheless, the average positions can usefully be described in terms of a radial distribution function and the same quantity can be determined indirectly from X-ray and neutron-scattering experiments. The radial distribution functions are easily obtained by either MC or MD. Figure 4 shows the g_{++}, g_{+-} and g_{--}

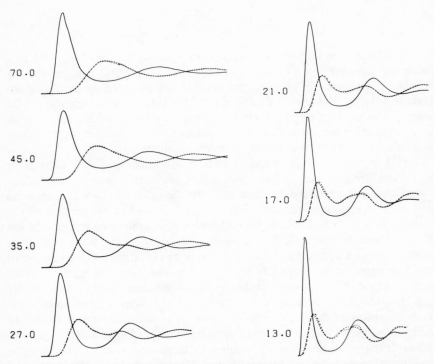

FIGURE 4. Radial distribution functions of simulated KCl obtained by MC at $T = 1700$ K over a wide density range. The numbers are the molar volumes in cm^3 mol^{-1}. The continuous line is g_{+-}, long dashes g_{++}, and short dashes g_{--}.

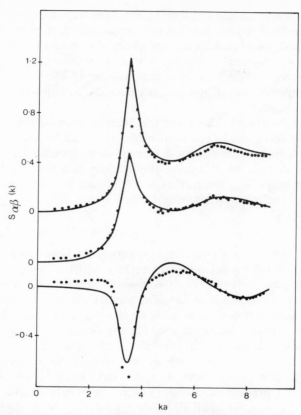

FIGURE 5. Static structure factors of NaCl. The points are MD results; the curves are from the neutron-scattering experiments of F. G. Edwards, J. E. Enderby, R. A. Howe, and D. I. Page, *J. Phys. C* **8**, 3483 (1976). The vertical scale has the same magnitude in every case, but each has a different origin. From top to bottom $S_{--}(k)$, $S_{++}(k)$, and $S_{+-}(k)$.

distributions of simulated KCl obtained by MC at $T = 1700$ K over a wide density range.[12] The effects of increasing pressure on the structure are clearly shown. The highest density shown had a computed pressure of ~ 2 M bar. The experimentalist cannot obtain the radial distribution functions directly, but by X-ray or neutron scattering it is possible to obtain the static structure factors, the Fourier transform of the radial distribution function. By the use of isotope substitution it is possible to separate the three contributions to the total structure factor and obtain S_{++}, S_{+-} and S_{--}. Figure 5 compares the partial static structure factors of real liquid NaCl with those obtained by MD.[13] The importance of such comparisons cannot be overstated. Often the simulator is not concerned to model specific liquids but is concerned with some simplified system of wider theoretical interest, such as the hard-sphere fluid. However, when the intention is to model a specific liquid then comparison of the properties of the simulated fluid with those of the actual liquid is essential. In this respect comparison of structural properties is at least as important as comparison of thermodynamic properties. It is also possible to use the comparison of real and simulated fluid structures systematically to improve the model on a cut-and-try basis.

It is worth noting at this point that there is not a complete freedom of choice in the form of the potential model that can be used in computer simulation, or indeed in statistical mechanics in general. As noted at the beginning of this chapter it is presently necessary to assume that the potential energy is pairwise additive, that is, that it can be written as a sum over all pairs of particles or molecules:

$$U(\mathbf{r}^N) = \sum_{i=2}^{N} \sum_{j<i} \phi_{ij}(\mathbf{r}_{ij}) \tag{5}$$

For most liquids this is a good approximation. At worst the empirical pair potential, ϕ_{ij}, is an "effective" pair potential which models the properties of the liquid very well but is not necessarily close to the interaction between particles i and j in isolation. The simulations illustrated in Figures 3 and 4 both used a potential of the form of equation (5) and this has been found reasonably satisfactory at least for the molten alkali halides.

The main deficiency of equation (5) is that polarization, the induction of a dipole moment in a molecule or ion by the other molecules or ions, cannot be written in this form. Sangster and Dixon[11] have discussed in great detail this problem with respect to the alkali halides, solid and liquid. The dynamic properties of alkali halide crystals cannot be explained without taking into account the effects of polarization. It is not yet clear how polarization affects the dynamic properties of molten salts but the pairwise additive approximation seems adequate to explain their equilibrium properties. It is possible to include polarization into an MD calculation, but only at the cost of even more computer time. It is necessary to make the calculation of the polarization at each time step iterative, for the polarization of one ion will depend on the polarization of all others. The polarization has thus to be found by successive approximation until an adequately self-consistent answer is found. Fortunately, only a few, sometimes as little as one or two iterations, are necessary and the arrival of the new generation of vector-processing computers will make the inclusion of polarization in both ionic and dipolar liquids a practical proposition.

5. Dynamic Properties

The difficulty of calculating such quantities as shear viscosity by MD has already been mentioned. Fortunately, there is another approach which makes the calculation of such N-body properties, as they are called, much easier. Unlike MC, MD is not restricted to the equilibrium fluid. Gosling, McDonald, and Singer[14] calculated the viscosity of the Lennard-Jones fluid by the direct approach of applying a shear force and observing the steady-state shear strain, the velocity gradient. This is, of course, the normal method of measuring the viscosity experimentally. In a computer simulation, however, the experiment is less straightforward. The applied force has to be consistent with the periodic boundary conditions and Gosling *et al.* used a force that varied sinusoidally across the periodic cell. In a real experiment the flow of a large quantity of fluid can be observed over a long period but in a compu'ter experiment one can only observe a tiny volume for a short perio~ of time. It is thus necessary to apply an extremely large shear force, otherwise the resulting flow will be too small and will be completely swamped by the thermal motion of the particles. Normally, the thermal motion of the particles is much greater than the shear flow, and in computer simulation this is a limiting factor. Gosling, McDonald, and Singer found it necessary to apply such a large force that the average velocities in the direction of the flow varied by ~ 80 m/sec in the space of a less than a nanometer. The difficulty with such large forces is that the simulated liquid rapidly heats up, and this limits the accuracy attainable. The results obtained were in good agreement both with the viscosity calculated from an equilibrium simulation and the measured viscosity of real argon. This method is likely to be limited to spherical molecules, such as the Lennard-Jones potential. Nonspherical molecules are likely to exhibit a nonlinear streaming effect in the very large velocity gradients.

A more satisfactory method has been devised by Ciccotti and Jacucci[15] and is likely to prove a powerful tool in computer simulation. They argue that the noise of thermal motion could be effectively suppressed by performing two calculations, one with and one without the applied force. The net effect of the force can then be obtained by subtracting the zero-force trajectory from that with the applied force. The main advantage of the method is that forces orders of magnitude smaller than necessary for use with the direct method are sufficient and there is no problem with nonlinear effects. The calculation is comparatively short and good statistics are obtained from averaging over a few hundred pairs of trajectories. An important bonus of the method is that it produces not merely the steady-state response to an applied force but the average time-dependent response, so that it becomes possible to obtain such properties as the frequency-dependent shear viscosity. The only limitation to the method, which in principle can be used for a wide variety of properties, is that it is only capable of providing the short-time response of the fluid. The two trajectories rapidly diverge from each other and after, say, 100 time steps they are sufficiently dissimilar that the subtraction to remove thermal

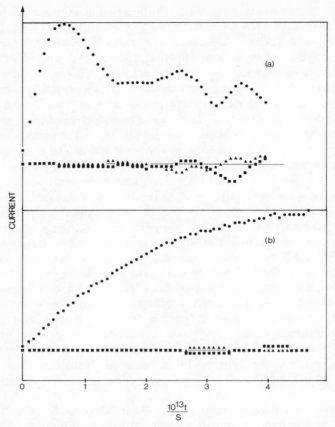

FIGURE 6. Computed responses, in arbitrary units, in the calculation of (a) electrical conductivity and (b) shear viscosity of liquid rubidium iodide.

motion no longer works. If, by this time, the response has not reached the steady-state plateau, then the method cannot provide the zero-frequency property.

The Ciccotti–Jacucci method is a direct application of linear-response theory[15] and can always be used to study at least the short-time part of the various correlation functions of interest. Figure 6, from a paper by Ciccotti, Jacucci, and McDonald,[16] shows the use of the method to obtain electrical conductivity and shear viscosity. The top curve is the average electric current in molten rubidium iodide when an external electric field is suddenly applied. The curves immediately below are the two cross terms, the currents perpendicular to the applied field which should be identically zero. The growth of these curves at large time shows the onset of noise as the subtraction of thermal motion fails. Before this occurs the current in the direction of the applied field has passed through a maximum and reached a plateau from which the steady-state conductivity can be obtained. The lower curves show the response to a shear force both in the direction of the force and perpendicular to it. In this case the response rises asymptotically to a steady state. The authors used a sinusoidal force similar to the one used by Gosling *et al.*[14] and they point out that the quantity actually computed is the time-dependent response to a short-wavelength shear wave. The extrapolation to a shear wave of infinite wavelength and zero frequency, to obtain the ordinary coefficient of viscosity, can only be made tentatively.

6. Polar Systems

One of the problems of simulating molten salts is that the Coulombic potential falls only as $1/r$ and its effective range is always larger than the dimensions of the periodic cell. The summation of forces or energies over all pairs of ions poses a special problem. The method commonly used is one devised by Ewald for the calculation of the Madelung constant of crystals. The ion–ion interactions are summed out to infinity by making use of the periodicity imposed on the simulation. The implementation of the Ewald summation has been discussed at length by Sangster and Dixon.[11] All the molten salt calculations mentioned in this chapter used this summation method, but it has received some censure. It has been attacked for introducing a spurious lattice-like structure to the ionic melt and compared unfavorably with another method based, rather loosely, on the Evjen summation.[7] However, we are satisfied that the Ewald summation is in no way defective but is the most satisfactory method for simulating ionic systems.[17]

The interaction potential between permanent dipoles falls as $1/r^3$, and this is also sufficiently slow to give problems with the method of summation over all dipole–dipole forces and energies. The matter of which method is the most appropriate is far from settled and the simulation of polar liquids to study dielectric properties has been held back. However, if only the thermodynamic properties are required then the method of summation has little effect and some results have been obtained for simple models of polar fluids. The simplest model potential for a polar molecule is the hard sphere plus central point dipole and this is being studied as a first step in understanding polar fluids and how best to simulate them.[18] It is being used by MC only as this potential is a particularly difficult one to use with MD because it combines a continuous force (the dipole–dipole interaction) with an impulsive force (the hard-sphere interaction). Another

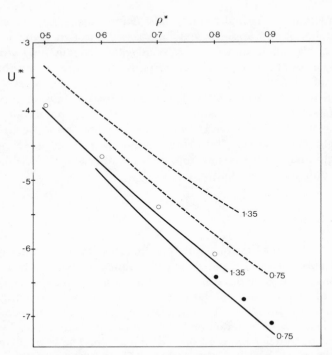

FIGURE 7. Internal energy of the Lennard-Jones and Stockmayer fluids at $T^* = 0.75$ and 1.35. The full lines show the results for the Stockmayer system for the cases $\mu^{*4} = \frac{1}{2}$ (at $T^* = 0.75$) and $\mu^{*4} = 1$ (at $T^* = 1.35$) based on the Padé approximate to the free energy and the broken lines are the results of L. Verlet and J. J. Weis [*Phys. Rev. A* **5**, 939–952 (1972)] for the Lennard-Jones potential. The full and open circles are the Monte Carlo results for the Stockmayer fluid for $T^* = 0.75$ and 1.35, respectively.

simple potential is the Lennard-Jones potential plus a central point dipole, the so-called Stockmayer potential. Figure 7 shows the use of MC results[19] to test theories of the thermodynamic properties of the Stockmayer fluid. Often simulation results for a simple potential, in this case the Lennard-Jones, are used as input data for theories of fluids with a more complicated potential, in this case the Stockmayer potential.

If a Lennard-Jones center plus dipole can be used as a simple model for a polar molecule, then a Lennard-Jones center plus a point charge can be used as a simple model of an ion. Electrolyte solutions can easily be modeled as mixtures of the two. Only a few tentative calculations have been reported so far, probably because the problems of pure polar liquids are far from settled. Adams and Rasaiah[20] used MC to study one and two ions in a Stockmayer solvent. The main aim of their calculations was to obtain the solvent-averaged potential of mean force between a pair of oppositely charged ions. Their results for this were rather disappointing because the mean force calculated remained large and repulsive at large ion separations, contrary to all reasonable expectation. This calculation, incidently, is one of the few cases (apart from the location of phase transitions) where MC should be superior to MD. They were able to bias the MC sampling in favor of the more important solvent molecules close to the ions in an effort to improve the statistics. This biasing method is capable of considerable refinement and a second attempt at this important calculation can be expected. What did emerge from Adams and Rasaiah's calculation was information on the structure of the solvation layer around an isolated ion and around a pair of ions. Figure 8 compares the density distribution of solvent molecules around a solvent molecule in the pure solvent with the density distribution of solvent molecules around a solitary ion. In the latter case there is a much more pronounced first peak corresponding to the formation of a solvation layer around the ion. Figure 9 shows, in rather compressed form, contours of the density distribution around one ion (labeled 0.0) and around pairs of oppositely charged ions at different separations expressed in units of σ, the Lennard-Jones length parameter, in this case the diameter of the ion and solvent molecule, which are the same. The contours are plotted in a reduced form using the fact that not only is the

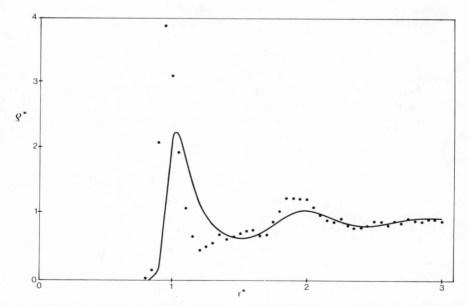

FIGURE 8. Density distribution functions for a simple model of an ionic solution. The line is $\rho^*(r^*)$ for the pure fluid and the points are $\rho^*(r^*)$ of solvent particles around a single ion.

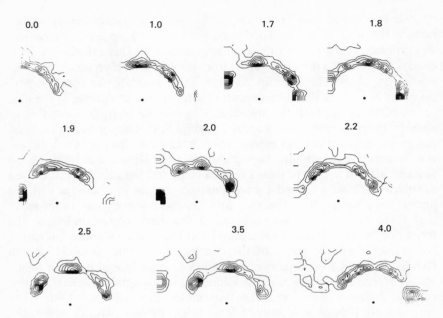

FIGURE 9. Solvent particle density distribution contours around one ion (labeled 0.0) and around pairs of oppositely charged ions at different separations expressed in units of σ, the Lennard-Jones length parameter, in this case the diameter of the ion and solvent molecules, which are the same.

average solvent distribution around a pair of ions cylindrically symmetric but that also the two ions differ only in the sign of their charges and so for density there is a mirror plane of symmetry normal to the line joining them. Figure 10 is a more graphic illustration of solvent structure. It shows

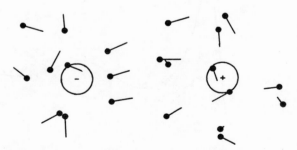

FIGURE 10. An illustration of the configuration of solvent dipoles in the neighborhood of two oppositely charged ions.

just one configuration generated for an ion separation of 2.5σ and the cause of rings of high density shown in Figure 9 is clearly visible. Figure 10 shows only the nearest 22 solvent molecules and the lack of symmetry emphasizes the difference between the *average* structure and the range of structures that actually occur. Though such results are of interest, it must be remembered that they are for a comparatively crude model that does not correspond to any real electrolyte solution, though the calculations were intended to model singly charged ions in acetonitrile.

7. Ions in Water

Water is a polar liquid of great importance, and a few simulation studies have been published, the most notable being the MD calculations of Rahman and Stillinger.[21] Water is particularly difficult to simulate in this way because the very low moment of inertia of the H_2O molecule requires a very small time step, $\sim 10^{-16}$ sec, if the equations of motion are to be solved accurately. This is unfortunate, because the time scale of observable events in associated liquids such as water is long compared to that for simple liquids. Thus the experimental dielectric relaxation time of water is $\sim 10^{-11}$ sec. This would require at least 10^5 time steps to simulate, which makes the computer study of the dielectric response of water prohibitively expensive in computer time. The appropriate pairpotential for modeling water is as yet by no means settled though some simple rigid-molecule potentials give good agreement between simulated and real liquid structures. Stillinger has reviewed the modeling of water in considerable detail,[22] but no model of polarizable water molecules has yet been evaluated.

There seem to be two ways around the very small time step needed for water. One way is to study instead a pseudo-water model, which differs from ordinary water in that its moment of inertia is much higher. This has not yet been done, probably because the effects on the long-time dynamic properties are difficult to assess. However, even the simulation of heavy water would be easier. The second method is to use a much larger time step regardless and to accept that the calculation will only be an approximate one. Too large a time step will cause the total energy to rise and this can be countered by rescaling the velocities as necessary. Another cause of upward energy drift which can also be suppressed in this way is that due to the truncation of the long-range charge–charge interactions between water molecules. This truncation produces a comparatively small upward rise in energy but it also removes any possibility of the simulation being used to study the dielectric properties of water, which arise from the long-range interactions. The effect on the thermodynamic and structural properties of simulated water of this truncation of the potential, which has so far been used in all large-scale studies, is probably small but this is not known for certain. It seems possible that the use of an infinite summation method, such as the Ewald summation, will lead to results that differ from the truncated-potential calculations as much as the results from rival model potentials for liquid water differ from each other.

Little has yet been published on the simulation of ions in water, probably because the full nature of the ion–H_2O interactions is as yet far from clear. Watts[23] has studied the structure of water molecules around pairs of ions, the pair potentials used being obtained from quantum mechanical calculations. His MC calculations are a rare case where periodic boundary conditions have not been used; instead he studied isolated clusters of 50 water molecules and two ions. This eliminates two problems: (i) the difficulty of correctly summing the long-range interactions and (ii) the problem of the periodic boundary conditions upsetting the structure of solvent molecules around the ions. Adams and Rasaiah[20] attempted to avoid this problem by using an enormous periodic system of 4000 dipoles. Examples of Watts' results[23] are given in Figures 11 and 12.

The most ambitious simulations of ions in water so far are those of Vogel and Heinzinger.[24] They used MD and, because they were interested in the equilibrium properties of the solution rather than the solvent, they also allowed the ions to move. The radial distributions they obtained for cesium chloride solution are shown in Figures 13–15. They used a periodic cell of 16 ions (eight of each sign) and 200 water molecules corresponding to $\sim 2.2\ M$ solution. Despite using a time step of only $\sim 10^{-16}$ sec they

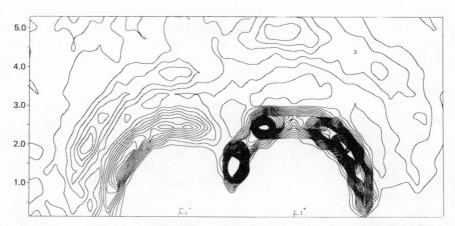

FIGURE 11. Hydrogen density distribution contours from a MC simulation of water molecules around a pair of Li^+ and Cl^- ions fixed at 4 Å apart ($T = 298$ K).

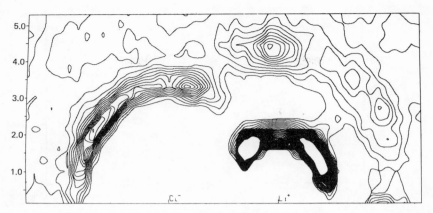

FIGURE 12. Oxygen density distribution contours from a MC simulation of water molecules around a pair of Li^+ and Cl^- ions fixed at 4 Å apart ($T = 298$ K).

still found it necessary to constantly adjust the velocities in a way which suppressed large fluctuations in the kinetic energies of the ions and water molecules. Such alteration of the equations of motion probably improves the statistics of the equilibrium properties computed but casts doubts on the worth of any time-dependent results.

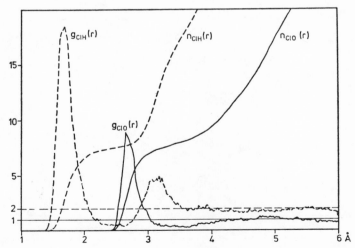

FIGURE 13. Radial pair correlation functions and running integration numbers for chlorine–oxygen and chlorine–hydrogen in the computer simulation of aqueous cesium chloride solution.

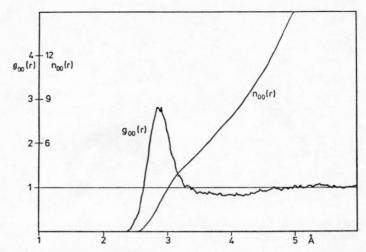

FIGURE 14. Radial pair correlation function and running integration number for oxygen-oxygen in aqueous cesium chloride solution.

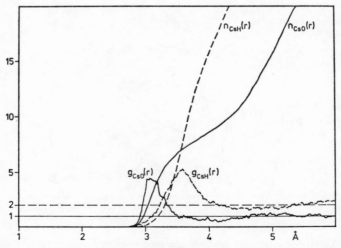

FIGURE 15. Radial pair correlation functions and running integration numbers for cesium-oxygen and cesium-hydrogen in aqueous cesium chloride solution.

8. Interfaces

Computer simulations have not been confined only to the bulk fluids but are turning increasingly to the study of interfaces. The simplest case, the vapor–liquid interface of a simple liquid, has received the most attention. The primary quantity of interest has been the density profile through the interface. The study of a phase boundary in a small periodic cell is obviously not straightforward. There have been two ways of tackling this, one is to start the simulation as a thin ribbon of liquid with vapor on both sides.[25] This is easily accommodated within the periodic boundary conditions in the normal way but has the disadvantage that the two phase boundaries may affect each other because the bulk fluid between them is less than 1 nm thick. The justification of the method is that the center portion of the liquid ribbon can be shown to have much the same density and internal energy as the ordinary bulk liquid. Unless very unfavorable conditions have been chosen then the system will always reach an equilibrium with two phases present. The starting configuration influences the simulation sufficiently that two phases in equilibrium with each other are achieved, whereas if the simulation were started from a homogeneous configuration of the same density then phase separation would not take place but a single, possibly metastable phase would remain. The ribbon method has also been used to study the surface of crystalline potassium chloride,[26] the only interface simulation known to us involving long-range forces.

The alternative method used to study the liquid–vapor interface is to dispense with the periodic boundary conditions in one of the three dimensions. Instead, the cell is bounded on one pair of opposite sides with "walls," which approximate the bulk fluid with a continuum. The vapor phase wall need be nothing more than a reflecting surface to keep particles from leaving the cell. The liquid phase wall will be a smooth attractive surface, with a well-defined interaction potential between it and the particles in the cell. This wall produces a rather strange distribution of particles close to it and the periodic cell has to be sufficiently large that the density fluctuations produced by it have disappeared before the phase boundary has been reached. Figure 16 shows a density profile obtained using this method for the Lennard-Jones potential fluid.[27] This figure illustrates one of the great problems that arise in the simulation of phase boundaries. Well beyond the region where the massive density fluctuations produced by the wall have died away the simulation shows density fluctuations in the liquid phase close to the phase boundary. It was once thought that these oscillations were real, but it is now largely accepted that they are spurious. In a sufficiently long calculation they disappear. This problem seems particularly acute for MC and runs of many millions of steps are

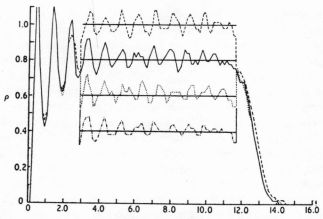

FIGURE 16. MC density profiles for a Lennard–Jones liquid–vapor interface. From the top are the first, second, and third 5×10^6 configurations; at the bottom is the total profile. The large density oscillations at the left are due to the presence of the "wall."

necessary to obtain reliable results. Fortunately, MD seems to produce reliable results more rapidly.

The future is always difficult to predict but it seems to us that computer simulation will provide a powerful means for studying many sorts of interfaces. The present restrictions on time scale and size are not likely to change greatly for many years but a variety of problems that can be squeezed into less than 10^{-9} sec and a few thousand particles will be tackled. For example, there does not seem to be any reason why at least some simple electrode–electrolyte solution interfaces should not be simulated and double-layer theory brought under detailed scrutiny.

9. Conclusions

However, before these developments will materialize much more remains to be done to improve the simulation of a range of electrolyte solutions. The technical problems of simulating pure polar liquids are being actively studied and the routine calculation of both the static dielectric constant and the characteristics of dielectric relaxation should be possible, we predict, by 1980. Much has already been published on the simulation of molten salts, and in that case the main technical problems were solved some years ago.[5, 11] The way is now open, therefore, for the study of electrolyte solutions and they compete for attention with the wider field of molecular liquids.

Computer simulation has two similar but slightly different purposes. One is to study simple models in order to test theories and to gain an insight into general properties of fluids and solutions in terms of molecular parameters. The other is the modeling of a specific substance in order to learn about that substance. In this latter case the main purpose of the calculation may be the testing of the model, i.e., to see how well the chosen potential predicts the experimentally observed properties. Occasionally, as when the Lennard-Jones potential is used, a calculation can achieve both aims by simulating a simple, theoretically interesting model and at the same time simulating liquid argon. With more complicated liquids it will not in general be possible to achieve both ends in the one calculation. Either the theoretically interesting hard-sphere-plus-point-dipole fluid is simulated or, say, water is simulated.

A great deal can be learned from simple potentials and their use is not to be decried. It was soon discovered that although thermodynamic properties are closely related to the interatomic potentials, the structures of a wide variety of simulated liquids with different potentials are very similar. Computer simulation can also be used to good effect with a deliberately contrived potential, for example, to determine the relative roles of the repulsive and attractive parts of a pair potential on the process of diffusion.[28]

In short, a simple model of a real liquid is sufficient to test simple ideas about it. However, the simulation of a particular liquid, with the confidence that the results obtained from the simulation are true also for that real liquid, because the model potential used is a good one, is also of great importance. The future of this type of simulation is less clear because there is a great lack of intermolecular potentials that can be used with confidence. Almost all simulation work on molten salts has been carried out on the alkali halides because the excellent Tosi–Fumi potentials are available.[29] For electrolyte solutions there are as yet no such models. The simulation of specific, "real" electrolyte solutions cannot be carried out without support from other fields of research. First, realistic model potentials for ion–ion, ion–solvent, and solvent–solvent interactions have to be obtained and we look to quantum mechanics for this. Secondly, computer simulation can be used to obtain the bulk structural properties from the potentials. Thirdly, experimental evidence must be brought to test the simulation and the potentials on which it is based. For this purpose we expect that the new high-flux neutron beam apparatus that will be in use by the early 1980s will provide the most useful data. Comparison of simulation studies with neutron diffraction data seems a very powerful tool for the study of electrolyte solutions, as it has already been for simple liquids. The scattering data will test the simulations and permit a methodical improvement on the model potentials, if only on a cut-and-try basis, and the simulation studies should be invaluable in the interpretation of the scattering data. We can expect then an explosion of fundamental knowledge on ionic solutions!

We have deliberately set out to make this chapter an essay on computer simulation rather than a review of the whole field. In many cases we have been content to refer to at most one or two papers because they struck us either as interesting or typical, when we could as well have catalogued a dozen. We must therefore apologize to the authors of many excellent papers that we have not cited. We must also apologize to any readers who would have preferred a more comprehensive introduction to the literature. For them there are many good reviews to turn to, including References 3, 7, 10, 11, 22, and 30–35.

References

1. R. O. Watts, Specialist Periodical Reports of the Chemical Society, *Statistical Mechanics*, Vol. 1, Chemical Society, London (1974).
2. N. Metropolis, A. W. Rosenbluth, M. N. Rosenbluth, A. H. Teller, and E. Teller, *J. Chem. Phys.* **21**, 1087 (1953).
3. C. W. Outhwaite, Specialist Periodical Reports of the Chemical Society, *Statistical Mechanics*, Vol. 2, Chemical Society, London (1975).
4. P. Turq, F. Lantelme, and H. L. Friedman, *J. Chem. Phys.* **66**, 3039 (1977).
5. L. V. Woodcock and K. Singer, *Trans. Faraday Soc.* **67**, 12 (1971).
6. D. J. Adams and I. R. McDonald, *J. Phys. C.* **7**, 2761 (1974).
7. J. P. Valleau and S. G. Whittington, in *Statistical Mechanics Part A. Equilibrium Methods*, Ed. B. J. Berne, Plenum, New York (1977), Chap. 4; J. P. Valleau and G. M. Torrie, in *Statistical Mechanics Part A. Equilibrium Methods*, Ed. B. J. Berne, Plenum, New York (1977), Chap. 5.
8. D. J. Adams, *Mol. Phys.* **32**, 647 (1976); *Mol. Phys.* **37**, 211 (1979).
9. W. G. Hoover and F. H. Ree, *J. Chem. Phys.* **49**, 3609 (1968).
10. W. W. Wood, *Physics of Simple Liquids*, Eds. J. N. V. Temperley, J. S. Rowlinson, and G. S. Rushbrooke, North-Holland, Amsterdam (1968), Chap. 5.
11. M. J. L. Sangster and M. Dixon, *Adv. Phys.* **25**, 247 (1976).
12. D. J. Adams, *J. Chem. Soc. Faraday Trans. II* **72**, 1372 (1976).
13. E. M. Adams, I. R. McDonald, and K. Singer, *Proc. R. Soc. London A* Ser. **357**, 37 (1977).
14. E. M. Gosling, I. R. McDonald, and K. Singer, *Mol. Phys.* **26**, 1475 (1973).
15. G. Ciccotti and G. Jacucci, *Phys. Rev. Lett.* **35**, 789 (1975).
16. G. Ciccotti, G. Jacucci, and I. R. McDonald, *Phys. Rev. A* **13**, 426 (1976).
17. D. J. Adams, *Chem. Phys. Lett.* **62**, 329 (1979).
18. D. Levesque, G. N. Patey, and J. J. Weis, *Mol. Phys.* **34**, 1077 (1977).
19. I. R. McDonald, *J. Phys. C.* **7**, 1225 (1974).
20. D. J. Adams and J. C. Rasaiah, *Faraday Discuss.* **64**, 22 (1978).
21. A. Rahman and F. H. Stillinger, *J. Chem. Phys.* **55**, 3336 (1971); F. H. Stillinger and A. Rahman, *J. Chem. Phys.* **60**, 1545 (1974).
22. F. H. Stillinger, *Adv. Chem. Phys.* **31**, 1 (1975).
23. R. O. Watts, *Mol. Phys.* **32**, 659 (1976).
24. P. C. Vogel and K. Heinzinger, *Z. Naturforsch.* **30a**, 789 (1975).

25. M. Rao and D. Levesque, *J. Chem. Phys.* **65**, 3233 (1976).
26. D. M. Heyes, M. Barber, and J. H. R. Clarke, *J. Chem. Soc. Faraday Trans. II* **73**, 1485 (1977).
27. G. A. Chapel, G. Saville, S. M. Thompson, and J. S. Rowlinson, *J. Chem. Soc. Faraday Trans. II* **73**, 1133 (1977).
28. P. Schofield, *Comput. Phys. Commun.* **5**, 17 (1973).
29. M. P. Tosi and F. G. Fumi, *J. Phys. Chem. Solids* **25**, 45 (1964).
30. B. J. Berne and G. D. Harp, *Adv. Chem. Phys.* **24**, 257 (1973).
31. J. R. D. Copley and S. W. Lovesey, *Rep. Prog. Phys.* **38**, 461 (1975).
32. A. Rahman, *Rivista Nuovo Cimento* **1**, *Numero Specialie 315* (1969).
33. W. W. Wood and J. L. Erpenbeck, *Ann. Rev. Phys. Chem.* **27**, 319 (1976).
34. W. B. Streett and K. E. Gubbins, *Ann. Rev. Phys. Chem.* **28**, 373 (1977).
35. J. A. Barker and D. Henderson, *Rev. Modern Phys.* **48**, 587 (1976).

5

Transport Properties in Concentrated Aqueous Electrolyte Solutions

M. Spiro and F. King

1. Introduction

Concentrated electrolyte solutions have become fashionable again. That they were unpopular for so long owes much to the advent of the Debye–Hückel–Onsager theory in the 1920s. For several decades thereafter the electrolyte solution spotlight was firmly fixed on the D–H–O model and ever more dilute solutions became the rage. It is true that in the late 19th and early 20th centuries the study of concentrated ionic solutions had been quite respectable but for various reasons much of the data then gathered is now of limited value. A selection of more recent experimental data is presented in Section 3 of this chapter and may serve as a guide to future work. We shall consider here only two-component systems, i.e., one electrolyte plus water. Fortunately the revival of interest in highly concentrated electrolyte solutions has transformed our knowledge not only empirically but also theoretically. A good illustration of the present thriving nature of the field is shown by the fact that one recent year, 1977, saw the appearance of two new theories, one dealing with conductance, transference and diffusion and the other concerned with viscosities. The state of the theoretical scene up to the beginning of 1979 is reviewed in the following section.

M. Spiro and F. King • Department of Chemistry, Imperial College of Science and Technology, London SW7 2AY, England.

2. Theories of Transport Properties

2.1. Extensions of the Debye and Hückel Approach

The Debye–Hückel model pictures hard-sphere ions surrounded by their ionic atmospheres in a dielectric continuum. The resulting equations have proved so successful for low concentrations of electrolyte $(< \sim 0.01\ M)$ that considerable efforts have been made to try to widen their domain of applicability. The most obvious way was to add extra terms which sometimes had a theoretical origin and in other cases were frankly empirical. The best established of these is the Jones and Dole[1] equation for viscosity η (or fluidity φ),

$$\eta/\eta_0 = \varphi_0/\varphi = 1 + Ac^{1/2} + B_\eta c \qquad (1)$$

where η_0 is the viscosity and φ_0 the fluidity of the pure solvent at the given temperature, c the molarity of the electrolyte, and A the small Falkenhagen ionic-atmosphere parameter. The tiny $Ac^{1/2}$ term is normally swamped by the theoretically based $B_\eta c$ term, and the introduction of the latter extends the validity of the equation up to a few tenths molar in aqueous solutions. The B_η values have turned out to be additive solute properties and useful accordingly. They are, however, manifestations of ion–solvent rather than ion–ion interaction[2] and this explains the breakdown of equation (1) at higher electrolyte concentrations.

For molar conductances even the latest theoretical extensions of the Debye–Hückel–Onsager theory hold precisely only at very low molarities.[3] The most successful modification, which applies for 1 : 1 electrolytes to concentrations of $1\ M$ $(M = \text{mol dm}^{-3})$ and beyond, is that of Wishaw and Stokes,[4]

$$\Lambda = \left(\Lambda^\circ - \frac{B_2 c^{1/2}}{1 + Bac^{1/2}} \right)\left(1 - \frac{B_1 c^{1/2} \cdot F'}{1 + Bac^{1/2}} \right)\frac{\eta_0}{\eta} \qquad (2)$$

where Λ° is the limiting molar conductance, B_2 and B_1 are the Onsager electrophoretic and relaxation parameters, respectively,[5] B the Debye–Hückel parameter[5] associated with the distance of closest approach of the ions (a), and F' a factor proposed by Falkenhagen. Wishaw and Stokes argued that the semiempirical viscosity correction factor should in principle exclude the small $Ac^{1/2}$ contribution of equation (1) but left it in for simplicity of expression; later evidence suggests[6] that the viscosity effect is appreciably overcorrected. Equation (2), which can be suitably modified to allow for ion association, contains only the one adjustable parameter a. The range of applicability of the equation varies from electrolyte to electrolyte[5,7] and so, therefore, does its usefulness.

The transference number of the ion constituent i of a completely dissociated $1:1$ electrolyte in dilute solutions is given to $\sim 0.1\%$ by the
Debye–Hückel–Onsager formula[8]

$$t_i = t_i^\circ + \frac{(t_i - \frac{1}{2})B_2 c^{1/2}}{\Lambda^\circ (1 + Bac^{1/2})} \tag{3}$$

At concentrations of a few tenths molar (or higher in favorable cases like
KCl) the limiting t_i° values calculated in this way are no longer constant
but vary linearly with concentration as expressed by Longsworth's empirical relation[9]

$$t_i^\circ = t_i^{\circ'} + bc \tag{4}$$

Although formally this equation incorporates the two adjustable parameters a and b, the $t_i - c$ relation can often be reduced to a one-
parameter form by omitting the $Bac^{1/2}$ term and appropriately modifying
the value of b. No viscosity correction factor is required since transference
numbers are essentially ratios of ionic conductances.

A rather more complicated relationship was found necessary[4, 5] to
account for the variation with molality m of the mutual (salt) diffusion
coefficient D of $1:1$ electrolytes:

$$D = (D^\circ + \Delta_1 + \Delta_2)\left(1 + \frac{m \, d\ln \gamma_\pm}{dm}\right)\left[1 + 0.036m\left(\frac{D_w}{D^\circ} - h\right)\right]\frac{\eta_0}{\eta} \tag{5}$$

In the first bracket Δ_1 and Δ_2 are small electrophoretic correction terms to
the limiting diffusion coefficient D° and the second bracket allows for the
thermodynamic nonideality of the solution through the mean molal activity coefficient $\gamma_\pm$. In the third bracket h is the "hydration number" of the
electrolyte and D_w the self-diffusion coefficient of water. If association is
allowed for where necessary, equation (5) represents to within $\sim 2\%$ the
variation of D up to molalities ranging from 2 to 7.[4] Even without the
viscosity factor, agreement to $\sim \frac{1}{2}\%$ is obtained up to $1 \, M$.[5] The only
empirical parameter here is the hydration number, and the h values
required for fitting are not unreasonable. The fit is less good for asymmetri-
cal electrolytes.[5] As regards the self-diffusion coefficients D_j of various ions
j, these decrease approximately linearly with concentration c of the supporting electrolyte up to ~ 2–$4 \, M$.[10] In a given supporting electrolyte
solution, values of D_j/D_j° of different trace ions j show a wide variation
which can clearly not be attributed to a η/η_0 factor.

2.2. Quasi-lattice and Transition State Models

A highly concentrated solution may profitably be viewed as a quasicrystalline electrolyte to which a small amount of water has been added.

Stokes and Robinson[11, 12] treated the water as adsorbed onto the ions and in this way successfully accounted for the activity of water in solutions more concentrated than 10 M. In a later stepwise hydration model[13] they were able to fit osmotic coefficients of several solutions up to 30 m. Other attempts to calculate the activity coefficients of the electrolyte by means of some type of irregular ionic lattice theory have been somewhat less successful.[14, 15] So far, with one exception,[16] lattice theories have been applied only to the prediction of thermodynamic properties, and transport properties have been tackled differently.

The most acceptable starting point for transport properties lies in transition state theory.[17, 18] Picture a particle oscillating with a frequency v_i° inside a cage formed by its neighbors. In order to escape from this cage to another quasi-equilibrium state nearby it must acquire a certain minimum Gibbs free energy of activation, $\Delta G_i^\ddagger$. Thus the frequency with which the particle escapes or "jumps" is given by

$$v_i = v_i^\circ \exp[-\Delta G_i^\ddagger/RT] = v_i^\circ \exp[\Delta S_i^\ddagger/R]\exp[-\Delta H_i^\ddagger/RT] \qquad (6)$$

In highly concentrated solutions of low fluidity it is more likely that transport occurs through the cooperative movement of all the particles in relatively independent microscopic regions in the liquid.[18, 19] Adam and Gibbs[19] have treated such a model in statistical thermodynamic terms and derived an equation for the average probability of rearrangement of the cooperative regions:

$$W = A \exp[-z^* \Delta\mu/kT] \qquad (7)$$

This bears a strong resemblance to the preceding equation. Here z^* is the minimum number of molecules (ions) in a cooperative region and $z^* \Delta\mu$ the potential energy hindering the cooperative rearrangement of the region. Adam and Gibbs go on to show that if a cooperatively rearranging region of the critical size z^* possesses a critical configurational entropy s_c^* then

$$s_c^*/z^* = S_c/L \qquad (8)$$

where L is the Avogadro number $(= R/k)$ and S_c the configurational entropy per mole of solution. Hence

$$W = A \exp[-s_c^* L \Delta\mu/kTS_c] = A \exp[-C/TS_c] \qquad (9)$$

The parameter C is a constant on the assumption that both s_c^* and $\Delta\mu$ are essentially independent of temperature. S_c, on the other hand, varies with temperature according to equation (10):

$$S_c = \int_{T_0}^{T} \frac{\Delta C_p}{T} \, dT \qquad (10)$$

T_0 is the ideal (extrapolated) glass transition temperature at which the configurational entropy of the whole assembly of particles has vanished; typically, T_0 lies 10–20°C below the experimental glass transition temperature T_g.[20] The difference ΔC_p between C_p of the solution at T and at T_0 therefore represents the configurational heat capacity. Experiment indicates[21] that ΔC_p is inversely proportional to T. Integration of equation (10) on this basis and insertion of S_c into equation (9) then yields

$$W = A \exp[-C'T_0/(T - T_0)] = A \exp[-B/(T - T_0)] \qquad (11)$$

where C' and B are constants. The same equation has been derived by Angell and Rao[16] from a "bond-lattice" model.

It is evident that the fluidity φ as well as the transport properties λ_i and D_i of the ions are directly proportional to the probability of cooperative rearrangements W or, on the transition state model, to the jump frequency v_i. Where opinions have differed is in connection with the temperature dependence of the pre-exponential terms, as summarized in Table 1. Equation (11) has been most frequently employed by Angell and his group who have usually opted for the $T^{\pm 1/2}$ formulations for A given by the free-volume theory. Unfortunately this theory has been under suspicion

TABLE 1. Temperature Dependence of Pre-exponential Factors of Various Transport Properties [Equations (6) and (11)]

	Property		
Source	φ	λ^a	D
Transition state theory[18]	v_φ° independent of T	v_λ° independent of T	$v_D^\circ \propto T$
Adam and Gibbs[19]	A_φ independent of T		
Adam and Gibbs plus Stokes equation[b]		A_λ independent of T	
Adam and Gibbs plus Stokes–Einstein[18] equation[b]			$A_D \propto T$
Angell (1966)[22]	$A_\varphi \propto T^{-1}$	$A_\lambda \propto T^{-1}$	A_D independent of T
Free-volume theory[18, 23, 24]	$A_\varphi \propto T^{-1/2}$	$A_\lambda \propto T^{-1/2}$	$A_D \propto T^{1/2}$

[a] λ is the molar ionic conductance.
[b] The Stokes ($\lambda_i \propto 1/\eta \propto \varphi$) and Stokes–Einstein ($D_i \propto T\varphi$) relations are derived from a hydrodynamic model in which the particles of species i are large compared to the surrounding particles and are of both constant radius and constant slip/stick ratio.[5] Although the systems in question are unlikely to meet these requirements, the equations do offer guidance on approximate behavior and have frequently been applied to molten salt systems[18] despite certain theoretical objections.[25]

for some time[18] and recently Williams and Angell[26] obtained volume expansion data for LiOAc · 10 H_2O that were incompatible with a free-volume interpretation. We therefore recommend for future analyses of data the TS/Adam–Gibbs–Stokes (TS = transition state) consensus by which only A_D includes a T factor. Luckily the temperature variation of W depends almost entirely on the exponential term and is relatively unaffected by the choices in Table 1.

The enormous effects of changing temperature are well illustrated in Figure 1, which is taken from Angell and Bressel's classic work with $Ca(NO_3)_2$ solutions.[21] The molar conductance is seen to increase by some three powers of ten when the temperature is raised by 100°C, the increase

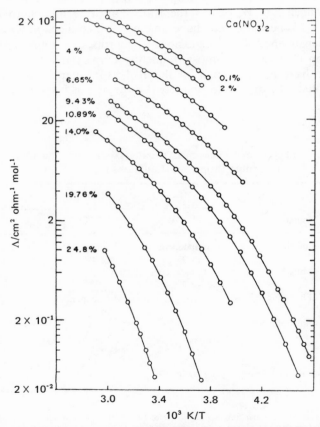

FIGURE 1. Arrhenius plots of molar conductance (log scale) versus reciprocal temperature for $Ca(NO_3)_2$ solutions of various compositions (shown as mol % salt for each curve), from the paper by Angell and Bressel.[21]

being steeper the more concentrated the solution. It is particularly noticeable that the plots of log Λ vs. $1/T$ display pronounced curvature. Only at the extreme ends of the temperature range can the transport behavior be described by an Arrhenius equation with approximately constant activation energy $E^{\ddagger}$. Over most of the range $E^{\ddagger}$ rises in marked fashion as the temperature falls. This cannot be explained by the standard transition state theory but is just what would be expected from equation (11). Indeed, an equation of the form of (11) was proposed independently by Vogel, Tammann, and Fulcher in the 1920s on purely empirical grounds to describe the fluidities of a variety of glass-forming liquids, and Angell has therefore termed it the VTF equation. It predicts for the Arrhenius activation energy

$$E^{\ddagger} = -R\frac{d \ln W}{d(1/T)} = \frac{BRT^2}{(T - T_0)^2} \tag{12}$$

so that $E^{\ddagger}$ should increase as T decreases, tending to infinity as T approaches T_0. (For this reason T_0 has been called the temperature of zero mobility.) However, experiments have shown[27] that equation (11) loses validity at low temperatures close to T_0 where, for reasons unknown, Arrhenius behavior returns. Over the large temperature range where equation (11) does apply, its usefulness is much enhanced by the fact that in the cases tested [e.g., $Ca(NO_3)_2$ solutions and $Na_2S_2O_3 \cdot 5H_2O$ (References 21 and 28, respectively)] the T_0 values derived from fluidity data agree to within a few degrees with those obtained from conductances, and both are $\sim 10\%$ lower than the calorimetric glass transition temperatures T_g. Furthermore, it has proved possible to describe the variation of conductance with pressure by a simple pressure dependence of T_0 alone.[29]

The concentration variation of transport properties is equally vast. Thus at 298 K both Λ (see Figure 1) and fluidity φ decrease by about four powers of ten when the $Ca(NO_3)_2$ concentration is raised to 25 mol %. A modified form of equation (11) can accommodate this great range. Angell and Bressel[21] found empirically that the parameter T_0, obtained from the temperature variations, increased linearly with increasing mole fraction x of the salt. This may be expressed by

$$T_0 = T_0^{\circ} + Qx \tag{13}$$

or

$$T - T_0 = Q(x_0 - x) \tag{14}$$

where Q is a constant and x_0 is defined as the mole fraction at which T_0 equals the temperature T of the solution. Substitution into (11) gives

$$W = A \exp\left[\frac{-C'T}{Q(x_0 - x)} + C'\right] = A' \exp\left[-\frac{C'T}{Q(x_0 - x)}\right]$$

$$= A' \exp[-B'/(x_0 - x)] \qquad (T = \text{const}) \tag{15}$$

where A' is constant and so is B' at any given temperature. This simple equation fits the concentration dependence of Λ and φ very well from about 5 mol % up to the highest concentrations studied. Equation (15) also explains why the specific conductivity κ passes through a maximum.[21] The actual value of x corresponding to κ_{max} predicted by the equation is somewhat larger than the experimental value because the maximum falls in the low concentration region where equation (15) is no longer valid. To sum up, equations (11) and (15) have proved remarkably successful in accounting for the way in which conductances and viscosities of a few systems vary over several orders of magnitude with changing T, P, and c. There is now a considerable need to test these equations further. Reliable data are required not only for other chemical systems but also for the complementary properties of transference number and diffusion coefficient over wide ranges of temperature and composition.

A different transition state treatment of viscosities has recently been published by Goldsack and Franchetto.[30, 31] They begin with the basic equation (6) as applied specifically to viscous flow:[17]

$$\eta = \frac{1}{\varphi} = \frac{hL}{V} \exp\left[\frac{\Delta G^{\ddagger}}{RT}\right] \tag{16}$$

where h is Planck's constant, and $\Delta G^{\ddagger}$ is the molar Gibbs free energy of activation most of which is required to form the holes for the particle to flow into.[17] In the original treatise by Glasstone, Laidler, and Eyring[17] V is the molar volume of the molecules in the liquid, whereas Goldsack and Franchetto, quoting this book, state that V is the molar volume of the holes in the liquid.[30] This does not affect the subsequent algebraic derivation in which the main premise is that V and $\Delta G^{\ddagger}$ be additive properties. Thus

$$V = x_0 V_0 + x_+ V_+ + x_- V_- \tag{17}$$

$$\Delta G^{\ddagger} = x_0 \Delta G_0^{\ddagger} + x_+ \Delta G_+^{\ddagger} + x_- \Delta G_-^{\ddagger} \tag{18}$$

where x is mole fraction and subscripts 0, $+$, and $-$ denote solvent, cation, and anion, respectively. All the V_i and $\Delta G_i^{\ddagger}$ terms are implicitly assumed to be independent of concentration. For a $1:1$ electrolyte $x_+ = x_- = x$, $x_0 = 1 - x_+ - x_- = 1 - 2x$, and hence

$$V = V_0\left[1 + x\left(\frac{V_+ + V_-}{V_0} - 2\right)\right] = V_0[1 + xv] \tag{19}$$

$$\Delta G^{\ddagger} = \Delta G_0^{\ddagger} + x(\Delta G_+^{\ddagger} + \Delta G_-^{\ddagger} - 2\,\Delta G_0^{\ddagger})$$

$$= \Delta G_0^{\ddagger} + xRTE \tag{20}$$

where v and E serve as convenient abbreviations. Insertion of (19) and (20) into equation (16), and division by equation (16) in the form appropriate for the pure solvent ($\eta = \eta_0$, $V = V_0$, $\Delta G^{\ddagger} = \Delta G_0^{\ddagger}$), finally yields

$$\eta = \frac{\eta_0 \exp[xE]}{1 + xv} \tag{21}$$

Similar two-parameter equations emerge for strong electrolytes of different valence types.[30] Goldsack and Franchetto demonstrate that several other empirical viscosity equations proposed in the literature become special cases of equation (21). In particular, for dilute solutions the Jones–Dole B_η coefficient in equation (1) is approximately equal to $(E - v)/55.5$.

Figure 2 illustrates the goodness of fit obtained by equation (21) irrespective of whether the viscosity rises or falls with increasing concentration.

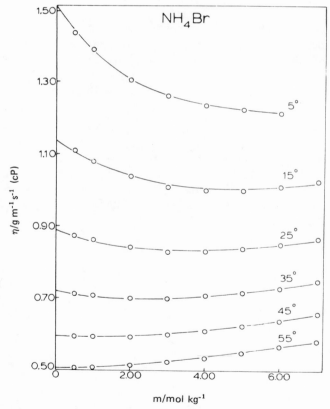

FIGURE 2. Variation of the viscosity of NH_4Br solutions with molality at several temperatures, from Goldsack and Franchetto.[31] The solid lines have been calculated by equation (21) with appropriate E and v parameters.

The equation fares rather less well by the second criterion of acceptability: the reasonableness of the parameters required for fitting. The ratios $(V_+ + V_-)/V_0$ are of the order of 10 for the alkali halides, which cannot be explained[32] on Eyring's interpretation of V_i as partial molar volumes[18] and is not easy to understand even if V represents molar hole volumes. Neither the $(V_+ + V_-)/V_0$ nor the E parameters are additive; they do not exhibit any systematic trend from salt to salt[30] or even vary smoothly over a temperature range.[31] Salt trends are shown, however, by the differences $(E - v)$ which are proportional to the additive B_n parameter.

To end this section, we cannot but express surprise that two equations of such disparate form as (15) and (21) appear capable of representing viscosities over wide ranges of concentration.

2.3. Representation by Phenomenological Coefficients

The transport behavior of electrolytes has traditionally been described by the properties of conductance, transference, and diffusion. An alternative method of representation is provided by the thermodynamics of irreversible processes.[25, 33, 34] For an electrolyte $M_{\nu_1}^{z_1} Y_{\nu_2}^{z_2}$ in a neutral solvent at constant temperature the basic linear equations are

$$J_1 = l_{11} X_1 + l_{12} X_2 \tag{22a}$$

$$J_2 = l_{21} X_1 + l_{22} X_2 \tag{22b}$$

where J_i are the flows of the ions (or more correctly, of the ion constituents) relative to the solvent and X_i the driving forces in terms of the electrochemical potential gradients of the ions. J and X are connected by the linear "Onsager" phenomenological coefficients, or ionic transport

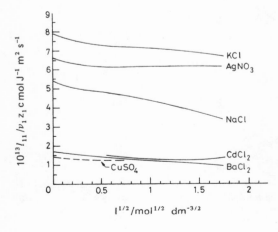

FIGURE 3. Variation of cation intrinsic mobilities ($l_{11}/\nu_1 z_1 c$) with the square root of the ionic strength I at 25°C. Above $I = 0.5$, the curve for $CuSO_4$ is indistinguishable from that for $BaCl_2$. From the review by Miller.[33]

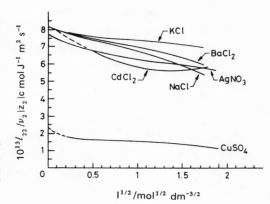

FIGURE 4. Variation of anion intrinsic mobilities $(l_{22}/v_2|z_2|c)$ with the square root of the ionic strength I at 25°C. From the review by Miller.[33]

coefficients,[35] l_{ij}. Subscript 1 conventionally refers to the cation constituent and subscript 2 to the anion one. These l_{ij} coefficients are of general and fundamental importance and provide a new and different insight into ionic transport properties. To obtain numerical values of the four l_{ij} coefficients at any given molarity c we require four independent experimental quantities. Those normally chosen are the molar conductance, the transference number t^h determined by the Hittorf or moving boundary method and that (t^e) calculated from the emfs of cells with transference,[36] and the electrolyte diffusion coefficient. (The trace diffusion coefficient has been treated by Paterson.[34, 88]) Miller[25] has shown that, on a solvent frame of reference,

$$\Lambda = \frac{F^2}{c}[z_1^2 l_{11} + z_1 z_2(l_{12} + l_{21}) + z_2^2 l_{22}] = \frac{F^2}{c}\alpha \tag{23}$$

$$t_1^h = (z_1^2 l_{11} + z_1 z_2 l_{12})/\alpha \tag{24}$$

$$t_1^e = (z_1^2 l_{11} + z_1 z_2 l_{21})/\alpha \tag{25}$$

$$D = \frac{RTv(1 + cd\ln y_\pm/dc)}{c}\frac{z_1 z_2}{v_1 v_2}\frac{(l_{11} l_{22} - l_{12} l_{21})}{\alpha} \tag{26}$$

where α is defined by equation (23), F is Faraday's constant, $y_\pm$ the mean molar activity coefficient, z the algebraic charge number (so that $v_1 z_1 = -v_2 z_2$), and $v = v_1 + v_2$. Equations (23)–(26) can now be solved to yield expressions for the l_{ij} coefficients:[25]

$$\frac{l_{ij}}{c} = \frac{t_i^h t_j^e \Lambda}{F^2 z_i z_j} + \frac{v_i v_j D}{RTv(1 + cd\ln y_\pm/dc)} \tag{27}$$

Values of l_{11}, l_{22}, and l_{12} derived from equation (27) for a number of electrolytes are plotted in Figures 3–5.

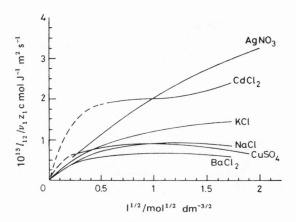

FIGURE 5. Variation of interaction mobilities $l_{12}/v_1 z_1 c$ with the square root of the ionic strength I at 25°C. From the review by Miller.[33]

The four l_{ij} coefficients can be reduced to three if the Onsager reciprocal relations (ORR) hold:

$$l_{ij} = l_{ji} \tag{28}$$

The validity of these relations is in fact of crucial importance to the whole subject of irreversible thermodynamics. The most sensitive test of (28) is offered by transference measurements in concentrated electrolyte solutions. From equations (24) and (25),

$$\frac{t_1^h - t_1^e}{t_1^e} = \left(\frac{l_{12}}{l_{21}} - 1\right) \frac{z_2(l_{21}/c)}{z_1(l_{11}/c) + z_2(l_{21}/c)} \tag{29}$$

Any deviations between l_{12}/l_{21} and unity will clearly be reflected in corresponding differences between transference numbers obtained by the Hittorf and emf methods, $(t_1^h - t_1^e)$ being larger the greater (l_{21}/c) and the smaller (l_{11}/c). Inspection of Figures 3–5 indicates that these conditions are best fulfilled by working with highly concentrated solutions and by choosing somewhat associated electrolytes. Within the last decade three definitive studies have been carried out along these lines. Miller and Pikal's results with AgNO$_3$ solutions up to 11 m,[37, 35] McQuillan's experiments with CdCl$_2$ up to 6 m (calculations up to 1 m),[38] and Agnew and Paterson's work[39] with ZnCl$_2$ from 2–3.4 M have shown conclusively that the ORR are valid within the experimental uncertainties of $\sim 1\%$–2%. Onsager's original postulate has thus been vindicated by transport experiments in concentrated salt solutions.

The $(l_{ii}/v_i|z_i|c)$ parameters plotted in Figures 3 and 4 are often called intrinsic mobilities, and they represent essentially interactions of the ion i with the water and with itself.[33] It is therefore reasonable that the intrinsic mobilities of most ions should vary only gradually with concentration. Only for H^+ does (l_{11}/c) decline steeply with increasing c,[25] for the proton-jumping ability is markedly impaired when all the water molecules are tightly held in hydration shells. The general cation trend[25, 33] $La^{3+} < Ba^{2+} < Li^+ < Na^+ < K^+ < H^+$ (partly shown in Figure 3) seems sensible: ions that interact strongly with the solvent by hydration will be bulky and possess lower intrinsic mobilities.[25] As regards anions, Figure 4 shows that the curves for the chloride ion constituent of various salts tend to diverge at higher concentrations. This arises indirectly through interactions of the solvent with the different cations and, in the case of $CdCl_2$, through the interactions resulting from ion association. Quite different behavior is exhibited in Figure 5 by the $(l_{12}/v_1 z_1 c)$ parameters. These have been termed interaction mobilities and are primarily a measure of the degree of coupling between the motions of cation and anion. It follows that (l_{12}/c) should be zero at infinitesimal concentration where the ions are an infinite distance apart; as $c \to 0$, the right-hand side of equation (27) becomes zero because the two terms in it are then of equal magnitude and opposite sign. (Equating these two terms leads to the Nernst–Hartley diffusion equation.[25]) With rising concentration the degree of cation–anion interaction must increase and Figure 5 shows that the interaction mobilities follow suit. This increase is specific, especially at the higher concentrations.[25] Cations and anions that associate by coulombic or covalent forces to form ion-pairs or complexes should possess larger interaction mobilities than ions of the same charge type that do not associate. Figure 5 demonstrates that this is true for $AgNO_3$ vis-à-vis the alkali halides and for $CdCl_2$ as compared with $BaCl_2$. The cases of $AgNO_3$ and of the highly complexed CdI_2 (References 35 and 34, respectively) have recently been discussed in some detail.

Two alternative methods of representation exist. One is the inverse description using friction coefficients which is gradually being displaced by the more popular l_{ij} formalism. The other and latest method has been put forward by Hertz.[40, 41] It is based on velocity correlation coefficients f defined in terms of time integrals over velocity correlation functions of the type[40, 42]

$$f_{ij} = \frac{v_j cLV}{3} \int_0^\infty \langle v_i(0) \cdot v_j(t) \rangle \, dt \tag{30}$$

where V is the volume, v the velocity of the subscripted ion, and where $\langle \ \rangle$ stands for the ensemble average. Use of linear response theory then enables the conventional transport properties to be expressed as functions of the velocity correlation coefficients. The molar conductance Λ and the transference numbers can be written in the symmetrical form[42]

$$\Lambda = \frac{v_1 z_1^2 F^2}{RT} \left[D_1 + f_{11} - f_{21} + \frac{|z_2|}{z_1}(D_2 + f_{22} - f_{12}) \right] \tag{31}$$

$$t_1 = 1 - t_2 = \frac{v_1 z_1^2 F^2}{RT\Lambda}(D_1 + f_{11} - f_{21}) \tag{32}$$

where $f_{12}|z_2| = f_{21} z_1$ (Reference 42b) and where D_1 and D_2 are the self(tracer)-diffusion coefficients of the cation and anion, respectively. The equation for the salt diffusion coefficient is more complicated. It suffices here to say that by a suitable rearrangement f_{11}, f_{22}, and f_{12} can be expressed in terms of, and calculated from, the measured values of Λ, t, D, D_1, and D_2. Woolf and Harris[43] have pointed out that the f coefficients were, originally defined on a mass-fixed reference frame whereas experimental transference numbers are based on a solvent-fixed reference frame, and they corrected for this by an appropriate conversion formula. Application to several 1:1 and 1:2 chlorides shows[41–43] that the anion–cation velocity correlation coefficients f_{21} all rise from zero (at $c = 0$) to a maximum of $1–2 \times 10^{-10}$ m^2 sec^{-1} (at $c < 0.5$ M) and then decline, sometimes to negative values. The initial rise has been attributed to ion association (which seems strange for salts like the alkali halides), the later fall to mutual averaging of the electrostatic forces around the ions.[41] The cation–cation and anion–anion coefficients f_{11} and f_{22}, on the other hand, all decline from zero to negative values and in some cases pass through a minimum. The negative signs of f_{ii} indicate that it takes a comparatively long time for an ion to diffuse into the disturbed surroundings recently vacated by another ion of the same kind.[41] This microdynamic theory is still being developed and has not yet been applied to solutions of concentration higher than 4 M.

3. Selection of Experimental Data

In certain respects it is easier to work with concentrated than with very dilute electrolyte solutions. To begin with, weighing errors in preparing solutions are negligible. Viscosities are often far higher and can consequently be determined more accurately. In conductance experiments one is relieved of the sometimes awkward solvent correction, and precise results can be obtained with cells of large cell constant. In transference measurements, worry about the solvent correction is replaced in moving boundary work by worry about the big volume correction. This is the main reason why indirect moving boundary, Hittorf, and emf determinations have predominated for high concentrations. The viscous nature of concentrated solutions leads to correspondingly slow diffusion, and in measuring diffusion coefficients care must be taken to avoid convective mixing over the long periods required.

Good experimental data are now quite extensive but widely scattered. Tables 2–5 have therefore been compiled to provide an up-to-date summary of data published since the appearance of Landolt–Börnstein[44] and other cited collections. Excluded from Tables 2–5 have been all electrolytes that hydrolyze or are liable to decompose and all nonbinary electrolytes

TABLE 2. Recent Viscosity Measurements in Concentrated Aqueous Solutions

Earlier Data May Be Found in the Compilations of Landolt–Börnstein[45] and of Stokes and Mills[2]

Solute	Max. concentration[a]	Temp./°C	Ref.
HCl	8.5 m	25	46
LiClO$_3$	100%	25, 131.8	47
NaCl	5.8 m	25–50	48
NaCl	5.0 M	25	49
NaBr	7.0 m	5–55	30, 31
NaI	3.0 m	5–45	30, 31
NaSCN, NaClO$_4$	9.0 M	25	49
NaNO$_3$	7.0 M	25	49
NaNO$_3$	7.8 M	25	50
KF	7.0 m	5–55	30, 31
KCl	4.6 m	20–50	48
KBr	5.0 m	5–55	30, 31
KI	3.0 m	5–45	30, 31
NH$_4$Cl	7.0 m	10–60	31
NH$_4$Br	7.0 m	5–55	30, 31
(NH$_4$)$_2$SO$_4$	4.0 m	25	46
R$_4$NBr (R = Me, Et, Pr, Bu)	4 m	15–35	51
AgClO$_4$	5.6 M	25	52
CuSO$_4$	1.4 M	25	53
NiSO$_4\cdot$6H$_2$O	50%–60%	20–60	54
AlBr$_3$	1.3 M	25	55
InBr$_3$	1.6 M	25	55
In(NO$_3$)$_3$	2.6 M	25	55
MCl$_3$ (M = Pr, Eu, Gd, Tm, Yb, Lu)	~ 4 m	25	56
M(ClO$_4$)$_3$ (M = La, Pr, Nd, Sm, Eu, Gd, Tb, Dy, Ho, Er, Tm, Yb, Lu)	~ 4.6 m	25	57
M(NO$_3$)$_3$ (M = as above, without Eu, Tm)	4.3–6.8 m	25	58

[a] $M = $ mol dm^{-3}, $m = $ mol kg^{-1}, $\% = $ wt %.

TABLE 3. Recent   Conductance   Measurements   in   Concentrated   Aqueous
Solutions
Earlier Data May Be Found in the Compilations of Landolt–Börnstein [59] *and of Robinson and Stokes* [60]

Solute	Max. concentration	Temp./°C	Ref.
LiCl	12 M	25	61, 62
LiCl	12 M	-122–(-102)	27
LiCl	9.0 m	23–60	63
LiBr	9.9 M	25	61
LiI	7.6 M	25	61, 62
$LiNO_3$	6.0 M	25	62
$LiClO_3$	100%	131.8	47
$LiClO_4$	4.0 M	25	62
Li_2SO_4	2.9 M	25–75	64, 65
NaCl	5.0 M	25	49, 62, 63
NaBr	6.1 M	25	61, 62
NaBr	8.0 m	0–55	63
NaI	7.5 M	25	61, 62
$NaNO_3$	7.8 M	25	49, 50, 62
$NaClO_4$, NaSCN	9.0 M	25	49, 62
Na_2SO_4	3.1 M	25–75	64
$Na_2S_2O_8$	2.4 M	15–25	66
KF	5.0 M	25	62
KCl	4.0 M	25	62, 63
KBr	4.3 M	25	61, 62
KBr	5.0 m	0–55	63
KI	5.8 M	25	61, 62
KI	3.0 M	0–45	63
K_2SO_4	1.1 M	25–75	64
RbCl	5.6 M	25	61, 67
RbCl	7.0 m	23–60	63
RbBr, RbI	4.8–5.0 M	25	61
Rb_2SO_4	2.4 M	25–85	64
CsCl	7.0 M	25	61
CsCl	9.0 m	23–60	63
CsBr	4.1 M	25	61
CsI	2.4 M	25	61
Cs_2SO_4	3.9 M	25–75	64
NH_4Cl	5.2 M	25	61, 62, 63
NH_4Cl	7.0 m	23–60	63
NH_4Br	5.5 M	25	61
NH_4Br	7.0 m	0–55	63
NH_4I	6.8 M	25	61
$(NH_4)_2S_2O_8$	2.6 M	15–25	66
$(CH_3)_xNH_{4-x}Cl$ $(x = 1$–$4)$	5.6–8.7 M	25	61
$AgNO_3$	8.7 M	25	52
$AgClO_4$	5.6 M	25	52

TABLE 3 (continued)

Solute	Max. concentration	Temp./°C	Ref.
$MgCl_2$	2.5 M	25	65
$Mg(NO_3)_2$, $Mg(ClO_4)_2$	3.0 M	25	65
$CaCl_2$, $Ca(NO_3)_2$	4.0 M	25	65
$Ca(NO_3)_2$	20 m	-20–$60+$	21, 29, 73
$BaCl_2$	1.2 M	25	65
$NiCl_2$	3.9 M	25	68
$CdCl_2$	6.0 m	25	38
$ZnCl_2$	3.5 M	25	39
$Zn(ClO_4)_2$	3.0 M	25	69
$AlBr_3$	1.9 M	25	55
$Al(NO_3)_3$, $LaCl_3$	1.3, 1.0 M	25	65
$InBr_3$	1.6 M	25	55
$In(NO_3)_3$	2.6 M	25	55
MCl_3 (M = La, Pr, Nd, Sm, Eu, Gd, Tb, Dy, Ho, Er, Tm, Yb, Lu)	~ 4 m	25	70
$M(NO_3)_3$ (as for MCl_3, but no Eu, Tm)	4.3–6.8 m	25	71
$M(ClO_4)_3$ (as for MCl_3, but no Eu, Yb)	~ 4.6 m	25	72

TABLE 4. Recent Transference Number Measurements in Concentrated Aqueous Solutions

Earlier Data May Be Found in the Compilations of Landolt–Börnstein,[74] Kaimakov and Varshavskaya,[75] and Spiro[76]

Solute	Method	Max. concentration	Temp./°C	Ref.
HBr	emf	5.4 m	20–40	77
NaCl	emf	5.0 m	25–55	78, 79
NaOH, KOH	emf	17.0 m	20	80
RbCl	Indirect moving boundary	4.7 m	25	81
CsCl	Indirect moving boundary	3.5 m	25	81
$AgClO_4$	Hittorf	4.2 M	25	52, 82
$CaCl_2$	Tracer migration	5.3 M	25	83
$CuSO_4$	Hittorf	1.4 m	25	84
$NiCl_2$	Indirect moving boundary, Hittorf	3.9 M	25	68
$CdCl_2$	Hittorf, emf	6.0 M	25	38
$ZnCl_2$	Hittorf, emf	4.4 M	25	39
$Zn(ClO_4)_2$	Hittorf	2.4 M	25	69

TABLE 5. Recent Diffusion Coefficient Measurements in Concentrated Aqueous Solutions

Earlier Data May Be Found in the Compilations of Landolt–Börnstein[85]

Solute	D of species[a]	Max. concentration	Temp./°C	Ref.
HNO_3	Salt	9.2 M	25	86
LiCl	Li^+	7.0 M	25?	87
LiCl	Cl^-	2.0 M	25?	87
LiCl	H_2O	2.5 M	25	88
$LiClO_3$	Salt	14.6 M	25	89
NaCl	Salt	4.0 M	25	49
NaCl	Na^+, H_2O	2.2, 2.9 M	25	88
NaCl	Na^+, Cl^-	5.0 M	10	90
$NaClO_3$	Salt	6.3 M	25	89
$NaClO_4$, NaSCN	Salt	9.0 M	25	49
$NaNO_3$	Salt	7.0 M	25	49
KCl	K^+	4.0 M	25	91
KCl	Cl^-	4.0 M	25	87, 92
KCl	H_2O	4.0 M	25	88, 92, 93
KCl	K^+, Cl^-, H_2O	3.0 M	10	92
KBr	H_2O	3.5 M	25	93
KI	K^+, I^-	6.0 M	25	91
KI	H_2O	6.0 M	25	92, 93
KI	K^+, I^-	5.0 M	10	91
KI	H_2O	5.0 M	10	92
RbCl	Salt	2.9 M	25	67
RbCl	Cl^-, H_2O	2.7 M	25	88
RbCl	H_2O	5.6 M	25	93
RbBr, RbI	H_2O	3.8, 4.6 M	25	93
CsCl	Cs^+, Cl^-, H_2O	2.4–2.8 M	25	88
CsCl	Cs^+, Cl^-, H_2O	4.0 M	10, 25	92
CsCl	H_2O	5.1 M	25	93
CsBr	H_2O	3.8 M	25	93
CsI	H_2O	1.7 M	25	93
$MgCl_2$	Mg^{2+}, Cl^-	3.9 M	10, 25	90
$CaCl_2$	Ca^{2+}, Cl^-	4.0 M	25	42
$BaCl_2$	Ba^{2+}, Cl^-, H_2O	1.5 M	25	42
$CuSO_4$	Salt	1.4 M	25	53
$NiCl_2$	Ni^{2+}, Cl^-, H_2O	4.0 M	25	68
$ZnCl_2$	Salt	3.9 M	25	39
$Zn(ClO_4)_2$	Salt	3.2 M	25	69
$ZnSO_4$	Salt	3.2 M	25	94

[a] "Salt" refers to the mutual diffusion coefficient of the electrolyte, even when the latter is an acid. The other symbols refer to tracer (self-) diffusion coefficients.

such as Na_2HPO_4; excluded also have been references in which data have been displayed graphically and not tabulated, in which only one or two concentrations have been investigated, or in which the concentrations do not significantly exceed 1 *M*. In a few cases, data of low accuracy have been included if no better high concentration data exist for the electrolyte in question. In addition to these tables, useful references may be found to the transport properties of alkali and some alkaline earth chlorides and certain other electrolytes like $AgNO_3$ in the recent phenomenological analyses of Miller[25, 33] and of Hertz.[41, 42]

References

1. G. Jones and M. Dole, *J. Am. Chem. Soc.* **51**, 2950 (1929).
2. R. H. Stokes and R. Mills, *Viscosity of Electrolytes and Related Properties*, Pergamon, Oxford (1965).
3. J.-C. Justice, *J. Solut. Chem.*, **7**, 859 (1978).
4. B. F. Wishaw and R. H. Stokes, *J. Am. Chem. Soc.* **76**, 2065 (1954).
5. R. A. Robinson and R. H. Stokes, *Electrolyte Solutions*, 2nd ed., Butterworths, London (1959).
6. R. H. Stokes, in *The Structure of Electrolytic Solutions*, Ed. W. J. Hamer, Wiley, New York (1959), Chap. 20.
7. J. Salvinien, B. Brun, and J. Molenat, *J. Chim. Phys. Suppl.*, October, 19 (1969).
8. D. P. Sidebottom and M. Spiro, *J. Chem. Soc. Faraday Trans. 1* **69**, 1287 (1973).
9. L. G. Longsworth, *J. Am. Chem. Soc.* **54**, 2741 (1932).
10. R. A. Robinson and R. H. Stokes, *Electrolyte Solutions*, 2nd ed., Butterworths, London (1959), Chap. 11, Table 11.7.
11. R. H. Stokes and R. A. Robinson, *J. Am. Chem. Soc.* **70**, 1870 (1948).
12. J. Braunstein, in *Ionic Interactions: From Dilute Solutions to Fused Salts*, Vol. I, Ed. S. Petrucci, Academic, New York (1971), Chap. 4.
13. R. H. Stokes and R. A. Robinson, *J. Solut. Chem.* **2**, 173 (1973).
14. M. H. Lietzke, R. W. Stoughton, and R. M. Fuoss, *Proc. Natl. Acad. Sci. USA* **59**, 39 (1968).
15. I. Ruff, *J. Chem. Soc. Faraday Trans. II* **73**, 1858 (1977).
16. C. A. Angell and K. J. Rao, *J. Chem. Phys.* **57**, 470 (1972).
17. S. Glasstone, K. J. Laidler, and H. Eyring, *The Theory of Rate Processes*, McGraw-Hill, New York (1941).
18. C. T. Moynihan, in *Ionic Interactions: From Dilute Solutions to Fused Salts*, Vol. I, Ed. S. Petrucci, Academic, New York (1971), Chap. 5.
19. G. Adam and J. H. Gibbs, *J. Chem. Phys.* **43**, 139 (1965).
20. C. A. Angell and E. J. Sare, *J. Chem. Phys.* **52**, 1058 (1970).
21. C. A. Angell and R. D. Bressel, *J. Phys. Chem.* **76**, 3244 (1972) (data in microfilm ed.).
22. C. A. Angell, *J. Phys. Chem.* **70**, 2793 (1966).
23. C. A. Angell, *J. Phys. Chem.* **68**, 218, 1917 (1964).
24. C. A. Angell, *J. Phys. Chem.* **69**, 2137 (1965).
25. D. G. Miller, *J. Phys. Chem.* **70**, 2639 (1966).
26. E. Williams and C. A. Angell, *J. Phys. Chem.* **81**, 232 (1977).
27. C. T. Moynihan, R. D. Bressel, and C. A. Angell, *J. Chem. Phys.* **55**, 4414 (1971).
28. C. T. Moynihan, *J. Phys. Chem.* **70**, 3399 (1966).

29. C. A. Angell, L. J. Pollard, and W. Strauss, *J. Solut. Chem.*, **1**, 517 (1972).
30. D. E. Goldsack and R. Franchetto, *Can. J. Chem.* **55**, 1062 (1977).
31. D. E. Goldsack and R. Franchetto, *Can. J. Chem.* **56**, 1442 (1978).
32. F. J. Millero, *Chem. Rev.* **71**, 147 (1971).
33. D. G. Miller, *Faraday Discuss. Chem. Soc.* **64**, 295 (1977).
34. R. Paterson, *Faraday Discuss. Chem. Soc.* **64**, 304 (1977).
35. D. G. Miller and M. J. Pikal, *J. Solut. Chem.* **1**, 111 (1972).
36. M. Spiro, in *Physical Methods of Chemistry*, Vol. I, Eds. A. Weissberger and B. W. Rossiter, Wiley Interscience, New York (1971), Part IIA, Chap. 4.
37. M. J. Pikal and D. G. Miller, *J. Phys. Chem.* **74**, 1337 (1970).
38. A. J. McQuillan, *J. Chem. Soc. Faraday Trans. 1* **70**, 1558 (1974).
39. A. Agnew and R. Paterson, *J. Chem. Soc. Faraday Trans. 1* **74**, 2896 (1978).
40. H. G. Hertz, *Ber. Bunsenges.* **81**, 656 (1977).
41. H. G. Hertz, K. R. Harris, R. Mills, and L. A. Woolf, *Ber. Bunsenges.* **81**, 664 (1977).
42. (a) H. G. Hertz and R. Mills, *J. Phys. Chem.* **82**, 952 (1978); (b) H. G. Hertz, private communication (1979).
43. L. A. Woolf and K. R. Harris, *J. Chem. Soc. Faraday Trans. 1* **74**, 933 (1978).
44. Landolt–Börnstein, *Zahlenwerte und Funktionen*, 6th ed., Vol. II, Springer, Berlin (1959–1971).
45. Landolt–Börnstein, *Zahlenwerte und Funktionen*, 6th ed., Vol. II, Springer, Berlin (1969), Part 5a, p. 305.
46. D. E. Goldsack and A. A. Franchetto, *Electrochim. Acta* **22**, 1287 (1977).
47. A. N. Campbell and W. G. Paterson, *Can. J. Chem.* **36**, 1004 (1958); A. N. Campbell and D. F. Williams, *Can. J. Chem.* **42**, 1778, 1984 (1964).
48. F. A. Gonçalves and J. Kestin, *Ber. Bunsenges.* **81**, 1156 (1977).
49. G. J. Janz, B. G. Oliver, G. R. Lakshminarayanan, and G. E. Mayer, *J. Phys. Chem.* **74**, 1285 (1970).
50. E. M. Kartzmark, *Can. J. Chem.* **50**, 2845 (1972).
51. D. Eagland and G. Pilling, *J. Phys. Chem.* **76**, 1902 (1972) (microfilm ed.).
52. A. N. Campbell and K. P. Singh, *Can. J. Chem.* **37**, 1959 (1959).
53. L. A. Woolf and A. W. Hoveling, *J. Phys. Chem.* **74**, 2406 (1970).
54. V. R. Phillips, *J. Chem. Eng. Data* **17**, 357 (1972).
55. A. N. Campbell, *Can. J. Chem.* **54**, 3732 (1976).
56. F. H. Spedding, D. L. Witte, L. E. Shiers, and J. A. Rard, *J. Chem. Eng. Data* **19**, 369 (1974).
57. F. H. Spedding, L. E. Shiers, and J. A. Rard, *J. Chem. Eng. Data* **20**, 66 (1975).
58. F. H. Spedding, L. E. Shiers, and J. A. Rard, *J. Chem. Eng. Data* **20**, 88 (1975).
59. Landolt–Börnstein, *Zahlenwerte und Funktionen*, 6th ed., Vol. II, Springer, Berlin (1960), Part 7II, p. 33.
60. R. A. Robinson and R. H. Stokes, *Electrolyte Solutions*, 2nd ed., Butterworths, London (1959), Appendix 6.3.
61. J. Molenat, *J. Chim. Phys.* **66**, 825 (1969).
62. M. Postler, *Collect. Czech. Chem. Commun.* **35**, 535 (1970).
63. D. E. Goldsack, R. Franchetto, and A. A. Franchetto, *Can. J. Chem.* **54**, 2953 (1976).
64. V. M. Valyashko and A. A. Ivanov, *Russ. J. Inorg. Chem.* **19**, 1628 (1974).
65. M. Postler, *Collect. Czech. Chem. Commun.*, **35**, 2244 (1970).
66. J. Balej and A. Kitzingerova, *Collect. Czech. Chem. Commun.*, **39**, 49 (1974).
67. S. K. Jalota and R. Paterson, *J. Chem. Soc. Faraday Trans. 1* **69**, 1510 (1973).
68. R. H. Stokes, S. Phang, and R. Mills, *J. Solut. Chem.*, **8**, 489 (1979).
69. A. Agnew and R. Paterson, *J. Chem. Soc. Faraday Trans. 1* **74**, 2885 (1978).
70. F. H. Spedding, J. A. Rard, and V. W. Saeger, *J. Chem. Eng. Data* **19**, 373 (1974) (microfilm ed.).

71. J. A. Rard and F. H. Spedding, *J. Phys. Chem.* **79**, 257 (1975) (microfilm ed.).
72. F. H. Spedding and J. A. Rard, *J. Phys. Chem.* **78**, 1435 (1974) (microfilm ed.).
73. C. A. Angell, *J. Phys. Chem.* **70**, 3988 (1966).
74. Landolt–Börnstein, *Zahenwerte und Funktioner*, 6th ed., Vol. II, Springer, Berlin (1960), Part 7II, p. 237.
75. E. A. Kaimakov and N. L. Varshavskaya, *Russ. Chem. Revs.* **35**, 89 (1966).
76. M. Spiro, in *Physical Methods of Chemistry*, Vol. I, Eds. A. Weissberger and B. W. Rossiter, Wiley Interscience, New York (1971), Part IIA, Appendix, Table 4A.1.
77. J. W. Augustynski, G. Faita, and T. Mussini, *J. Chem. Eng. Data* **12**, 369 (1967).
78. T. Mussini and A. Pagella, *Chim. Ind.* (*Milan*) **52**, 1187 (1970).
79. L. J. M. Smits and E. M. Duyvis, *J. Phys. Chem.* **70**, 2747 (1966), and references therein.
80. S. Lengyel, J. Giber, G. Beke, and A. Vértes, *Acta Chim. Acad. Sci. Hung.* **39**, 357 (1963).
81. J. Tamás, O. Kaposi, and P. Scheiber, *Acta Chim. Acad. Sci. Hung.* **48**, 309 (1966).
82. W. V. Childs and E. S. Amis, *J. Inorg. Nucl. Chem.* **16**, 114 (1960).
83. M. Fromon, F. Lantelme, and M. Chemla, *Bull. Soc. Chim. France*, 3388 (1970).
84. M. J. Pikal and D. G. Miller, *J. Chem. Eng. Data* **16**, 226 (1971).
85. Landolt–Börnstein, *Zahlenwerte und Funktionen*, 6th ed., Vol. II, Springer, Berlin (1969), Part 5a, p. 612.
86. K. Nisancioglu and J. Newman, *Am. Inst. Chem. Eng. J.* **19**, 797 (1973).
87. P. Turq, F. Lantelme, Y. Roumegous, and M. Chemla, *J. Chim. Phys.* **68**, 527 (1971).
88. J. Anderson and R. Paterson, *J. Chem. Soc. Faraday Trans. 1* **71**, 1335 (1975).
89. A. N. Campbell and B. G. Oliver, *Can. J. Chem.* **47**, 2681 (1969).
90. K. R. Harris, H. G. Hertz, and R. Mills, *J. Chim. Phys.* **75**, 391 (1978).
91. H. G. Hertz, M. Holz, and R. Mills, *J. Chim. Phys.*, **71**, 1355 (1974).
92. H. G. Hertz and R. Mills, *J. Chim. Phys.* **73**, 499 (1976).
93. B. Brun, M. Servent, and J. Salvinien, *C. R. Acad. Sci. Ser. C* **269**, 1 (1969).
94. J. G. Albright and D. G. Miller, *J. Solut. Chem.* **4**, 809 (1975).

6

A Direct Correspondence between Spectroscopic Measurements and Electrochemical Data and Theories

A Linear Relationship up to 10 mol dm^{-3}, Precise Madelung Constants from Spectral Shift Data, and Correlations with Thermodynamic Parameters

Trevor R. Griffiths and Ranmuthu H. Wijayanayake

1. Introduction

Spectroscopic measurements of aqueous electrolyte solutions have been made for nearly 50 years, and electrochemical measurements for very much longer. Although some relationships between infrared measurements of vibrational bands and electrochemical data have at times been noted, no relationships have, to our knowledge, so far been reported for charge-transfer-to-solvent (CTTS) spectra. Further, all the previous relationships have not sought to interrelate the theoretical understanding of the nature of the transition with that of electrochemical measurements. In this chapter we report and discuss such a relationship, and show that an all-embracing

Trevor R. Griffiths and Ranmuthu H. Wijayanayake • Department of Inorganic and Structural Chemistry, The University, Leeds LS2 9JT, England.   R. H. Wijayanayake's present address: Department of Chemistry, University of Sri Lanka, Ruhuna Campus, Matara, Sri Lanka.

account is now available, which extends from infinitely dilute, through saturated salt solutions, to, in principle, the anhydrous salts.

We commence by considering the nature of CTTS spectra, how they are affected on addition of an electrolyte, and then on to the relationship between the resulting salt shifts and mean activity coefficient data from electrochemical measurements as a function of concentration. We conclude with a discussion of the results and implications of the relationship.

2. Charge-Transfer-To-Solvent Spectra

There are several species which, in solution, exhibit CTTS spectra, but the most well known are the free halide ions, and of these the iodide ion has been most studied. It has an ionization potential of about 280 kJ mol^{-1}, and in aqueous solution at 20°C has an ultraviolet absorption spectrum which shows two well-resolved peaks at much higher energy, 531.7 and 620.8 kJ mol^{-1} approximately.[1,2] The energy separation of these band maxima is close to the energy difference between the $^2P_{1/2}$ and $^2P_{3/2}$ states of the iodine atom. Hence it is concluded that the process of light absorption yields an iodine atom and an electron, i.e.,

$$I_{aq}^- \rightarrow (I + e^-)_{aq}$$

Since there are no stable binding atomic levels for halide ions in the gas phase above the ground state, the energy associated with a (for iodide) $5p^6 \rightarrow 5p^56s^1$ transition being greater than the ionization potential, the electron is bound in a stationary state approximately 200–250 kJ mol^{-1} above the gas phase ionization energy, and the excited state is consequently defined by the environment of the ion. It is for this reason that this class of spectra is called charge-transfer-to-solvent spectra, to differentiate it from other electron transfer spectra.

The intense CTTS bands of iodide are smooth and structureless with no indication of vibrational fine structure. These bands obey Beer's law, with the low-energy aqueous iodide band having a molar absorbance of $1.35 \times 10^4 \ M^{-1} \ cm^{-1}$ at 20°C and an oscillator strength of 0.25.[3]

The energy of maximum absorption (E_{max}) of iodide is consequently extremely sensitive to changes in the environment around the iodide ion. Thus E_{max} is greatly affected by added electrolytes (and nonelectrolytes), temperature, pressure, and change in solvent composition, as well as being subject to an isotope effect on replacing water by deuterium oxide. Further, E_{max} of iodide in water as solvent is independent of the cation present at low salt concentrations ($< 10^{-3}$ mol dm^{-3}), indicating no ion pairing.

If we are going to relate this sensitivity to the concentration of added electrolyte we must first consider the theoretical models for CTTS spectra, particularly those which employ thermodynamic cycles.

3. Quantitative Treatments of CTTS Spectra

Quantitative treatments of CTTS spectra have, in the majority of cases, attempted to correlate the variation in E_{max} with the above-mentioned constraints. Since E_{max} measures the difference between electronic excited and ground state energies, a change in E_{max} brought about by a change in the environment of iodide may reflect changes in the energies of the ground state, the excited state, or both. Quantitative analysis of CTTS spectra is made more manageable if the Franck–Condon principle is extended to include the solvent molecules contiguous to the anion. Thus, the arrangement of solvent molecules in the excited state is identical to that in the ground state. Information concerning the ground state ion–solvent interactions can therefore be utilized in formulating the nature of the excited state.

The first important model proposed to explain quantitatively CTTS spectra was the "polaron" model of Platzmann and Franck,[4] which provided the basis for later treatments. The important feature of this model is the treatment of the excited state; the discrete orbital for the electron in the excited state is spherically symmetric with its center coinciding with the center of the cavity containing the iodide ion. The solvent was treated as a dielectric continuum (a common feature of electrochemical treatments). Their essential feature was the recognition that an electron in a polar medium can be trapped by self-polarization of the medium. Within this shallow potential energy well there exist a number of stationary states. The CTTS excited states can be treated in this manner such that the excited states are hydrogenlike. Although this approach was able to predict the E_{max} value of iodide in water at 25°C it did not account for the marked variation of E_{max} with, for example, temperature.

From this approach two major models have been developed[5, 6] which use the same thermodynamic cycle as their starting point, although the final expressions for E_{max} are quite different. This cycle may be represented as

$$I^-(H_2O)_x \xrightarrow{\;E_1\;} (H_2O)_x + I^-_{gas} \xrightarrow{\;E_2\;} (H_2O)_x + I_{gas} + e^-_{gas}$$

$$\downarrow E_{max} \qquad\qquad\qquad\qquad\qquad\qquad\qquad\qquad \downarrow E_3$$

$$(I + e^-)(H_2O)_x \xleftarrow{\hspace{3cm} E_4 \hspace{3cm}} I(H_2O)_x + e^-_{gas}$$

Unlike the Platzmann and Franck model,[4] the cavity of oriented solvent molecules, $(H_2O)_x$, was regarded as remaining intact throughout the cycle, in accordance with the Franck–Condon principle. It follows directly from the cycle that

$$E_{max} = E_1 + E_2 + E_3 + E_4 \tag{1}$$

4. The Stein–Treinin Model

This model[6] is a slightly modified version of that of Platzmann and Franck.[4] The solvent was again regarded as a dielectric continuum with constant dielectric properties, and the electron as polarizing the solvent medium. The iodide ion was viewed as situated in a spherical cavity of radius r_d, and the electron contained in a diffuse excited state. For aqueous solutions at 25°C their final expression was

$$E_{max} = E_x - L_x + 0.77e^2/r_d - 1.58 \tag{2}$$

where E_x is the ionization potential of the iodide ion, L_x, the heat of solvation of the iodine atom, and r_d the radius of the spherical cavity. The extent of penetration of the excited electron into the solvent medium, r_d, was calculated to be about 580 pm at 25°C, and since this model used a diffuse excited state, it is termed the "diffuse" model.

5. The Smith–Symons Model

This model[1, 5, 7] envisages the excited electron as being in an orbital within a potential energy well with infinitely steep sides. E_1 is the energy required to remove an iodide ion from the original cavity, E_2 is the vertical ionization potential of iodide, and E_3 is the energy of solvation of the iodine atom. In contrast to the Stein–Treinin approach, where E_3 was approximated to the solvation energy of the corresponding rare gas atom, the Smith–Symons model assumes E_3 negligible. The term E_4 embraces two factors: (i) the energy required to place the electron into the lowest energy level of the cavity, and (ii) the energy required to place the iodine atom in this cavity. The first contribution to E_4 was derived from the Schrödinger wave equation, and the second was equated to $-E_1$, on the grounds that the electrostatic energies involved in both cases are of the same magnitude. Hence

$$E_4 = \frac{h^2}{8mr_c^2} - E_1 \tag{3}$$

and the final expression for E_{max} is

$$E_{max} = E_x + \frac{h^2}{8mr_c^2} \tag{4}$$

where E_x is again the ionization potential of the iodide ion, h is Planck's constant, m the mass of the electron, and r_c is the radius of the potential energy well.

This model predicts that the penetration of the excited electron into the water medium is 390 pm at 25°C. Thus in contrast to the larger penetration calculated by Stein and Treinin,[6] this model is termed the "confined" model.

6. Comparison of the Two Models

The confined model has a very simple expression for E_{max} for the first (low-energy) CTTS band of iodide, in contrast to the diffuse model. In the former model all changes in E_{max} are attributed to changes in the radius of the cavity, r_c. In the latter, however, changes in E_{max} are attributed to changes in the optical and static dielectric constants, as well as to changes in the radius parameter, r_d.

The two treatments also differ basically in two other respects:

a. The extent of penetration of the excited electron into the solvent medium is greater in the diffuse than in the confined model.

b. Environmental changes in E_{max} are dominated by changes in the energy of the excited state according to the confined model but in the ground state according to the diffuse model.

Although the effects of constraints on E_{max} for iodide have been explained by both these models in terms of consequent changes in the radius parameter of the cavity, it is now generally argued and agreed that the confined model is the better of the two. Among the main reasons advanced, in addition to its simplicity, are the following: (i) Its r_c parameter corresponds to the near inside edge of solvent molecules of the cavity, the positively charged region of these molecules, due to the permanent dipole and induced dipole on these molecules; the r_d parameter value would place the excited electron on the lone pairs of the oxygen atoms of the contiguous water molecules, a region of high electron density, and hence intuitively less likely. (ii) The excited states of the solvated halide ions have been successfully compared with the ground states of F centers in ionic crystals,[8-10] and with that of the solvated electron in liquid ammonia and similar solvents.[10-13]

7. Other Models

Jorgensen has proposed that the orbital of the excited state is predominantly the $(n + 1)s$ orbital of the *halogen* with the radical function somewhat changed by its environment.[14-17] The model has not been developed in detail and comparison with the confined and diffuse models is difficult.

Siano and Metzler[18] have proposed a "configuration-coordinate" model, based on the configuration-coordinate model commonly used in the theory of color centers in alkali halide crystals. Their model does not have a radius parameter. They suggest that their model is useful for analyzing the CTTS spectrum of the iodide ion and that it yields expressions for the higher moments of the spectrum upon which the polaron, diffuse, and confined models are silent. It has not yet been used to evaluate the effects of added electrolytes.

8. The Iodide Ion as a Probe

The integral role that the solvent medium plays in determining the excited state of the CTTS spectrum of iodide, coupled with the extreme sensitivity of E_{max} to environmental changes in the solvent, enables information about the structure of the solvent, ion–solvent, and ion–ion interactions to be obtained. For example, a detectable and reproducible shift in the position of E_{max} for iodide (10^{-5} mol dm^{-3}) is observed on addition of an electrolyte, say, KCl, at a concentration of 2.5×10^{-2} mol dm^{-3}. Such a shift would be around 10 cm^{-1}, around 400 cm^{-1} for LiCl at 4 mol dm^{-3}, and about double that for CsCl. Thus the CTTS spectrum of the iodide ion is a very sensitive probe of the environment of an anion in an electrolyte solution; similar conclusions can therefore also be drawn for cations in solution.

9. The Effects of Added Electrolyte on CTTS Spectra

The effects of added salt on CTTS spectra have been extensively investigated, but generally only explained qualitatively. The basic observations are that in nearly all the systems investigated the addition of a solute shifts the iodide band to higher energies. The most notable feature is the relative insensitivity of the band shape to the added salt. For example, the molar absorbance ε_{max} and bandwidth of iodide ion in water is unaffected on the addition of LiCl up to 12 mol dm^{-3}, further justifying the use of iodide CTTS spectra as a probe, and indicating, at high electrolyte concentrations, the absence of any (or significant) amounts of discrete ion pairing.

These effects have been explained qualitatively as E_{max} increasing as water molecules are removed from the solvent shell of the anion since, in the first limit, E_{max} is governed by the radius, and hence the number of molecules in this shell.[1, 19] The approach of considering the effect of added salt on individual water molecules is a considerable oversimplification: we take a more realistic average approach, by using the calculated average separation between ions, assuming no ion pairing, obtained as a function of concentration. The increasing charge cloud around iodide (and all other ions) with increasing electrolyte concentration will reduce the radius parameter, and this approach can be extended to the limit where all water molecules are removed and a truly ionic liquid, a molten salt is obtained.

So far there has been no direct evidence for the consequent implication, that, in the limit, the solvent cavity radius decreases with increase in salt concentration to the crystallographic radius for the confined model (or to a somewhat greater value for the diffuse model). The trend has been confirmed by computations using a modified Hepler equation derived from the partial molar volume variation of both potassium chloride and bromide with added sodium chloride,[20–22] but because an iodide ion was not investigated, quantitative comparisons cannot be made. However, this is an example of the way in which spectroscopic and thermodynamic data can be linked. A quantitative relationship between the shift of the CTTS spectrum of iodide (as a probe) in mixed aqueous electrolyte solutions, up to 1 mol dm^{-3}, with the cross–square relationship for ΔH_m, the heat of mixing of binary mixtures, has been published.[23] We now extend this approach to consider the shift relationship with the Debye–Hückel charge cloud[24] and the expanded lattice (sometimes termed the quasi- or diffuse lattice) approach for more concentrated solutions.[25–28]

10. Relating CTTS Spectra and Coulombic Interactions

The observed shift in the band maximum E_{max} of iodide to higher energies with increasing electrolyte concentration is now generally understood[10] as a Coulombic interaction upon the ground state energy of the reference iodide ion. The properties of electrolytic solutions at low concentrations are those theoretically predicted for a model system in which the effects of all other ions on a reference ion can be treated as those of a continuous space charge, the familiar ion atmosphere proposed by Debye and Hückel. At high concentrations, on the other hand, the effects of near neighbors on a reference ion become far more significant than those of more distant ions; interionic electrostatic energies no longer are small compared to kT and thereby the necessary condition for the existence of a smoothed charge density disappears, and a different model must be chosen as the basis of theory. In concentrated solutions, average interionic distances will vary as the cube root of concentration. As the solution is diluted, we know that the $c^{1/3}$ behavior must give way to the $c^{1/2}$ properties of the Debye–Hückel charge cloud. There is, of course, no critical concentration at which a solution suddenly stops acting like an expanded crystal lattice and rearranges its ions to the distribution characteristic of the ion atmosphere. Rather, a smooth transition from one model to the other must occur. A stochastic description can be formulated[29] as

$$P(c) = P(0)f(c) + P(\infty)[1 - f(c)] \tag{5}$$

where P is the value of a given property of the solution at concentration c, $P(0)$ is the Debye–Hückel description of that property, $P(\infty)$ is the concentrated electrolyte model description, and $f(c)$ satisfies the conditions

$$f(0) = 1, \qquad f(\infty) = 0$$

the symbols 0 and ∞ implying solute–solvent ratios of $0 : 1$ and $1 : 0$. In effect, the Coulombic interactions can be treated as a mixture of two components, the contribution of each being weighed by a partition function $f(c)$.

Expanded lattice models have long been considered for electrolytic solutions, commencing with Ghosh[25] and Bjerrum[30, 31] and now recently by Ruff;[27] these lead to $c^{1/3}$ laws, but they have generally been limited to concentrations of 1 mol dm^{-3} and occasionally up to 2 mol dm^{-1}. Dependence on $c^{1/3}$ in properties such as logarithm of activity coefficient, equivalent conductance, and diffusion coefficients has also been noted.[32, 33] But in dilute solutions these properties go as the square root and not as the cube root of concentration. This effect is also now noted in CTTS spectral shift. Figure 1 shows a plot of E_{max} for iodide against the square root of the concentration of various alkali chlorides, and Figure 2 the shift in E_{max} as a function of the cube root of the concentration. The square root plot (Figure 1) shows a linear plot with E_{max} between 1 and 4 mol dm^{-3} approximately, and the cube root plot (Figure 2) is linear above 4 mol dm^{-3}.

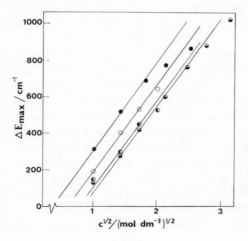

FIGURE 1. Plot of ΔE_{max}, the shift in the position of the first CTTS band of iodide in water, upon the addition of chloride against the square root of the chloride concentration. The parallel lines drawn through the data have the Debye–Hückel limiting slope, k. ◓, lithium; ◑, sodium; ○, potassium; ●, cesium.

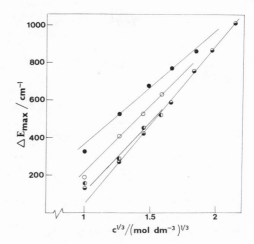

FIGURE 2. Plot of ΔE_{max}, the shift in the position of the first CTTS band of iodide in water, upon the addition of chloride against the cube root of the chloride concentration. ◒, lithium; ◓, sodium; ○, potassium; ●, cesium.

Thus qualitatively there exists a parallelism between E_{max} and the logarithm of activity coefficient as a function of concentration.

The next stage is to relate the spectral shifts to equation (5). The crucial problem is the choice of the partition (or transition) function $f(c)$. There are some accounts in the literature for activity coefficient data.[28, 34–36] For example, Mitra et al.[36] used the molarity of a saturated solution for $f(c)$. The expression they derived for $\log \gamma_\pm$ comprised the electrical (ion cloud) term and the quasilattice term, but also included the Glueckauf expression[37] for the statistical part, $\log \gamma^s$, of $\log \gamma_\pm$. Within their equation there was an adjustable parameter h, the hydration number, and their approach can be further criticized since they did not recognize supersaturated solutions, or that the limit of the quasilattice theory is the anhydrous crystal. Leitzke et al.[28] have used a Debye–Hückel plus cell model and for $f(c)$ an exponential,

$$f(c) = e^{-a\varphi} \qquad (6)$$

where a is a constant, φ is the volume fraction of solute,

$$\varphi = Vc/1000 \qquad (7)$$

computed on the basis of the molar volume V of the dry salt, and c is concentration in mol dm^{-3}. Their cell model is based on the upper limit of the concentration scale corresponding to a randomized[38, 39] fused salt, where a solute particle may be anywhere in its cell, and empty cells matched by cells containing two ions (ion pairs) are not excluded. They found equation (5) a valid description of activity coefficients up to a concentration of 2 mol dm^{-3}.

We have taken a similar approach, and avoiding any arbitrary constants, we relate $f(c)$ to the molar concentration of the salt in the solid state c_s. The interionic distance in solution l can vary between infinity, for infinitely dilute solutions, to that of the sum of the anion and cation radii in the anhydrous salt. Thus l may be obtained from

$$l = \alpha c^{-1/3} \tag{8}$$

where c is the concentration in mol dm^{-3}, and α is $(1000/2N)^{1/3}$, N being Avogadro's number. The limit, in the anhydrous salt, is therefore l_s, and may be calculated from

$$l_s = \alpha c_s^{-1/3} \tag{9}$$

From X-ray data the sum of the anion and cation radii for LiCl and KCl are 257.1 and 315.0 pm, respectively. From the above equation l_s is calculated as 241 and 314 pm, respectively.

The ratio of l and l_s is our $f(c)$ function and we define l_s/l as l^*, and, as required by equation (5), it is zero for infinitely dilute solutions and unity for the solid salt (where the lattice no longer has water molecules between the ions). This ratio is also related to the electrolyte concentration by

$$l^* = (c/c_s)^{1/3} \tag{10}$$

A plot of l^* against concentration for various alkali metal chlorides is shown in Figure 3. This plot also shows the relationship with average

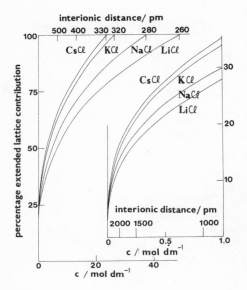

FIGURE 3. Plot relating average interionic distances and concentration with the percentage of the expanded lattice component of the ionic interactions for the chlorides of Li, K, and Cs. Note that this component only begins to become significant above 1 mol dm^{-3}.

interionic distance. The inset further shows that l^* only begins to become significant above about 1 mol dm^{-3}, that is, the expanded lattice approach, and its contribution to the Coulombic interactions around an ion in solution only become significant above 1 mol dm^{-3}.

The contribution of the expanded lattice to the overall Coulombic interactions is by definition related to the cube root of the concentration. Thus the ordinate of Figure 3 also defines the fraction of this contribution to the Coulombic interaction absolutely, without the use of any of the adjustable parameters of previous workers.[28, 36]

The remaining Coulombic interactions arising from the Debye–Hückel charge cloud[24] can also be obtained absolutely, by difference.

11. Quantifying the Coulombic Interaction Components

According to expanded lattice theory the Coulombic interaction is proportional to $(\beta/\alpha)c^{1/3}$, where β is $2NMz^2e^2/\varepsilon$, and M is the Madelung constant, and z, e, and ε have their usual meanings. The expanded lattice contribution to the Coulombic interaction is thus $l^*(\beta/\alpha)c^{1/3}$. The Debye–Hückel charge cloud component, related to the square root of the concentration, is quantified as $(1 - l^*)kc^{1/2}$. It is important to note that k is not some arbitrary constant, but is the Debye–Hückel limiting slope. We therefore express the change in Coulombic interactions upon an ion on going from an infinitely dilute solution to one of concentration c mol dm^{-3} as

$$\Delta E_c = l^*(\beta/\alpha)c^{1/3} + (1 - l^*)kc^{1/2} \tag{11}$$

12. Comparison with CTTS Spectral Shifts

The observed spectral shift ΔE_{max} for an infinitely dilute solution of iodide (in practice 10^{-5} mol dm^{-3}) as an electrolyte such as KCl is added may be expressed as

$$\Delta E_{max} = \Delta E_c + S \tag{12}$$

where S includes the remaining concentration-dependent ground state energy components, essentially the changes in the solvation of the ions. ΔE_c cannot be equated with changes in the radius parameter of CTTS theory models, but it will later be seen that at high concentrations it is directly related to the expanded lattice component of ΔE_c.

13. Relationship with Mean Activity Coefficient

The activity coefficient is also a measure of ionic interaction, and its relationship with ΔE_C is

$$RT \ln \gamma_\pm = \Delta E_C + S' \tag{13}$$

where $\gamma_\pm$ is the mean activity coefficient (since single-ion coefficients, for more precise comparisons with iodide shifts, are not yet reliable). S' is not identical with S as expected, but will subsequently be shown to be of comparable value.

We therefore now have two essentially identical expressions for the ionic interactions in electrolyte solutions, but from two very different and previously unrelated techniques, one based on spectroscopic band shifts and the other on activity coefficients from conductivity data, viz.,

$$\Delta E_{max} - (1 - l^*)kc^{1/2} = l^*(\beta/\alpha)c^{1/3} + S \tag{14}$$

and

$$RT \ln \gamma_\pm - (1 - l^*)kc^{1/2} = l^*(\beta/\alpha)c^{1/3} + S' \tag{15}$$

To check the validity of our procedure for obtaining $f(c)$ as $(c/c_s)^{1/3}$ in determining the relative proportions of charge cloud to expanded lattice contributions within the Coulombic interactions, plots of the left-hand side of these two equations against $l^*c^{1/3}$ should both have identical slopes, (β/α). Figure 4 shows a composite plot of our spectral shift data, and mean activity data, taken from the standard literature.[40] Identical correspon-

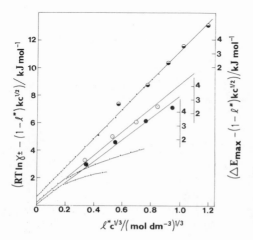

FIGURE 4. Display of the identical slopes of the electrochemical (left-hand ordinate) and spectral shift (right-hand ordinate) data against $l^*c^{1/3}$ for LiCl, KCl, and CsCl.

dence in slopes (β/α) for the data from these two different techniques, within experimental error, is evident. Although the data for Li, K, and Cs are shown the correspondence applies to all the alkali chlorides. The ordinates for the spectral shift data have been adjusted to show the correspondence clearly; S and S' are also thereby shown not to be identical. The difference between S and S' is small and approximately 8, 4, and 2 kJ mol^{-1} for LiCl, KCl, and CsCl, respectively.

We now consider the implications of this figure.

14. Linear Relationships up to 10 mol dm^{-3}

It is possible to measure the position of E_{max} for iodide in alkali metal chlorides up to their saturation concentration at room temperature. The most concentrated solution here measured was that for LiCl of 10 mol dm^{-3}, and this point is firmly on the lithium chloride line. It is in principle possible to derive E_{max} values well beyond the saturation concentration for all the alkali chlorides: there is a linear relationship between E_{max} and temperature, at all concentrations of added electrolyte, and therefore by measuring E_{max} at several elevated temperatures, and extrapolating back to 25°C the data in Figure 4 can be extended. This we have not yet done. We do, however, have the value for KI in single-crystal KCl at room temperature. At the present time it cannot be incorporated into Figure 4 because in the crystal the iodide ions are forced into the smaller cavity normally occupied by a chloride ion. The spectrum of iodide, which is now technically no longer CTTS, is thus experiencing two constraints, the maximum salt shift plus a pressure effect. When this latter can be accurately evaluated then the maximum concentration value can be incorporated into Figure 4. We have also the spectra of iodide in molten KCl, but because the temperature range studied was limited, extrapolation back to 25°C from over 800°C was not realistic.

The mean activity data reported in the literature does not, of course, extend to such high concentrations, but the linear portions of the plots in Figure 4 clearly have the same slope as the spectral shift data. We note with interest that the lithium chloride data are linear over the entire concentration range. It is generally the case that for plots involving the cube root of the concentration (e.g., Ruff[27]) all the alkali chlorides except lithium exhibit linear relationships up to the maximum concentration investigated, usually between 1 and 2 mol dm^{-3}. Using our function, $l^*c^{1/3}$, all the alkali chlorides, including lithium, are linear up to around 1.3 mol dm^{-3}, and lithium remains linear up to 4 mol dm^{-3}. It is therefore now possible to calculate mean activity coefficients for lithium chloride up to at least 10 mol dm^{-3}.

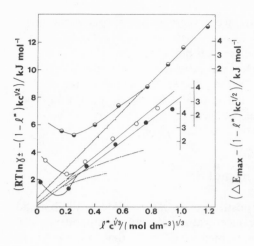

FIGURE 5. Display of the identical slopes of the electrochemical (left-hand ordinate) and spectral shift (right-hand ordinate) data against $l^*c^{1/3}$ for LiCl, KCl, and CsCl, including low-concentration spectral shift results.

The spectral shift data plotted on Figure 4 were deliberately limited to high concentrations, for reasons of clarity. Figure 5 now includes all the spectral shift data, and shows deviations at low concentrations. This arises from the nature of CTTS spectra. The term S in equation (12) is concentration dependent at low concentrations, and it is only strictly comparable with S' in equation (13) at high concentrations (Section 13). The nonlinearity at low concentrations arises from the very sensitive nature of the iodide probe. We have remarked (Section 8) that it can detect the effect upon the solvating molecules of iodide ions of the addition of an added electrolyte at a concentration of 2.5×10^{-2} mol dm^{-3}, where the average interionic distance is 2500 pm. Even at this distance the electrostatic field around iodide is sufficiently altered to affect the radius parameter. As concentration is increased the solvation of iodide will be changing at the same time as the Coulombic interaction is increasing. At around 4 mol dm^{-3} all the water molecules in the system are in contact with at least one ion, as the interionic distance is now around 600 pm. Consequently there are no longer any "free" water molecules in solution which could play a role in secondary solvation. Ion–solvent interactions thus dominate ΔE_{max} at low added electrolyte concentrations; above 1 mol dm^{-3} the ion–ion interactions due to the Debye–Hückel charge cloud dominate, and above about 4 mol dm^{-3} the expanded lattice contribution becomes significant. Returning to Figures 1 and 2, the plots of ΔE_{max} against $c^{1/2}$ not only are linear between 1 and 4 mol dm^{-3}, but in this region all have the *same* slope, and this slope is the Debye–Hückel limiting slope. The plots of ΔE_{max} against $c^{1/3}$ are not all colinear above 4 mol dm^{-3}, but are all linear up to our measured maximum concentration.

We are currently examining the shape and magnitude of the deviations in Figure 5 for qualitative and quantitative information on solvation energies.

15. Precise Madelung Constants from Spectral Shift Data

Some attempts have been made to obtain Madelung constants from conductivity data because it is obtainable from essentially activity-coefficient–concentration plots. Using a two-structure model for electrolyte solutions Lietzke *et al.*[28] (see Section 10) obtained a value for a constant A, described as analogous to the Madelung constant, of 0.88, compared to a Madelung constant of 1.74756 for the bcc NaCl crystal. This was attributed to their use of a cell model, based on the randomized[38, 39] fused salt, and the discrepancy to the lack of longer-range structure due to thermal motion and diffusion. Glueckauf,[33] using a treatment somewhat different from that of Lietzke *et al.*,[28] obtained values a little closer to the Madelung constants of crystals. Mitra *et al.*[36] obtained values in excess of 2.0. To date there are no reported values close to the theoretical value.

In our treatment [equation (14)] of iodide spectral shifts (Figure 4) the slope of the lines (β/α) includes the Madelung constant. The slope is evaluated as $-2^{4/3}N^{4/3}Mz^2e^2/10\varepsilon$. Thus only one value of the Madelung constant should be found for all the alkali metal chlorides. However, because of experimental error, when the best straight line is drawn through each individual chloride spectral data set, small variations between individual Madelung constants emerge. The results are, from the slope of each plot, 2.123 for LiCl, 1.760 for NaCl, 1.720 for KCl, and 1.740 for CsCl. These compare very favorably with the theoretical value of 1.74756. We note two points, firstly, that for previously determined Madelung constants the value for LiCl has generally been significantly higher than the other chlorides (and the average of our remaining values is 1.740). Secondly, the CsCl value corresponds to the theoretical bcc (NaCl) structure value: the theoretical value for fcc (CsCl) is 1.76267. Very accurate spectral shift measurements at higher concentrations in the future may detect a slight change in slope as the expanded lattice becomes slightly reorganized.

The above experimental values of the Madelung constant are, to our knowledge, the first determinations from absorption spectra, and the most accurate to date from spectroscopic or electrochemical measurements. The values calculated from the mean activity coefficient data are the same within experimental error as the spectra-based data: the reduced range in Figure 4 over which the former are linear is compensated by greater precision in the measured activity coefficients.

We therefore contend that these *a priori* Madelung constant values lend strong support to the validity of our approach, interpretation, and the particular partition function $f(c)$ chosen, equation (10), for the two-structure model for electrolyte solutions over the complete concentration range.

16. Correlations with Thermodynamic Parameters

From Figures 1 and 2 it may be seen that a negative value of ΔE_{max} is obtained upon extrapolating the linear parts of these plots to infinite dilution. The values obtained for any particular chloride are not identical from the $c^{1/2}$ and $c^{1/3}$ plots, but the infinite dilution values correspond to the ionic solvation effects, since the charge cloud and expanded lattice contributions are now zero. This negative value is therefore expected to correlate with the free energy of solvation at infinite dilution (ΔG_{sol}^{∞}). Figure 6 shows plots of ΔG_{sol}^{∞} against the intercepts for the alkali halides; a linear correlation obtains. We thus now have the basis for a direct and simple spectroscopic method for determining thermodynamic parameters of electrolyte solutions which we are currently investigating.

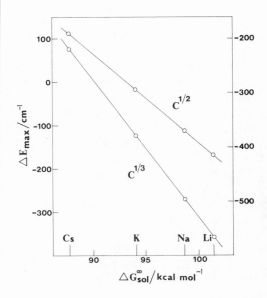

FIGURE 6. Plots of the intercepts at infinite dilution from the square root concentration-dependent spectral shifts of Figure 1 (right-hand ordinate) and the cube root concentration-dependent spectral shifts of Figure 2 (left-hand ordinate) against ΔG_{sol}^{∞}, the free energy of solvation at infinite dilution, for the various alkali metal chlorides.

17. Experimental

The absorption spectra of the aqueous iodide solutions, and with various added alkali metal chlorides, were recorded on an Applied Physics Cary 14H, a Unicam SP 700C, a Unicam SP 700A, and a Beckmann DK2A spectrophotometer. Special care was taken to maintain the solutions at constant, and accurately known, temperature, that thermal equilibrium had been attained, and that possible evaporation from the sample and reference solutions was eliminated. The average error in the estimation of E_{max} of iodide in all solutions was ± 3.0 cm^{-1}. Previous results in our laboratory[41] have been incorporated into our data. All chemicals used were of the highest purity available.

ACKNOWLEDGMENTS

We thank Professor M. C. R. Symons and several of our colleagues at Leeds University for valuable and stimulating discussions. We also thank the S.R.C. for a grant for the purchase of the Cary 14H spectrophotometer, and the Association of Commonwealth Universities for financial assistance and a University Academic Staff Fellowship to R.H.W.

References

1. M. Smith and M. C. R. Symons, *Trans. Faraday Soc.* **54**, 338 (1958).
2. J. Jortner, B. Raz, and G. Stein, *Trans. Faraday Soc.* **56**, 1273 (1960).
3. D. Myerstein and A. Treinin, *J. Phys. Chem.* **66**, 446 (1962).
4. R. Platzmann and J. Franck, *Z. Phys.* **138**, 411 (1954).
5. M. Smith and M. C. R. Symons, *Trans. Faraday Soc.* **54**, 346 (1958).
6. G. Stein and A. Treinin, *Trans. Faraday Soc.* **55**, 1086, 1091 (1959).
7. M. Smith and M. C. R. Symons, *Discuss Faraday Soc.* **24**, 206 (1957).
8. B. S. Gourary and F. J. Adrian, *Solid State Phys.* **10**, 127 (1960).
9. M. C. R. Symons and W. T. Doyle, *Quart. Rev.* **14**, 62 (1960).
10. M. J. Blandamer and M. F. Fox, *Chem. Rev.* **70**, 59 (1970).
11. M. Anbar and E. J. Hart, *J. Phys. Chem.* **69**, 1244 (1965).
12. W. Gottschall and E. J. Hart, *J. Phys. Chem.* **71**, 2102 (1967).
13. M. C. R. Symons, *Chem. Soc. Rev.* **5**, 337 (1976).
14. C. K. Jørgensen, *Solid State Phys.* **13**, 375 (1962).
15. C. K. Jørgensen, *Adv. Chem. Phys.* **5**, 33 (1963).
16. C. K. Jørgensen, *Orbitals in Atoms and Molecules*, Academic, London (1962).
17. C. K. Jørgensen, *International Review of Halogen Chemistry*, Vol. 1, Academic, London (1967).
18. D. B. Siano and D. E. Metzler, *J. Chem. Soc. Faraday Trans. 2* **68**, 2042 (1972).
19. G. Stein and A. Treinin, *Trans. Faraday Soc.* **56**, 1393 (1960).
20. H. E. Wirth, *J. Am. Chem. Soc.* **39**, 2549 (1937).

21. H. E. Wirth, *J. Am. Chem. Soc.* **42**, 1128 (1940).
22. H. E. Wirth and H. E. Collier, *J. Am. Chem. Soc.* **72**, 5292 (1950).
23. M. J. Blandamer, T. R. Griffiths, and K. J. Wood, *J. Chem. Soc. Chem. Commun.* 933, (1969).
24. P. Debye and E. Hückel, *Phys. Z.* **24**, 185 (1923).
25. J. C. Ghosh, *J. Chem. Soc.* **113**, 449 (1918).
26. L. W. Bahe, *J. Phys. Chem.* **76**, 1062, 1608 (1972).
27. I. Ruff, *J. Chem. Soc. Faraday Trans. 1* **73**, 1858 (1977).
28. M. H. Lietzke, R. W. Stoughton, and R. M. Fuoss, *Proc. Natl. Acad. Sci. USA* **59**, 39, (1968).
29. R. M. Fuoss and L. Onsager, *Proc. Natl. Acad. Sci. USA* **47**, 818 (1961).
30. N. Bjerrum, *Z. Elektrochem.* **24**, 321 (1918).
31. N. Bjerrum, *Z. Anorg. Chem.* **109**, 275 (1920).
32. H. S. Frank and P. T. Thompson, in *The Structure of Electrolytic Solutions*, Ed. W. J. Hamer, Wiley, New York (1959), pp. 113–134.
33. E. Glueckauf, in *The Structure of Electrolytic Solutions*, Ed. W. J. Hamer, Wiley, New York (1959), Part 7, pp. 97–112.
34. G. W. Murphy and E. W. Smith, *J. Chem. Phys.* **31**, 1086 (1959).
35. M. A. Devanathan, *J. Sci. Ind. Res. India B* **20**, 256 (1961).
36. P. Mitra, D. V. S. Jain, and M. H. Kapoor, *Indian J. Chem.* **6**, 391 (1968).
37. E. Glueckauf, *Trans. Faraday Soc.* **51**, 1235 (1955).
38. C. A. Angell, *J. Phys. Chem.* **70**, 3988 (1966).
39. C. A. Angell, E. J. Sare, and R. D. Bressel, *J. Phys. Chem.* **71**, 2759 (1967).
40. R. A. Robinson and R. H. Stokes, *Electrolyte Solutions*, 2nd ed., Butterworths, London (1959), p. 491.
41. K. J. Wood, Ph.D. thesis, Leeds University (1970).

7

Dead Sea Brines: Natural Highly Concentrated Salt Solutions

Y. Marcus

1. Introduction

The Dead Sea is located in the lowest basin on the surface of the earth, near the northern end of the Great Syro-African Rift. Its surface is at a mean elevation of -395 m, with an approximate length of 80 km and width of 15 km, its surface area being about 1000 km^2. It consists of a deep northern basin having a maximum depth of about 400 m, and a shallow southern basin, of average depth of only 3 m. This latter basin has suffered considerable change in recent years through both climatic and man-made causes; its eastern half has almost dried up completely, while its western half is a series of solar salt pans. Recent aerial photographs from satellites show a southern contour line quite different from those appearing in common maps (Figure 1).

The Dead Sea is stratified: it consists of an upper water mass approximately 35 m deep, a transition layer with strong density and salinity gradients, approximately 40 m deep, and a lower water mass.[1, 2] The latter two exist only in the northern basin. The minerals which precipitate out naturally from the Dead Sea are halite, gypsum, and calcite, but the precipitation processes are very complicated, and not near equilibrium. On the other hand, in the salt pans, where halite and a mixture of halite and carnallite are precipitated, the conditions are near equilibrium. The temperature of the surface water, both of the lake and of the pans, is nearly constant[1, 2] at the standard temperature familiar to solution chemists:

Y. Marcus • Department of Inorganic and Analytical Chemistry, The Hebrew University of Jerusalem, Jerusalem, Israel.

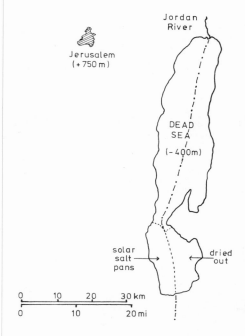

FIGURE 1. Map of the Dead Sea (before the recent changes in the southern basin).

25°C. This fortuitously convenient fact is due to a balance of the absorption of intense solar radiation, surface water evaporation, replenishing by the Jordan river, and surface mixing by wind and waves.

Although the Dead Sea seems to be the remnant of a large primordial sea connected to the ocean, its recent geochemical history has led to changes in its salt composition and concentration with respect to ocean water. For instance, its salinity is some eight times that of ocean water, and indeed it is the saltiest large body of water on the surface of the earth. Then, its ionic composition is different from that of ocean water (Figure 2).

TABLE 1. Molalities m_i° and Molarities c_i° of the Components of Dead Sea Water and the Parameters[5] for the Osmotic and Activity Coefficient Equations (7) and (9), at 298.15 K

Salt	m_i°, mol kg^{-1}	c_i°, mol dm^{-3}	A_i	$10^2\,B_i$	$10^3\,C_i$	$10^4\,D_i$
NaCl	1.752	1.570	1.45397	2.23565	9.30838	−5.36209
KCl	0.174	0.156	1.30753	−0.35919	7.17091	−5.67500
MgCl$_2$	1.555	1.394	1.60067	6.63253	9.00292	−2.54526
CaCl$_2$	0.427	0.383	1.61291	4.56577	8.57310	−2.73800

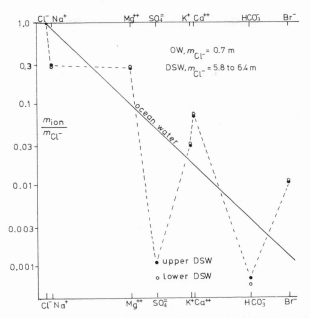

FIGURE 2. Concentrations of the constituent ions in ocean water (OW) and in Dead Sea water (DSW) in the upper water mass (top 35 m) and the lower water mass of the northern basin, relative to the chloride ion concentration, on a logarithmic scale. The positions of the ions on the abscissa have been arbitrarily selected so as to make the curve for OW a straight line. [After A. Lehrman, *Geochim, Cosmochim. Acta* **31**, 2309 (1967).]

The main differences are seen to be deficiencies in sulfate and bicarbonate, which are due to precipitation of gypsum and calcite, and an excess of bromide ions. Still, the concentration of the latter is quite low (1.02 equivalent per cent) and that of the sulfate and bicarbonate is even lower (0.36 equivalent per cent together). Dead Sea water can therefore be considered a quinquenary common ion aqueous electrolyte solution, composed of the chlorides of sodium, potassium, magnesium, and calcium. For present considerations, a solution of 1.752 m NaCl, 0.174 m KCl, 1.555 m MgCl$_2$, and 0.427 m CaCl$_2$ at 25.0°C will be taken as representing the surface water of the Dead Sea. It has a density of 1207 kg/m^3, from which the corresponding molar (mol dm^{-3}) concentrations can be calculated (Table 1).

2. The Activities of the Components of Dead Sea Water

Consider now Dead Sea water as a system of five components: water and the four salts with the common anion, chloride. It is a concentrated solution, the ionic strength is $I = 0.5 \sum z_i(z_i + 1)m_i = 7.874$ m, the osmolality is $\sum (z_i + 1)m_i = 9.799$ m, and the total chloride concentration is $\sum z_i m_i = 5.891$ m. The summation extends over the four salts $i = 1, 2, 3$, and 4, and z_i is the charge on the cation.

It is usual to record the activities of the salts in their anhydrous state, but for present purposes it will be convenient to calculate and present the activities of the hexahydrates of magnesium and calcium chlorides. On the one hand, these are the forms of these salts which are in equilibrium with the saturated solutions, on the other, this permits the calculation of the activity of carnallite, $KMgCl_3 \cdot 6H_2O$. Although not a component of the solution in the usual sense, carnallite is one of the salts which precipitate out of the solutions when concentrated by evaporation in the salt pans. We shall therefore calculate

$$a_{H_2O} = \exp\{-0.01801[\sum (z_i + 1)m_i]\phi\} \tag{1}$$

$$a_{NaCl} = m_{Na}(\sum z_i m_i)\gamma^2_{\pm NaCl} \tag{2}$$

$$a_{KCl} = m_K(\sum z_i m_i)\gamma^2_{\pm KCl} \tag{3}$$

$$a_{MgCl_2\ 6H_2O} = m_{Mg}(\sum z_i m_i)^2\gamma^3_{\pm MgCl_2} a^6_{H_2O} \tag{4}$$

$$a_{CaCl_2\ 6H_2O} = m_{Ca}(\sum z_i m_i)^2\gamma^3_{\pm CaCl_2} a^6_{H_2O} \tag{5}$$

$$a_{KMgCl_3\ 6H_2O} = m_K m_{Mg}(\sum m_i z_i)^3\gamma^2_{\pm KCl}\gamma^3_{\pm MgCl_2} a^6_{H_2O} \tag{6}$$

Here a_{H_2O} is the activity of water on the rational scale, ϕ is the osmotic coefficient of the solution, a_i are activities, relative to the usual standard state of the hypothetical 1 m ideal solution, and $\gamma_{\pm i}$ are the mean ionic activity coefficients of the indicated salts in the *mixture*.

Several approaches have been proposed for the calculation of the osmotic coefficient and the activity coefficients of the electrolytes in a multicomponent electrolyte solution. Their purpose is to express these quantities in terms of the corresponding quantities of the *binary* subsystems (i.e., those consisting of the solvent and one electrolyte), ϕ_i° and $\gamma_{\pm i}^\circ$, and of the *ternary* subsystems (i.e., those consisting of the solvent and two electrolytes with a common anion), where the required quantities are interaction parameters, g_{ij}, to be described further below. For the present type of systems, these approaches have recently been examined by the author.[3] Surprisingly, perhaps, it turned out that a good description of the multicomponent system in the present case can be obtained by a consideration of the binary subsystems alone, the terms involving the interaction parameters g_{ij} canceling out, or contributing negligibly to the osmotic coefficient ϕ and the activity coefficients $\gamma_{\pm i}$ in the mixture. A refinement involving the g_{ij}'s is certainly possible, but unnecessary, for the present purpose of obtaining a macroscopic description of the system. They will be required, however, if a microscopic description in terms of the chemical interactions is desired, as will be done in a later section.

We shall now follow the approach of Lietzke and Stoughton[4] for the mixtures, utilizing also their parametric descriptions for the binary solutions.[5] All we need as input data are the molalities m_i^o (i.e., those of Dead Sea water given above) and the parameters S (=1.1720, the Debye–Hückel parameter for 25°C, for natural logarithms of the activity coefficients), and A_i, B_i, C_i, and D_i, specific for each salt (Table 1). We then calculate in turn

$$\phi_i^o = 1 - z_i S A_i^{-3} I^{-1}[(1 + A_i I^{1/2}) - 2 \ln(1 + A_i I^{1/2}) - (1 + A_i I^{1/2})^{-1}]$$
$$+ B_i I + C_i I^2 + D_i I^3 \quad (7)$$

$$\phi = \sum (z_i + 1) m_i^o \phi_i^o / \sum (z_i + 1) m_i^o \quad (8)$$

$$\ln \gamma_{\pm i}^o = -z_i S I^{1/2} (1 + A_i I^{1/2})^{-1} + (2B_i)I + (3C_i/2)I^2 + (4D_i/3)I^3 \quad (9)$$

$$x_i = z_i(z_i + 1) m_i^o / \sum_j z_j(z_j + 1) m_j^o \quad (10)$$

$$\ln \gamma_{\pm i} = \ln \gamma_{\pm i}^o + 0.5 \sum_{j \neq i} x_j(z_i z_j^{-1} \ln \gamma_{\pm j}^o - \ln \gamma_{\pm i}^o) \quad (11)$$

Here x_i is the ionic strength fraction of salt i, $I = 0.5 \sum z_i(z_i + 1) m_i^o$ is the ionic strength, and all the other terms have already been defined.

It should be pointed out that the use of (7) and (9) at the ionic strength of Dead Sea water $I = 7.874 \ m$ is well within the range of experimental data for magnesium and calcium chlorides, but is outside that of the uni-univalent salts sodium and potassium chlorides, where the solubility is limited to 6.1 and 4.8 m, respectively, at 25°C. Thus an extrapolation is necessary, and great weight is placed on the higher terms in (7) and (9), involving the parameters C_i and D_i. It is therefore reassuring to note that the osmotic coefficient calculated from (7) and (8) leads to a water activity from (1) of 0.752, within 1% of the measured[3] quantity 0.760 ± 0.003. There are no activity coefficient data with which to compare the calculated values for the mixture, but they can be tested, when the equations are used to follow the changes in the salt solutions during evaporation, until precipitation occurs. They will be seen in the next section to predict correctly the points of precipitation of halite (NaCl) and of carnallite ($KMgCl_3 \cdot 6H_2O$).

Let us now consider briefly some alternative approaches, which do consider interactions between the cations of the different salts. The excess Gibbs free energy of a ternary mixture of salts i and j is available experimentally[6, 9] from solvent vapor pressure measurements as a function of the ionic strength I and the ionic strength fraction $x_i \equiv x$ (i.e., $x_j \equiv 1 - x$), that is, $\phi(x, I)$. At a given composition x,

$$G^E(x, I) = -4RTI \int_0^I [1 - \phi(x, I')](I')^{1/2} \, d(I')^{1/2} \quad (12)$$

The excess Gibbs free energy of a binary solution is given by

$$G_i^{E^\circ}(I) = RTm_i^\circ(1 - \phi_i^\circ + \ln \gamma_{\pm i}^\circ) \tag{13}$$

Hence, for an amount of ternary mixture containing 1 kg of solvent, $G^E(x, I)$ can be expressed in terms of the interaction parameter g_{ij} as

$$G^E(x, I) = xG_i^{E^\circ}(I) + (1 - x)G_j^{E^\circ}(I) + RTx(1 - x)I^2 g_{ij} \tag{14}$$

from which g_{ij} can be readily obtained. This expression is an approximation, in that it considers the deviations from linearity of G^E with composition to be symmetrical, that is, g_{ij} to be composition independent. Formally, higher terms $g_{ijq}(1 - 2x)^q$, $q = 1, 2, \ldots$, can be added to take account of unsymmetrical deviations, but for present purposes all g_{ijq} for $q > 0$ will be disregarded. We have then the following alternative equations for the osmotic and activity coefficients in the multicomponent mixture.

According to Reilley, Wood, and Robinson,[7]

$$\phi = \sum (z_i + 1)m_i\phi_i^\circ/\sum (z_i + 1)m_i^\circ$$
$$+ [I^2/\sum (z_i + 1)m_i^\circ] \sum\sum x_ix_j(g_{ij} + dg_{ij}/d \ln I) \tag{8a}$$

$$\ln \gamma_{\pm i} = \ln \gamma_{\pm i}^\circ + (1 - x_i)(1 - \phi_i^\circ) - \sum_{j \neq i} z_j x_j(1 - \phi_j^\circ) + 0.5I \sum x_i g_{ij}$$
$$+ 0.5I \sum_{j \neq i}\sum x_jx_i(dg_{ij}/d \ln I) \tag{11a}$$

For the present type of systems, common-ion mixtures, the neutral electrolyte approach of Scatchard, improved by Rush and Johnson,[9] is simpler than the ion component approach of Scatchard, Rush, and Johnson,[8] which has advantages only for reciprocal systems. According to this neutral electrolyte approach

$$\phi = 1 - [I/\sum (z_i + 1)m_i^\circ] \sum x_i(\phi_i^\circ - 1)$$
$$+ [I^2/\sum (z_i + 1)m_i^\circ] \sum_{j \neq i}\sum x_jx_i g_{ij} \tag{8b}$$

$$\ln \gamma_{\pm i} = 0.5 \ln \gamma_{\pm i}^\circ + 0.5 \sum_{j \neq i} x_j(\phi_j^\circ - \phi_i^\circ) + 0.5I \sum x_j g_{ij}$$
$$+ 0.5 \sum_{j \neq i} x_j\left(\int_0^I g_{ij}\, dI - Ig_{ij}\right) \tag{11b}$$

Equations (8b) and (11b) differ from (8a) and (11a) by a different weighting of the terms involving g_{ij} and $dg_{ij}/d \ln I$ and those involving ϕ_i°. The required values of the interaction parameters for the salt pairs i and j at the relevant ionic strength $I = 7.874\ m$ are shown in Table 2. Still more elaborate equations have been presented by these authors,[7-9] involving triple-ion

TABLE 2. Interaction Parameters g_{ij} in Common-Anion (Chloride) Aqueous Salt Mixtures from Equation (14) and Harned Coefficients α_{ij} from Equation (15), at 298.15 K, for $I = 7.874 \ m^3$

Salt i	Salt j	g_{ij}	α_{ij}	Interactions of cations
NaCl	KCl	-0.043	-0.0107	Attractive
NaCl	MgCl$_2$	$+0.024$	$+0.0131$	Repulsive
NaCl	CaCl$_2$	$+0.018$	$+0.0076$	Repulsive
KCl	MgCl$_2$	$+0.022$	$+0.0376$	Repulsive
KCl	CaCl$_2$	-0.048	$+0.0420$	Attractive
MgCl$_2$	CaCl$_2$	$+0.003$	-0.0120	Hardly any
(Na + K)Cl	MgCl$_2$	$+0.024$		Repulsive
(Na + K)Cl	CaCl$_2$	$+0.015$		Repulsive

interactions, but no advantage can be seen in going beyond the two-ion interaction terms used above, in view of the limited accuracy of the experimental data, good as they are.

From quite different theoretical premises, Pitzer and Kim[10] have derived a semiempirical set of equations,

$$\phi = 1 + \left[\sum (z_i + 1)m_i^\circ\right]^{-1}\{2If^\phi + 2(\sum z_i m_i^\circ) \sum m_i^\circ[B_i^\phi + 2(\sum z_j m_i^\circ)C_i']\}$$

(8c)

$$\ln \gamma_{\pm i} = z_i f^\gamma + 2(z_i + 1)^{-1}(\sum z_j m_j^\circ)\,[B_i' + (\sum z_j m_j^\circ)C_i']$$
$$+ 2z_i(z_i + 1)^{-1} \sum m_j^\circ[B_j' + (\sum z_j m_j^\circ)C_j']$$
$$+ (\sum z_j m_j^\circ) \sum m_j^\circ[z_j B_j'' + 2z_j(z_j + 1)^{-1}C_j']$$

(11c)

for common-ion mixtures. Here f^ϕ and f^γ are universal functions:

$$f^\phi = -(1/3)SI^{1/2}(1 + 1.2I^{1/2})^{-1}$$
$$f^\gamma = -(1/3)S[I^{1/2}(1 + 1.2I^{1/2})^{-1} + (5/3)\ln(1 + 1.2I^{1/2})]$$

with the same $S = 1.1720$ as before, while B_i^ϕ, B_i', B_i'', and C_i' are electrolyte-specific parameters, shown in Table 3 for $I = 7.874 \ m$. Again, in these equations terms involving triple-ion interactions with two different cations have been neglected, because of the empirical nature of the interaction parameters, and the relative success of a treatment without these terms.[10]

A final equation, based on the Harned coefficients α_{ij}, has been used by Lehman,[2] namely,

$$\ln \gamma_{\pm i} = \ln \gamma_{\pm i}^\circ - (\ln 10)I \sum_{j \neq i} x_j \alpha_{ij}$$

(11d)

TABLE 3. Parameters for Equations (8c) and
(11c) for $I = 7.874\ m$ at 298.15 K [12]

Salt	B^ϕ	B'	$10^4\ B''$	$10^3\ C'$
NaCl	0.0775	0.1685	1.598	+1.91
KCl	0.0491	0.1091	1.273	−1.26
MgCl$_2$	0.3585	0.8029	10.086	+7.79
CaCl$_2$	0.3218	0.7259	9.682	−0.51
$f^\phi = -0.2519$			$f^\gamma = -1.2150$	

The coefficients α_{ij} are concentration dependent, and the values for $I = 7.874\ m$ are shown in Table 2. It turns out[11] that the α's and g's are interrelated:

$$\alpha_{ij} = [(\phi_j^\circ - \phi_i^\circ)I^{-1} - \tfrac{1}{2}g_{ij}]/\ln 10 \tag{15}$$

but while g_{ij} is symmetric in the composition (hence $g_{ji} = g_{ij}$), α_{ji} is not equal to α_{ij} but equals $-[g_{ij}(2 \ln 10)^{-1} + \alpha_{ij}]$.

The application of these equations for the calculation of the osmotic coefficient of Dead Sea water and of the activity of potassium chloride therein has been carried out and discussed by the author.[3]

On the basis of the simplest equations, (7)–(11), the following activities are obtained for Dead Sea water: $a_{H_2O} = 0.7514$, $\log a_{NaCl} = 1.198$, $\log a_{KCl} = -0.102$, $\log a_{MgCl_2\ 6H_2O} = 1.591$, $\log a_{CaCl_2\ 6H_2O} = 0.788$, and $\log a_{KMgCl_3\ 6H_2O} = 1.489$. The activities of anhydrous magnesium and calcium chlorides can readily be obtained from these figures.

3. Precipitation of Salts from Dead Sea Brines

Dead Sea water (DSW) is undersaturated with respect to its chloride salt components (not so with respect to gypsum[2]). However, when it is concentrated by partial evaporation of the water in the solar salt pans, a point is reached when halite, that is, sodium chloride, starts to precipitate, the *halite point*. It is possible to predict quite accurately how much water must be evaporated isothermally, in order to reach this point, assuming a state of equilibrium between the solution and the precipitating salt.

At equilibrium between the solution and the solid salt, the activity of the salt in the solution equals that of the corresponding saturated binary solution. This, in turn, equals the exponent of $1/RT$ times the difference between the standard molar Gibbs free energy of the solid salt and that of its constituent aqueous ions. The values found in the literature[13-16] for the Gibbs free energy of formation of water, of the salts, and of the ions are summarized in Table 4. For a given anhydrous salt MCl_i

$$RT \ln a_{MCl_n}(\text{sat}) = -\Delta G_d^\circ(MCl_n)$$

$$= \Delta G_f^\circ(MCl_n(s)) - \Delta G_f^\circ(M^{n+}(aq)) - n\Delta G_f^\circ(Cl^-(aq)) \quad (16)$$

(where ΔG_d° and ΔG_f° are the Gibbs free energies of dissolution and formation, respectively) while for the hydrated salts

$$RT \ln a_{MCl_n \cdot 6H_2O}(\text{sat}) = -\Delta G_d^\circ(MCl_n \cdot 6H_2O) = \Delta G_f^\circ(MCl_n \cdot 6H_2O(s))$$

$$- \Delta G_f^\circ(M^{n+}(aq)) - n\Delta G_f^\circ(Cl^-(aq)) - 6\Delta G_f^\circ(H_2O(l)) \quad (17)$$

The logarithms of the activities of the salts in saturated aqueous solutions are also shown in Table 4. The tolerances reflect the probable error in the combined Gibbs free energies of formation.

Starting from an amount of solution containing 1 kg of water, the isothermal (25°C) evaporation of p g water causes an upward concentration of all molalities to $1000(1000 - p)^{-1}m_i^\circ$. When 192 ± 3 g water are evaporated, the solubility product of sodium chloride is just reached, and halite starts to precipitate, the *halite point* (Figure 3). The uncertainty in p corresponds to the tolerance of ± 0.008 units in $\log a_{NaCl}(\text{sat})$ from Table 3, assuming equations (7)–(11) to be exact, with the initial molalities m_i° and the parameters given in Table 1. The prediction of the halite point in DSW agrees well with the published data. Another test can be made on similar brines containing only sodium and magnesium chlorides, where the solubilities are again known, although with rather poor accuracy (Figure 4).

TABLE 4. Standard Molar Gibbs Free Energies of Formation and of Dissolution at 298.15 K

Ion, salt	$-\Delta G_f^\circ$, kJ mol^{-1}	$-\Delta G_d^\circ$, kJ mol^{-1}	$\log_{10} a_{salt}(\text{sat})$
Na$^+$(aq)[16]	261.80		
K$^+$(aq)[16]	282.68		
Mg^{2+}(aq)[16]	456.01		
Ca^{2+}(aq)[13]	553.54		
Cl$^-$(aq)[13]	131.26		
H$_2$O(l)[13]	237.18		
NaCl(s)[14]	384.04	9.02	1.580 ± 0.008
KCl(s)[14]	408.78	5.16	0.903 ± 0.008
MgCl$_2 \cdot$6H$_2$O(s)[13]	2114.97	26.64	4.46 ± 0.01
CaCl$_2 \cdot$6H$_2$O(s)[15]	2216.39	22.74	3.97 ± 0.01
KMgCl$_3 \cdot$6H$_2$O(s)[2]		22.8	4.00 ± 0.05

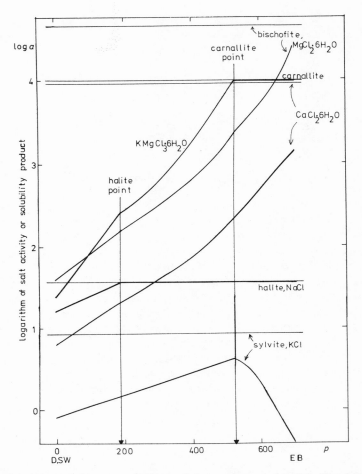

FIGURE 3. The changes of the activities of the component salts of the Dead Sea, and of carnallite, and the precipitation of salts from the brines, on isothermal evaporation of the water at 25°C. Here p represents grams of water evaporated from 1000 g water originally present in a given quantity of solution. The solubility products of sylvite (KCl), halite (NaCl), calcium chloride hexahydrate, carnallite ($KMgCl_3 \cdot 6H_2O$) and bischofite ($MgCl_2 \cdot 6H_2O$) and the halite and carnallite points are marked by appropriate straight lines. The composition at $p = 0$ corresponds to DSW (upper water mass) and at $p \sim 670$ to EB, end brine.

Beyond the halite point, the removal of sodium chloride from the system must be taken into account in following its changes when more water is evaporated isothermally. Since this changes the concentration of the chloride ions and of all the activity coefficients, and hence the activity of sodium chloride, an iterative computer method is required in order to

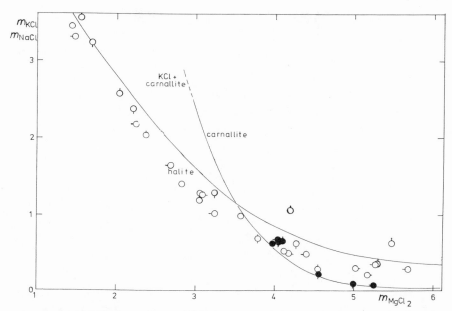

FIGURE 4. Calculated (full lines) and experimental molalities of sodium chloride in equilibrium with halite (empty circles) and of potassium chloride in equilibrium with carnallite (black circles) in the presence of the molalities of magnesium chloride given on the abscissa, at 25°C. Sources of data indicated by positions of dashes appended to circles. NaCl—none, Reference 17; left, 18; up, 19; right, 20; down, 21; KCl—none, Reference 22; left, 23; up, 24; down, 25.

calculate the amount of halite that must precipitate, for a given amount of water removed by evaporation, so that the activity of sodium chloride agrees with (16). For p g water evaporated, the molality of sodium chloride in the $(n + 1)$th iterative step is

$$\log m_{\text{NaCl}}(n + 1) = 1.58 - \log m_{\text{Cl}}(n) - 2 \log \gamma_{\pm \text{NaCl}}(n) \qquad (18)$$

where $1.58 = \log a_{\text{NaCl}}(\text{sat})$ from Table 3. Iteration is continued until the difference between two consecutive $\log m_{\text{NaCl}}$ values is below the tolerance in $\log a_{\text{NaCl}}(\text{sat})$, ± 0.008. The evolution of the activities of the various components with increasing p is shown in Figure 3. In no case are the solubility products, i.e., $\log a_{\text{salt}}(\text{sat})$ exceeded for sylvite (KCl), bischofite ($\text{MgCl}_2 \cdot 6\text{H}_2\text{O}$), or calcium chloride hexahydrate.

However, when the value $p = 520 \pm 6$ g is reached, and $78.8 \pm 1.0\%$ of the halite has precipitated, the value of $\log a_{\text{KMgCl}_3 \cdot 6\text{H}_2\text{O}}$ exceeds the limit of 4.00 ± 0.05 set for the saturation with carnallite,[2] and this double salt starts to precipitate. This is the *carnallite point* (Figure 3), again in agreement with published data. Fortunately, there are a few data in the literature on the bivariant solubilities in aqueous potassium–magnesium chloride solutions in equilibrium with carnallite (Figure 4). Agreement between the calculated and experimental solubilities is good. Beyond this point a more complicated iteration is necessary since the solution is in equilibrium with both halite and carnallite, and both are removed from the solution by precipitation as the water is evaporated away. If from a quantity of solution initially containing 1 kg of water, y mol of carnallite are precipitated upon the evaporation of p g of the water, then the solubility product equation becomes the quadratic in y:

$$(m_{KCl}^\circ - y)(1000/p)(m_{MgCl_2}^\circ - y)(1000/p)m_{Cl^-}^3 \cdot \gamma_{\pm KCl}^2 \gamma_{\pm MgCl_2}^3 a_{H_2O}^6 = 10^{4.00} \quad (19)$$

where $4.00 = \log a_{carnallite}(\text{sat})$ from Table 4. The value of $y(n + 1)$ for the $(n + 1)$th iteration is obtained from m_{Cl^-}, $\gamma_{\pm KCl}$, $\gamma_{\pm MgCl_2}$ and a_{H_2O} from the previous, nth, iteration, also taking into account the change in m_{NaCl} from (18), which affects m_{Cl^-} and the activity coefficients. With this double iteration, the changes of the solution can be followed until the composition of the *end brine* (EB) is reached on evaporation of ~ 670 g of the water, when $\sim 92\%$ of the sodium chloride and 89% of the potassium chloride have precipitated. This end brine is a solution of mainly magnesium $(4.24\ m)$ and calcium $(1.30\ m)$ chlorides, with small remnants of sodium $(0.42\ m)$ and potassium $(0.06\ m)$ chlorides. It is an extremely concentrated solution, with an ionic strength of $17.10\ m$, a water activity of only 0.371, and a density of 1.342 kg dm^{-3}.

4. Structural Aspects in Dead Sea Brines

The structural aspects of the concentrated brines become clearer when we turn from the molal scale used hitherto to the molar scale, mol dm^{-3}. Consider the Dead Sea water (DSW) (Table 1): it has a density of 1.207 kg dm^{-3}, and contains 9.80 mol of ions, of total mass 0.311 kg per kg water, hence 8.78 mol ions dm^{-3} or 5.29 ions nm^{-3}. The mean distance between two ions in DSW is therefore 0.57 nm. On the other hand, the end brine (EB) has a density of 1.342 kg dm^{-3}, contains 17.58 mol ions of total mass 0.577 kg per kg water, hence 14.96 mol ions dm^{-3} or 9.01 ions nm^{-3}. The mean distance between two ions in EB is then 0.48 nm. Thus, although the molality of the ions in EB is 79% larger than in DSW, the mean distance apart of the ions has shrunk by only 16%. The diameters of the ions themselves are all considerably smaller than 0.57 or 0.48 nm, the largest being the chloride anion, which has a Pauling diameter of 0.362 nm only. Thus the solution is not close packed in terms of the bare ions.

Consider, however, the hydrated ions (Figure 5). If only primary hydration is taken into account, then 6 mol of water per mole of mag-

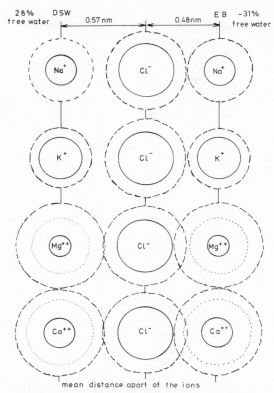

FIGURE 5. The ions in Dead Sea water (DSW, left-hand side) and in end brine (EB, right-hand side) at the mean distances apart of cations from anions. The primary (dotted lines) and secondary (dashed lines) hydration shells are indicated around the ions.

nesium or of calcium ions must be considered bound. In DSW these are altogether 11.9 mol of water, in EB 33.2 mol of water, out of the 55.5 mol present in 1 kg of water. The primary hydration, in the above sense, does not use up the "free" water in the solution. If, however, secondary hydration is also considered, the diameters of the ions[15] $(Cl \cdot 2.8H_2O)^-$, $(Mg \cdot 8.7H_2O)^{2+}$, and $(Ca \cdot 8.7H_2O)^{2+}$, 0.51, 0.62, and 0.62 nm are commensurate with the mean distance apart of the ions. The water bound in these ions, together with that in the ions $(Na \cdot 3.2H_2O)^+$ and $(K \cdot 2.2H_2O)^+$, becomes 39.7 mol in DSW and 82.1 mol in EB, for an amount of solution containing 1 kg = 55.5 mol of water. In the former there is still 28% "free" water, but in the latter there is a deficit of 31%, that is, the hydration numbers appearing in the formulas of the ions cannot be fully attained. Indeed, these are hydration numbers valid for infinitely dilute solutions,[11] and should not be expected to be valid in the very concentrated brines.

This decrease in the hydration numbers is illustrated in Fig. 5 as a penetration of the hydration shell of the chloride ions into the secondary hydration shells of the magnesium or the calcium ions (and even into the primary shell of the latter). This effect can be interpreted in terms of the formation of solvent-shared ion pairs, but too much weight should not be put on this deduction from an illustration. A further distinction that can be made between DSW and EB is that in the latter all the hydrated ions must be in intimate contact, so that EB is a *molten hydrated salt*, whereas in DSW the monovalent hydrated ions are not in contact, so that all the ions have more space available to them, in a continuous medium, the "free" water.

What kind of order, if any, prevails in these concentrated solutions? Several diffraction studies have been made of alkali metal and alkaline earth metal chlorides. In order to make comparisons, the results obtained in ~ 5 m solutions at room temperature, where the ions are on the average ~ 0.5 nm apart, have been assembled in Table 5. The interpretation of the radial distribution functions in terms of pair correlation functions is virtually impossible in the case of X-ray diffraction.[26-28] Six pairs of nearest neighbors: cation–cation, cation–anion, cation–solvent, anion–anion, anion–solvent, and solvent–solvent appear in X-ray diffraction, while in neutron diffraction, sensitive to hydrogen atoms, a further four correlations appear. However, recent neutron diffraction studies[29,30] that utilize second-order differences of data obtained with different isotopes, give unambiguous results as described by Enderby in Chapter 2. There is no conclusive evidence for ion–ion contacts or contact ion pairs (except for cesium chloride solutions), but solvent-shared ion pairs cannot, of course, be discounted, because of the general proximity of the ions. The "irregular" cation hydration noted in the alkaline earth chlorides, which is not further explained in the original paper,[28] may point to nonisotropic location of water molecules, hence to some kind of ion pairing.

TABLE 5. X-Ray Diffraction Studies of Aqueous Chlorides, ~ 5 m at Room Temperature; Ions ~ 0.5 nm Apart on Average

Salt and source	Qualitative findings
LiCl (Brady,[26] Lawrence,[27] Narten[30])	No ion–ion contacts
NaCl (Lawrence,[27] Enderby[29])	No ion–ion contacts
KCl (Brady[26])	
CsCl (Lawrence[27])	Ion–ion contacts found
MgCl$_2$ (Albright[28])	Irregular cation hydration
CaCl$_2$ (Albright,[28] Enderby[29])	Irregular cation hydration[28]

In all cases chloride ions interact with ~ 6–8 water molecules.

The pair correlation functions, when they become available, shed light on the interactions between the ions and the solvent and between the ions themselves. There is little hope, however, that diffraction methods will be able to give pair correlation functions in mixtures of electrolytes. Therefore these interactions must be obtained from other kinds of data, above all from the thermodynamic data. Here the interaction parameters g_{ij} play an important role. These have been given[31] the physical significance of describing the interactions between groups of ions: ... quadruplets, triplets, down to pairs. As the concentrations decrease, the pair interactions become dominant. If only these are considered, then the interaction parameter for the excess Gibbs free energy, g_{ij}, can be written for a ternary common anion solution:

$$\lim_{I \to 0} g_{ij} = 2g(i, j) - g(i, i) - g(j, j) \tag{20}$$

Here $g(i, i)$ pertains to interactions between two similar cations $i-i$ in the binary solution, and similarly for $g(j, j)$, whereas $g(i, j)$ is a new interaction $i-j$, present only in the ternary solutions. Note that no new pairwise cation–anion nor anion–anion interactions occur on mixing the two binary solutions. A positive value of g_{ij} thus can be interpreted as resulting from a conservation of the interactions in the binary solutions, with a low tendency for new ones to occur in the mixture. On the contrary, a negative value of g_{ij} points to a high probability for the occurrence of the new, $i-j$ interactions, and a decreased importance of the homoionic interactions.

At higher concentrations, triplet interactions also contribute to g_{ij}, and may even cause a change of its sign. Empirically, the expression

$$g_{ij} = A_{02} + (2A_{03}/3)I^{1/2} \tag{21}$$

has been found to account for the data in systems such as ours (up to quaternary systems) up to an ionic strength of 5 m. The coefficients A_{02} and A_{03}, describing the pair and triplet interactions, are shown in Table 6

TABLE 6. Coefficients A_{02} and A_{03} Contributing to the Interaction Parameter g_{ij}, Equation (21), for a Number of Ternary and Quaternary Aqueous Systems at 25°C

Mixture	A_{02}	A_{03}
$NaCl + KCl + H_2O$	−0.017	−0.010
$NaCl + MgCl_2 + H_2O$	0.072	−0.016
$KCl + MgCl_2 + H_2O$	0.091	−0.041
$(NaCl + KCl, 1:1) + MgCl_2 + H_2O$	0.077	−0.027
$(NaCl + KCl, 1:1) + CaCl_2 + H_2O$	0.081	−0.034

for some of the ternary and quaternary systems of interest. The former, A_{02}, is obviously given by the right-hand side of equation (20). The latter includes negligible contributions of three-cation interactions (these being rare on account of the coulombic repulsion), but interactions i–Cl–j should dominate the contribution to the excess function, again in comparison with i–Cl–i and j–Cl–j in the binary solutions.

Let us examine the values in Tables 2 and 6. The ternary NaCl + KCl mixture has a negative g_{ij}. When such a ternary mixture of fixed Na : K ratio is mixed with a magnesium or a calcium chloride solution, the new quaternary (pseudoternary) mixture exhibits a positive g_{ij}.[6] Thus, the old, preferred Na–K interactions are even more preferred over new divalent cation–univalent cation interactions. However, the negative values of A_{02} in these systems lead to an increasing importance of these latter interactions (as triplet interactions) with increasing ionic strength (Figure 6). The barrier (positive g_{ij} values) against K–Mg interactions, for instance, decreases enormously when the ionic strength of DSW is reached. Admittedly, this requires extrapolation beyond the range of known[6] validity of equation (21), but it is certain that the barrier against K–Mg interactions is less than that against (K + Na)–Mg interactions, since this is

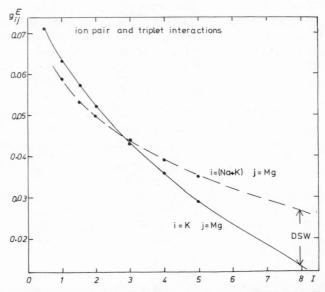

FIGURE 6. Ion pair and ion triplet interactions in the ternary KCl + MgCl$_2$ and in the pseudoternary (quaternary) (NaCl + KCl, 1 : 1) + MgCl$_2$ aqueous mixtures, as a function of the ionic strength. The black dots represent experimental data,[6] and the curves were obtained with equation (21) and the parameters in Table 6.

already true by $I = 3\,m$, where the data exist. In other words, since $|A_{03}(\text{K, Mg})| > |A_{03}(\text{K} + \text{Na, Mg})|$ (Table 6) and both are negative, K–Cl–Mg triplets have a higher probability of existence than (K, Na)–Cl–Mg triplets. Sodium ions are seen to interfere with the formation of the ion configurations necessary for carnallite precipitation. In fact, this double salt precipitates from DSW only after most of the sodium chloride has been removed. (See the previous section and Figure 3.)

The word "interactions" has been used in the above qualitative discussion without a clear definition. It is obvious that it does not refer to pure Coulombic interactions, since these cancel out when ions of the same charge type are compared, i.e., K with (K + Na) interacting with Mg, and since they cannot account for attractive cation–cation interactions (i.e., negative g_{ij} values). What is meant by "interaction" is a more subtle effect, in which the solvent is involved. The effects that the ions have on the structure of the solvent, breaking or enhancing it, via ion–dipole interactions and hydrogen bonding, etc., can lead to cooperative effects, and a lowering of the Gibbs free energy of the system, hence an "attractive interaction" between ions of the same sign. It will be of interest to follow this idea by means of molecular dynamics or Monte Carlo calculations, provided reasonable potential functions can be defined; the beginnings of such an approach are described by Hills and Adams in Chapter 4.

ACKNOWLEDGMENTS

Fruitful discussions with D. Saad and help with computer programming from S. Glikberg are gratefully acknowledged.

List of Symbols

A_i	coefficient in osmotic and activity coefficient equations (7) and (9)
A_{02}, A_{03}	coefficients in interaction parameter equation (21)
a_i	(relative) activity of component i in the mixture
B_i	coefficient in osmotic and activity coefficient equations (7) and (9)
B_i', B_i'', B_i^ϕ	coefficients in osmotic and activity coefficient equations (8c) and (11c)
C_i	coefficient in osmotic and activity coefficient equations (7) and (9)
C_i'	coefficient in osmotic and activity coefficient equations (8c) and (11c)

c_i°	molar concentration (mol dm^{-3}) of salt i
D_i	coefficient in osmotic and activity coefficient equations (7) and (9)
f^ϕ, f^γ	coefficients in osmotic and activity coefficient equations (8c) and (11c)
G^E	excess Gibbs free energy of ternary mixture with respect to its constituent binary mixtures
$G_i^{E\circ}$	excess Gibbs free energy of binary solution of salt i
ΔG_f°	Gibbs free energy of formation
ΔG_d°	Gibbs free energy of dissolution
g_{ij}	interaction parameter in common ion ternary mixtures, equation (14)
I	ionic strength, molal scale
i, j	salt components in solution
m_i	molality (mol/kg water) of salt i in the mixture
m_i°	molality of salt i in its pure (binary) solution
p	grams of water evaporated from a solution initially containing 1 kg water
q	order of interactions (pairs and higher for $q = 0$, which is implied but not written, triplets and higher for $q = 1$, etc.) in interaction parameter g_{ijq}
R	gas constant
S	coefficient in activity coefficient equations (7), (8c), (9), and (11c)
T	(absolute) temperature
x_i	ionic strength fraction
y	moles of carnallite precipitated from DSW initially containing 1 kg water
z_i	charge on cation of salt i
α_{ij}	Harned coefficient of activity coefficient of salt i in a ternary mixture containing also salt j, equation (15)
$\gamma_{\pm i}$	activity coefficient (molal scale) of salt i in a mixture
$\gamma_{\pm i}^\circ$	activity coefficient of salt i in its pure (binary) solution
ϕ	osmotic coefficient of mixture
ϕ_i°	osmotic coefficient of salt i in its pure (binary) solution

References

1. D. Neev and K. O. Emery, *Isr. Geol. Surv. Bull.* **41**, 1 (1967).
2. A. Lehrman, *Geochim. Cosmochim. Acta* **31**, 2309 (1967).
3. Y. Marcus, *Geochim. Cosmochim. Acta* **41**, 1739 (1977).
4. M. H. Lietzke and R. W. Stoughton, *J. Solution Chem.* **1**, 299 (1972); *J. Inorg. Nucl. Chem.* **36**, 1315 (1974).

5. M. H. Lietzke and R. W. Stoughton, *J. Phys. Chem.* **66**, 508 (1962).
6. D. Saad, Ph.D. thesis, Hebrew University of Jerusalem, 1978; D. Saad, J. Padova, and Y. Marcus, *J. Solution Chem.* **4**, 983 (1975); D. Saad and J. Padova, *J. Solution Chem.* **6**, 191 (1977); J. Padova, D. Saad, and D. Rosenzweig, *J. Solution Chem.* **6**, 309 (1977).
7. P. J. Reilley, R. H. Wood, and R. A. Robinson, *J. Phys. Chem.* **75**, 1305 (1971).
8. G. Scatchard, R. M. Rush, and J. S. Johnson, *J. Phys. Chem.* **74**, 3786 (1970).
9. G. Scatchard, *J. Am. Chem. Soc.* **83**, 2636 (1961); R. M. Rush and J. S. Johnson, *J. Phys. Chem.* **72**, 767 (1968).
10. K. S. Pitzer and J. J. Kim, *J. Am. Chem. Soc.* **96**, 5701 (1974).
11. Y. Marcus, *Introduction to Liquid State Chemistry*, Wiley, Chichester (1977), Chap. 6, pp. 247–249, and Appendix 6.2.
12. K. S. Pitzer and G. Mayoraga, *J. Phys. Chem.* **77**, 2300 (1973).
13. U.S. National Bureau of Standards, Technical Notes TN-270-3 (1968), TN-270-6 (1971).
14. JANAF Thermochem. Tables, 2nd ed., NSRDS-NBS-37 (1971).
15. V. I. Voznesenskaya, in *Vop. Fiz. Khim. Rastvorov Elektrolit.*, Ed. G. I. Mikulin, Khimiya, Leningrad (1968), pp. 172–201.
16. D. A. Johnson, *Some Thermodynamic Aspects of Inorganic Chemistry*, Cambridge University, Cambridge, England (1968).
17. S. Takegami, *Mem. Coll. Sci. Kyoto Imp. Univ.* **4**, 317 (1921).
18. N. S. Kurnakov and S. F. Zemkuznys, *Z. Anorg. Chem.* **140**, 153 (1924). (Quoted also in *Gmelins Handbuch der Anorganische Chemie*, 8th ed., Anhangsband K, 1942, Berlin, Verlag Chemie G.M.B.H., Ergänzungsband, 1970, Weinheim/Bergstz, Verlag Chemie G.M.B.H.)
19. Y. W. Lee, Master's thesis, University of Mississippi, 1969.
20. D. K. Reddy, quoted by J. C. Dhawan, Ph.D. thesis, University of Mississippi, 1974.
21. I. C. Juan, Master's thesis, University of Mississippi, 1966.
22. W. B. Lee and A. C. Egerton, *J. Chem. Soc.* **123**, 706 (1923).
23. N. S. Kurnakov, D. P. Manoev, and N. A. Osokoreva, *Kali* **2**, 25 (1932).
24. M. R. Bloch and J. Schnerb, *Bull. Res. Counc. Isr.* **3**, 151 (1953).
25. A. B. Zdanovskii, quoted in A. Lehrman, *Geochim. Cosmochim. Acta* **31**, 2317 (1967).
26. G. W. Brady and J. T. Krause, *J. Chem. Phys.* **27**, 304 (1957); G. W. Brady, *J. Chem. Phys.* **28**, 464 (1958).
27. R. M. Lawrence and R. F. Kruh, *J. Chem. Phys.* **47**, 4758 (1967).
28. J. N. Albright, *J. Chem. Phys.* **56**, 3783 (1972).
29. J. E. Enderby, R. A. Howe, and W. S. Howells, *Chem. Phys. Lett.* **21**, 105 (1973); J. E. Enderby, *Proc. R. Soc. London Ser.* A **345**, 107 (1975).
30. A. H. Narten, F. Vaslow, and H. A. Levy, *J. Chem. Phys.* **58**, 5017 (1973).
31. J. L. Friedman, *Ionic Solution Theory*, Wiley-Interscience, New York (1962).

8

The Metal–Electrolyte
Interface Problem

I. L. Cooper, J. A. Harrison, and J. Holloway

1. Introduction

The purpose of this chapter is to discuss the equilibrium data that can be obtained for the mercury–electrolyte and the solid-metal–electrolyte interfaces, together with their interpretation. Unfortunately, kinetic data cannot yet be obtained by electrochemical relaxation nor by spectroscopic methods for these systems. The thermodynamic data must, therefore, for the moment, stand alone. It is, however, sufficiently characteristic for us to be convinced that a satisfactory phenomenological explanation of the data is possible, which will incorporate the known basic properties of the species present at the interface. This model will be presented herein. A detailed molecular model is another matter and cannot be attempted without kinetic data.

The subject of the structure of the interface has a long history and is considered by many electrochemists and physical chemists to be settled. We hope in this chapter to undermine that illusion by adopting a different starting point to the problem. It must be said that although the modern view of the subject started with Grahame,[1] his original tentative suggestions have become elevated into an accepted theory. It was after reexamination of some of these ideas in a recent series of papers[2-7] that the need for a more realistic theory became, almost painfully, obvious.

I. L. Cooper, J. A. Harrison, and J. Holloway • School of Chemistry, University of Newcastle upon Tyne, Newcastle upon Tyne, England.

2. Experimental Data

The data must be discussed in a fragmented way because experimental techniques have not, until recently, been capable of completely analyzing one system under all conditions of potential.

2.1. Aqueous Solutions: Hg

The experiments are well documented for the Hg–electrolyte system (see, for example, the review by Payne[8]). The primary data consist of differential capacity(C)–potential(E) and interfacial tension γ–E data. In addition, the zero point of charge ($E_{e.c.m.}$) is measured independently. The C–E and γ–E curves can be measured as a function of concentration (or chemical potential of the salt, μ) temperature (T) and, rarely, pressure. Charge density on the metal q^m–E and surface excess $\Gamma_{\pm(\omega)}$–E curves are then obtained by the thermodynamic relations

$$q^M = \int_{E_{e.c.m.}}^{E} C \, dE \tag{1}$$

$$\Gamma_{\pm(\omega)} = -\left(\frac{\partial \gamma}{\partial \mu}\right)_{E^{\mp}} \tag{2}$$

2.2. Aqueous Solutions: Hg (Far Negative Potentials)

In the presence of a Faradaic reaction it is necessary to separate the double-layer contribution using standard electrochemical models. Although a systematic investigation of the dependence of the resulting C–E curves on ion type has not been carried out the following preliminary observations seem to appear from the data[9–13]:

1. C–E curves rise at negative-going potentials in a manner similar to the far positive-going region (which is associated predominantly with the anion). The order of the capacity at a potential with respect to a fixed reference electrode is

$$BaCl_2, BaI_2 > KCl, KOH > MgCl_2, LiCl$$

2. There does not seem to be a hump similar to the one on the positive potential side, although the accuracy of the measurements, with the measuring technique used, may not yet be sufficiently accurate to show it.

3. The C–E behavior seems to be determined by the cation and not the anion.

2.3. Aqueous Solutions: Solid Metals

A critical investigation of the *C–E* curves on solid metals has been carried out[14] recently. The *C–E* curves on various metals were measured during a linear potential sweep. Various potential pretreatments and chemical polishing methods were applied to the surface before the experiments. Typical results have been compared to measurements on Hg in the same solution. It has also been reported (see, for example, Reference 15) that on metal single crystals a large orientation effect exists. This has yet to be confirmed.

2.4. Molten Salts: Solid Metals

There is still some dispute as to the form of the *C–E* curves in metal–molten salt systems. In chloride melts Graves and Inman[16] have reviewed the situation. They pointed out the large influence of Faradaic currents at the high temperatures involved. Subsequent authors have given more attention to separating out the double-layer and Faradaic contributions. However, the methods still lag behind those routinely used in aqueous solution investigations. It now seems that the *C–E* curves in, for example, chloride melts are parabolic in shape about $E_{e.c.m.}$. The value of C at the C minimum seems to be dependent on the molten salt but less dependent on the metal. At, for example, 800°C in a chloride melt C at $E_{e.c.m.}$ is ~ 40 μF cm^{-2}. (But see remark above concerning Faradaic currents.) However, the surface roughness is not known as no detailed investigation has been made.

2.5. Experimental Methods for C–E

Most information in the literature has been obtained using bridge methods. The electrode is a dropping Hg electrode and the applied potential a dc potential with superimposed small-amplitude ac. The alternating current is balanced at a known time in the Hg drop life.

The bridge has largely been superseded by phase-sensitive detection methods in which the input ac potential and output current signals are compared. Although this can be operated at dropping Hg there is some advantage in using a standing Hg drop. If the dc potential is a linear potential sweep then *C–E* curves can be displayed directly, provided the ac frequency is low enough to make ohmic resistance corrections negligible. *C–E* curves can be obtained in this way without serious "poisoning" effects. The methods which use "purified" carbon to clean solutions over long periods of time seem to be discredited. In many cases the surface can be cleaned by an appropriate potential profile applied before the linear potential sweep in which the measurement is made.

2.6. Characteristics of the Data

A detailed description and critical assessment of the thermodynamic data and its interrelation has been published previously.[3] A few of the most important facts are picked out in this section, as these will be used later to formulate what factors are essential for a workable theory.

Of the available data the C–E, q^m–E, and $\Gamma_{\pm(\omega)}$–E curves are probably the most interesting. The C–E data have always attracted the most interest because, especially for the Hg–electrolyte system, of the precision with which they can be measured and their characteristic shape (see for example Reference 17). Unfortunately, the C–E curve, in spite of its attractions, is probably the most difficult to interpret qualitatively and even more difficult to interpret quantitatively. The $\Gamma_{\pm(\omega)}$–E data (Reference 18 gives some examples of experimental data) must be the simplest quantities to consider as they are directly concerned with the separate anion and cation distributions in the interfacial region. The measured $\Gamma_{\pm(\omega)}$–E curves for all common salts are similar and have the following important characteristics:

1. $E - E_{e.c.m.} < 0$: $\Gamma_{+(\omega)}$ increases with $E < E_{e.c.m.}$ approximately linearly. $\Gamma_{-(\omega)}$ reaches a limiting value.

2. $E - E_{e.c.m.} > 0$: $\Gamma_{+(\omega)}$ goes through a minimum then increases again, in the region of the "hump" in the C–E curve. $\Gamma_{-(\omega)}$ increases, possibly approximately linearly, in this region.

Even a crude theory must be able to provide an interpretation of these key facts before there is any hope of proceeding to a theory for the C–E curves.

Once an interpretation for $\Gamma_{\pm(\omega)}$–E is found then the q^m–E curve follows from the relation

$$q^s = -q^m = -F(\Gamma_{+(\omega)} - \Gamma_{-(\omega)}) \tag{3}$$

The C–E data, which are related to q^m and hence to $\Gamma_{+(\omega)}$ and $\Gamma_{-(\omega)}$ by the relation

$$C = \frac{dq^m}{dE} = F\left(\frac{d\Gamma_{-(\omega)}}{dE} - \frac{d\Gamma_{+(\omega)}}{dE}\right) \tag{4}$$

have the following characteristics:

1. $E - E_{e.c.m.} < 0$: C has a limiting value at $E < E_{e.c.m.}$ which eventually rises slowly, depending in the cation, at very negative potentials.

2. $E - E_{e.c.m.} \simeq 0$: C rises steeply depending on the anion.

3. $E - E_{e.c.m.} > 0$: The "hump" occurs.

4. $E - E_{e.c.m.} \gg 0$: A steep rise in C occurs.

The hump region is pronounced at low temperature ($\sim 20°C$) and tends to disappear as the temperature is raised.

3. The Prevailing View: Theory and Experiment

The "prevailing view," which appears in many research papers and textbooks (see, for example, Reference 19), is based on the following:

i. The distance of closest approach of nonadsorbed ions to the metal surface characterizes the outer Helmholtz plane (OHP), which serves to subdivide the double layer into inner and outer regions.

ii. Ions redistribute in the outer region according to diffuse layer theory, in response to the OHP potential rather than the external potential.

iii. $\Gamma_{\pm(\omega)}-E$ curves for KF can be rationalized (for negative values of E at least) on the basis of diffuse layer theory. It is assumed that the $\Gamma_{+(\omega)}-E$ curve represents the excess cation charge in the diffuse layer, and this is then used to calculate the $\Gamma_{-(\omega)}-E$ curve via the associated value of q^m. On this basis, F^- is said to be nonadsorbed.

iv. Similar arguments to those in (iii) have been applied to other systems, such as KCl, KI etc., and the "adsorbed charge" deduced on the assumption that $\Gamma_{+(\omega)}$ represents the excess cation charge in the diffuse layer.

v. Since all $C-E$ data are asymmetric with respect to $E_{e.c.m.}$, this characteristic asymmetry cannot be associated solely with adsorption, and so must be ascribed to the inner layer region itself.

vi. Since the OHP is situated a fixed distance from the metal surface, variations in its potential can only be ascribed to a variable inner layer permittivity. This has led to the belief that the shapes of the $C-E$ curves involving F^- are a consequence of orientational properties of a layer of solvent molecules lying within the inner region. A recent paper[20] invokes a four-state model for water molecules in the inner layer. No attempt has been made to correlate these models with the $\Gamma_{\pm(\omega)}-E$ curves.

4. Theoretical Treatment of the Double-Layer Problem

We have discussed elsewhere[3, 5] some of the internal inconsistencies associated with the prevailing view. Some of the more important points are as follows:

i. Theoretical models should be directed towards an explanation of the $\Gamma_{\pm(\omega)}-E$ data, from which the $C-E$ data follow.

ii. All $\Gamma_{\pm(\omega)}-E$ data show similar behavior, although shifted in magnitude and position. We have demonstrated[3] that humps in the capacity data are associated with minima in the $\Gamma_{\pm(\omega)}-E$ data.

iii. Anions and cations cannot be brought simultaneously into the interface through a dielectric modification of an external potential. Some other effect must be involved.

iv. Although different ions are known to undergo differing extents of solvation, no account is taken of variations in ion size.

v. Although the water layer adjacent to the electrode is assumed to be capable of undergoing orientational motion, ions are in some way prevented from displacing these solvent molecules.

We have previously demonstrated[5] that the double-layer problem can be reformulated in such a way that the mean distance from the metal of the net charge distribution in the interfacial region becomes a significant variable. Since diffuse layer theory treats ions as structureless point charges, all distance effects must be contained in the inner region, and it is here that differences in ion sizes must be of importance.

Although the double-layer problem is usually formulated in terms of q^m, it is instructive to invert the problem, treating the potential E as the independent variable, thereby bringing the theory into line with experimental procedure.

We shall consider a series of simple models for the interface.

4.1. Simple Diffuse Layer Theory

The model can be summarized as follows:

a. Assume the electrolyte remains homogeneous up to the boundary with the planar metal surface.

b. Assume the applied potential E represents that between the metal surface and the bulk electrolyte, such that $q^m = 0$ when $E = 0$.

c. Assume that E falls off linearly within the electrolyte, and is completely screened out over a distance $2L_D$, where L_D is the Debye length of the electrolyte.

d. The mean distance from the metal of the anion and cation distributions will be L_D, where the mean potential is $\frac{1}{2}E$.

e. Assume that each ion is capable of independent response to the potential, distributing classically according to Boltzmann statistics. Then

$$n_\pm(E) = n_0 \exp\left(\mp \frac{1}{2}\frac{e}{kT}E\right) \tag{5}$$

where $n_0 = 2L_D c_0$ is the total number of anions (cations) per unit area within a distance $2L_D$ of the metal, and c_0 is the bulk concentration. Then $n_\pm(E) = \Gamma_\pm(E)/N_A + n_0$, where N_A is Avogadro's number. Then

$$q^m(E) = -e[n_+(E) - n_-(E)] = 2en_0 \sinh\left(\frac{1}{2}\frac{e}{kT}E\right) \tag{6}$$

and

$$C(E) = \frac{dq^m(E)}{dE} = \frac{\varepsilon_0 \varepsilon_r}{L_D} \cosh\left(\frac{1}{2}\frac{e}{kT}E\right) \tag{7}$$

since $L_D = (kT\varepsilon_0\varepsilon_r/2c_0e^2)^{1/2}$, with ε_r the relative permittivity of the electrolyte.

This is the well-known result of diffuse layer theory, derived on the basis of certain well-defined physical assumptions. As is well known, this theory is unable to explain the characteristic shape of C–E curves since it yields a symmetric function of E.

4.2. Diffuse Layer Theory Plus Inner Region Dielectric

If we assume that a dielectric medium of width $2L_i$ and relative permittivity ε_i is interposed between the metal surface and the bulk electrolyte, we find from simple electrostatics that the mean effective potential acting at the mean position of the ion distribution (now situated distance $2L_i + L_D$ from the metal) is

$$E_{\text{eff}} = \gamma_0 E = \frac{\varepsilon_i/L_i}{\varepsilon_i/L_i + \varepsilon_r/L_D} E \tag{8}$$

Then

$$C(E) = \frac{C_i(0)C_0(0)}{C_i(0) + C_0(0)} \cosh\left(\frac{1}{2}\frac{e}{kT}\gamma_0 E\right) \tag{9}$$

where

$$C_i(0) = \varepsilon_0\varepsilon_i/L_i \quad \text{and} \quad C_0(0) = \varepsilon_0\varepsilon_r/L_D.$$

Note that the familiar series capacitor result is regained only when $E = 0$ i.e.,

$$C^{-1} = C_i^{-1} + C_0^{-1} \tag{10}$$

If we generalize this analysis to the case $C_i = C_i(E)$, then $\gamma_0 = \gamma_0(E)$, and we find

$$C(E) = \frac{C_i(E)C_0(0)}{C_i(E) + C_0(0)} \cosh\left[\frac{1}{2}\frac{e}{kT}\gamma_0(E)E\right]\left[1 + E\frac{C_0(0)}{C_i(E)}\frac{dC_i(E)/dE}{C_i(E) + C_0(0)}\right] \tag{11}$$

This result arises from a consistent theoretical treatment of the prevailing view, where the inner and outer regions are explicitly incorporated into the theory. We do not infer a separation simply by the incorporation of diffuse layer theory into the experimental data by means of a dual-capacitor model. From the above equation, we note that the assumption of a potential-dependent inner region capacity precludes the possibility of a formal separation of internal and external contributions, which forms the foundation of the prevailing view. It is interesting also to note that $C_i = C_i(E)$ includes not only the case $\varepsilon_i = \varepsilon_i(E)$, but also the possibility $L_i = L_i(E)$.

4.3. Incorporation of Ion Size

The simplest model to account for different anion and cation sizes is to allow anions and cations to have different distances of closest approach $(2L_\pm)$ from the metal surface, and to allow for solvation by the replacement of ε_i by ε_∞. Then we find the result

$$C(E) = C_0(0)\left[\gamma_- \exp\left(\frac{1}{2}\frac{e}{kT}\gamma_- E\right) + \gamma_+ \exp\left(-\frac{1}{2}\frac{e}{kT}\gamma_+ E\right)\right] \quad (12)$$

where $\gamma_\pm = (\varepsilon_\infty/L_\pm)/(\varepsilon_\infty/L_\pm + \varepsilon_r/L_D)$. This will exhibit the desired asymmetry about $E_{\text{e.c.m.}}$ (i.e., $E = 0$) if $L_+ > L_-$, but there is continuity between high- and low-concentration data (only L_D is a function of concentration) so that the overall shapes of the curves must be similar for different concentrations. This does not agree with the experimental facts.

The theory presented here can, and has been, developed further; see Reference 23.

5. New Experiments

More sophisticated instrumentation has recently opened the way to vastly improved experimental capability and precision in this field.[21, 22] The speed with which these measurements can be made has also been drastically reduced. The instrumentation is based on the control of E and in this case the simultaneous measurement of i–E and $Z(\omega)$–E curves using a minicomputer. The subsequent processing of the data is carried out by a

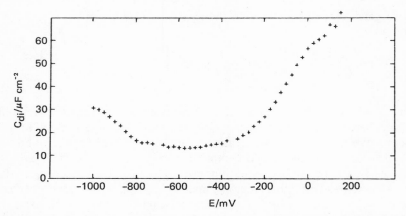

FIGURE 1. Differential capacity–potential curves obtained using the new instrumentation at a standing Hg drop in contact with 10^{-2} M HClO$_4$. The potential scale is relative to an S.C.E. (NaCl) reference electrode.

main-frame computer in contact with the minicomputer, by means of a communications interface. The measurements of double-layer capacity then become part of general electrode kinetic investigations. It is then possible as a matter of routine to make measurements which were practically impossible previously. These include the possibility of measuring $C–E$ curves under extremes of concentration and potential for Hg and solid metals in contact with aqueous, nonaqueous, and molten salt electrolytes. This data is almost completely lacking at present. An example is shown in Figure 1 where the $C–E$ data for Hg in contact with 10^{-4} M $HClO_4$ is given.

References

1. D. C. Grahame, *Chem. Rev.* **41**, 441 (1947).
2. I. L. Cooper and J. A. Harrison, *J. Electroanal. Chem.* **66**, 85 (1975).
3. I. L. Cooper and J. A. Harrison, *Electrochim. Acta* **22**, 1365 (1977).
4. I. L. Cooper and J. A. Harrison, *Electrochim. Acta* **22**, 519 (1977).
5. I. L. Cooper and J. A. Harrison, *Electrochim. Acta* **22**, 1361 (1977).
6. I. L. Cooper and J. A. Harrison, *Electrochim. Acta* **23**, 545 (1978).
7. I. L. Cooper and J. A. Harrison, *J. Electroanal. Chem.* **86**, 425 (1978).
8. R. Payne, in *Progress in Surface and Membrane Science*, Vol. 6, Ed. J. F. Danielli *et al.*, Academic, New York (1973), p. 51.
9. R. M. Reeves, M. Sluyters-Rehbach, and J. H. Sluyters, *J. Electroanal. Chem.* **34**, 55 (1972).
10. R. M. Reeves, M. Sluyters-Rehbach, and J. H. Sluyters, *J. Electroanal. Chem.* **34**, 69 (1972).
11. R. M. Reeves, M. Sluyters-Rehbach, and J. H. Sluyters, *J. Electroanal. Chem.* **36**, 101 (1972).
12. R. M. Reeves, M. Sluyters-Rehbach, and J. H. Sluyters, *J. Electroanal. Chem.* **36**, 287 (1972).
13. M. Sluyters-Rehbach, J. S. M. C. Brenkel, and J. H. Sluyters, *J. Electroanal. Chem.* **48**, 411 (1973).
14. I. L. Cooper, J. A. Harrison, and D. R. Sandbach, *Electrochim. Acta* **23**, 527 (1978).
15. A. Hamelin and J.-P. Bellier, *Surf. Sci.* **78**, 159 (1978).
16. A. D. Graves and D. Inman, *J. Electroanal. Chem.* **25**, 357 (1970).
17. J. A. Harrison, J. E. B. Randles, and D. J. Schiffrin, *J. Electroanal. Chem.* **48**, 359 (1973).
18. J. A. Harrison, J. E. B. Randles, and D. J. Schiffrin, *J. Electroanal. Chem.* **25**, 197 (1970).
19. J. Albery, *Electrode Kinetics*, Clarendon, Oxford (1975).
20. R. Parsons, *J. Electroanal. Chem.* **59**, 229 (1975).
21. J. A. Harrison, Third Symposium on Electrode Kinetics, Boston, 1979.
22. J. A. Harrison and C. E. Small, *Electrochim. Acta* **25**, 447 (1980).
23. I. L. Cooper, J. A. Harrison, and J. Holloway, A preliminary experimental and theoretical study of the differential-capacity potential curves, especially in low concentrations of electrolytes (submitted to *Electrochim. Acta*).

9

Acid–Base Properties of Concentrated Electrolyte Solutions

John A. Duffy and Malcolm D. Ingram

1. Introduction

Some metal salts are very soluble in water and it is possible to obtain solutions where the salt : water ratio is, for example, 1 : 6 or 1 : 4 or even lower. Indeed, it is sometimes possible to obtain such solutions simply by heating the solid salt hydrate, for example, $CaCl_2 \cdot 6H_2O$ melts to a clear liquid at 29.9°C. Some of these solutions possess remarkable chemical properties on account of their high acidity,[1-3] and it has been argued by the present authors[3] that this acidity can be predicted on the basis of calculations using the "optical basicity" approach. In this chapter we shall discuss in more detail why it is that concentrated salt solutions should exhibit acidic properties and attempt to interpret their behavior in the wider context of strong protonic acids.

The interaction between acids and bases can be approached from several viewpoints. According to the Brønsted–Lowry theory, an acidic substance can be regarded in terms of its tendency to transfer a proton from itself to a base. Writing the acid as HX and the base as B, we have the equilibrium

$$HX + B \rightleftharpoons BH^+ + X^- \tag{1}$$

Thus the stronger the acid HX, the more the equilibrium lies to the right (for a given base, B).

John A. Duffy and Malcolm D. Ingram • Department of Chemistry, The University, Old Aberdeen, Scotland.

If the base is a water molecule, we are concerned with the equilibrium

$$HX + H_2O \rightleftharpoons H_3O^+ + X^- \qquad (2)$$

In dilute aqueous solution, the ions H_3O^+ and X^- are hydrated, and although it is difficult to define the boundary of the hydration sheath, nevertheless (provided the solution is sufficiently dilute) there will be regions of space where the water molecules are unperturbed by ionic forces. Such molecules can be regarded as being in the bulk solvent and outside the sphere of hydration. Thus the solution consists of hydrated cations and hydrated anions separated by "inert" water molecules. It is this simple picture which allows one to adopt a second concept of acidity, which is in terms of the hydrogen ion concentration. Strong acids are those with large dissociation constants in equation (2) and vice versa.

It is possible to consider the strength of a protonic acid in terms of the nature of its conjugate base, that is, the anion to which it gives rise after losing a proton. Thus, it is expected that the strength of an acid would be related to the affinity of its conjugate base for protons. If the equilibrium, equation (2), is thought of in terms of the conjugate base and a water molecule competing for a proton,

$$X^- + H^+ \rightarrow HX \qquad (3)$$

$$H_2O + H^+ \rightarrow H_3O^+ \qquad (4)$$

then it follows that if the affinity of X^- for protons is greater than that of H_2O, then HX will be a weak acid, whereas if it is less, then HX will be a strong acid. An important factor influencing acid strength is therefore the *anion basicity*.

2. Anion Basicity

The approach to anion basicity is simplified very much if we restrict ourselves to oxide-containing systems. Most acids belong to this class and only a few, e.g., the hydrogen halide acids, are excluded from consideration. The restriction to oxidic systems means that the basic atom(s) to be considered, when comparing different anions (e.g., SO_4^{2-} with PO_4^{3-}), is always oxygen: there is no need to switch from oxygen to, say, chlorine, as would be necessary if SO_4^{2-} were being compared with Cl^-. As will become apparent, this simplification enables us to introduce the concept of *electron donor power* when discussing the basicity of anions. Ionic oxides, such as CaO, have a very great electron donor power and are very basic. Oxyanions, such as SO_4^{2-} or PO_4^{3-}, have much weaker electron donor powers owing to the polarizing influence of the sulfur or phosphorus on the oxygens.

TABLE 1. Basicity-Moderating
Parameters, γ, of Some Elements

Element	γ	Element	γ
K	0.73	Zn	1.82
Na	0.87	H	2.50
Li	1.00	P	2.50
Ca	1.00	S	3.04
Mg	1.28	Cl	3.73
Al	1.65		

Elsewhere[4] the authors have shown that the electron donor power of the oxygens can be probed by introducing small concentrations of metal ions which, by their acidic nature, accept negative charge. Metal ions such as Tl^+, Pb^{2+}, and Bi^{3+} appear willing to accept very large amounts of negative charge (from very basic environments) or more moderate amounts (from less basic environments). The amount of negative charge donated to the metal ion affects the frequency, v, of its $^1S_0 \to {}^3P_1$ absorption band, which occurs in the uv region. In the case of Pb^{2+}, for the condition where there is zero donation, $v = 60,700 \text{ cm}^{-1}$, while for the condition approaching maximum donation in CaO, $v = 29,700 \text{ cm}^{-1}$, representing a spectroscopic shift of 31,000 cm^{-1}. The $^1S_0 \to {}^3P_1$ transition involves transference of an electron from the $6s$ orbital of the Pb^{2+} to the $6p$. It is the increased screening of the nucleus, effected by negative charge being donated to the Pb^{2+} ion, which is responsible, in part, for the observed frequency decrease. The smaller electron donor power of an oxyanion produces a smaller frequency shift (to $v_{oxyanion}$, say), and this can be expressed, relative to the shift produced by an ionic oxide, such as CaO, as the ratio λ:

$$\lambda = \frac{60,700 - v_{oxyanion}}{31,000} \tag{5}$$

This numerical expression of electron donor power is termed the optical basicity.[†][4, 5]

Examination of the experimental optical basicity values for a wide range of oxidic materials has revealed that it is possible to assign a *basicity-moderating parameter*, γ, to each element.[6, 7] This parameter expresses the influence that the central atom in oxyanions such as PO_4^{3-}, SO_4^{2-}, etc. has upon the electron donor power of the oxygens. In other words, it expresses the ability of the central atom to contract the electron charge clouds of the

---

† A CNDO molecular orbital treatment of oxyanion basicity has shown that λ is proportional to the amount of negative charge borne by the oxygen atoms.[17]

TABLE 2. Values of λ for Some Oxidic
Species and of pK for Parent Acids

Oxidic species	λ	Parent acid	pK^a
H_2O	0.40	H_3O^+	-1.74
OH^-	0.70	H_2O	15.74
$H_2PO_4^-$	0.47	H_3PO_4	2.16
HPO_4^{2-}	0.55	$H_2PO_4^-$	7.21
PO_4^{3-}	0.625	HPO_4^{2-}	12.33
$H_3P_2O_7^-$	0.44	$H_4P_2O_7$	1.52
$H_2P_2O_7^{2-}$	0.49	$H_3P_2O_7^-$	2.36
$HP_2O_7^{3-}$	0.53	$H_2P_2O_7^{2-}$	6.60
$P_2O_7^{4-}$	0.57	$HP_2O_7^{3-}$	9.25
HSO_4^-	0.42	H_2SO_4	
SO_4^{2-}	0.50	HSO_4^-	1.96
ClO_4^-	0.37	$HClO_4$	(-3)

[a] Values of pK are taken from D. D. Perrin, *Dissociation Constants of Inorganic Acids and Bases in Aqueous Solutions*, Butterworths, London (1969).

oxygen atoms. The γ values of several elements have been determined and some are given in Table 1. From a knowledge of γ and the stoichiometry of the oxyanion, it is possible to calculate values of λ (see Appendix), and also to calculate changes brought about by such effects as protonation.[7–9] The λ values of several oxyanions and water, calculated in this way, are shown in Table 2.

3. Acids in Dilute Solution

In this section we shall consider both neutral and negatively charged oxyacids. Most of these are rather weak and, as can be seen from Table 2, they have pK values which lie between that for the dissociation

$$H_3O^+ \rightleftharpoons H^+ + H_2O \qquad (6)$$

which is -1.74, and the pK for the dissociation

$$H_2O \rightleftharpoons H^+ + OH^- \qquad (7)$$

which is 15.74. It can also be seen from Table 2 that the optical basicities of the anions to which the acids give rise (i.e., their conjugate bases) lie between the optical basicity values of the conjugate bases, H_2O and OH^-. These have λ values of 0.40 and 0.70, respectively, and the linear equation relating pK with λ, which satisfies the data for these species, is therefore[8]

$$pK = 58(\lambda - 0.40) - 1.74 \qquad (8)$$

This equation is shown in Figure 1 together with the plots of pK vs. λ for the oxyacids in Table 2. The points lie fairly close to the line; thus the idea that anions of low basicity are derived from the stronger acids and *vice versa* (see Section 1) is roughly true. Equation (8) should be regarded more as indicating a trend than a means of accurately ranking oxyacids. For example, the predicted pK for H_2SO_4 of -0.6 [substituting $\lambda = 0.42$ in equation (8)] is almost certainly too positive, but then a similar "error" is evident for HSO_4^- (Figure 1), the predicted pK value again being too positive by more than two units. Furthermore, acids of elements in oxidation states less than the maximum, e.g., $HClO_2$, have not been included in Figure 1 because it is not certain whether the γ values in Table 1 hold good for oxidation states other than the maximum value.

Probably the long-term significance of the pK–optical-basicity relationship is that it enables us to consider water as an *oxidic* solvent, and thus it establishes for the first time a direct link between acid–base reac-

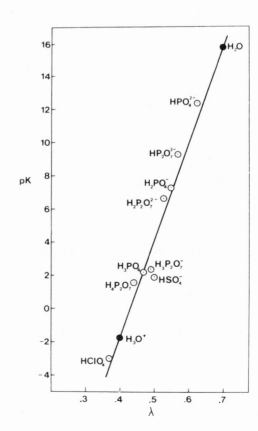

FIGURE 1. Plot of pK of the acids denoted versus the optical basicity, λ, of their conjugate bases. The straight line [equation (8) in text] has been drawn to pass through the points for H_3O^+ and H_2O.

tions in aqueous solutions and those in (normally anhydrous) molten salts and glasses. However, it must be emphasized that the link between optical basicity and acid strength is *entirely empirical*, and that the existence of such a relationship has neither been fully justified in terms of thermodynamic arguments nor of detailed molecular models.

One difficulty which arose in our first attempts to predict the strengths of acids in aqueous solution (prior to the appearance of Reference 8) was whether the optical basicity should be correlated with the enthalpy or the free energy of acid–base reactions. To simplify this problem we followed a "trial and error" approach which can, if necessary, be justified by reference to detailed arguments presented elsewhere by Larson and Hepler.[10] In dealing with the relative strengths of weak organic acids, they discussed the thermodynamics of reactions of the type

$$HA(aq) + R^-(aq) \rightarrow A^-(q) + HR(aq) \tag{9}$$

They considered the ΔH and ΔS values for this reaction as the sums of internal (intramolecular) and external (environmental) effects. Thus

$$\Delta H^\circ = \Delta H_{int} + \Delta H_{ext} \tag{10}$$

and

$$\Delta S^\circ = \Delta S_{int} + \Delta S_{ext} \tag{11}$$

To a good approximation, Larson and Hepler found that $\Delta S_{int} \simeq 0$, and that $\Delta H_{ext} \simeq T \Delta S_{ext}$. Thus, they propounded the very interesting thermodynamic relationship

$$\Delta G^\circ \simeq \Delta H_{int} \tag{12}$$

On this basis, the effects of substituents on the relative strengths of weak organic acids can be predicted in terms of very largely enthalpic effects in isolated or gas-phase acid–base reactions. They reasoned that it is this "principle of compensation" which underlies the success of the Hammett equation ($\log K = \rho\sigma$), where σ values can be assigned to substituent groups and then applied in a wide variety of organic acid–base reactions. Probably similar compensation effects are operative for *inorganic* oxyacids, and the (limited) success of the pK–λ relationship can be attributed to the same factors which have made the Hammett equation so useful in many aspects of physical organic chemistry.

4. Acidity of Metal Aquo Complexes

Traditionally, the oxides and hydroxides of the elements are classified as "basic" or "acidic." The former dissolve in water or dilute acids to give aquo complexes; for example, with the oxide of a bivalent metal, we may write

$$MO + 2H^+ \rightarrow M^{2+} + H_2O \tag{13}$$

and it is understood that M^{2+} denotes an (hydrated) aquo complex, usually of formula $[M(H_2O)_4]^{2+}$ or $[M(H_2O)_6]^{2+}$. Acidic oxides dissolve in water to give acid molecules in equilibrium with anions; with the oxide of a hexavalent element, for example, which can be a metal or nonmetal, we may write (assuming four coordination for M)

$$MO_3 + H_2O \rightarrow MO_2(OH)_2 \tag{14}$$

$$MO_2(OH)_2 \rightleftharpoons H^+ + [MO_3(OH)]^- \tag{15}$$

$$[MO_3(OH)]^- \rightleftharpoons H^+ + [MO_4]^{2-} \tag{16}$$

However, these two types of behavior, either "basic" or "acidic," sometimes merge, as illustrated by the amphoteric behavior of, say, $Al(OH)_3$, or the protonation of sulfuric acid in liquid HF to produce $[H_3SO_4]^+$. It is possible to envisage a scheme, embracing both these aspects of behavior, which involves a set of equilibria starting with the hydrolytic dissociation of the aquo complex. If four coordination for M is

TABLE 3. Values of pK for Metal Ions in Aqueous Solution and Relevant Optical Basicity Data

Metal ion	Experimental pK^a	λ value for conjugate base[b]	
		four coordination	six coordination
Li^+		0.475	0.45
Na^+		0.49	0.46
Mg^{2+}	11.4	0.42	0.41
Ca^{2+}	12.9	0.475	0.45
Zn^{2+}	9.0	0.36	0.375
Al^{3+}	5.0	0.33	0.35
$AlOH^{2+}$	9.9	0.40	0.40
$Al(OH)_2^+$	15.6	0.48	0.45
Mn^{2+}	10.5		
Fe^{2+}	9.5		
Fe^{3+}	2.7		
$FeOH^{2+}$	5.9		

[a] The hydrolysis constants pK are taken from C. F. Baes and R. E. Mesmer, *The Hydrolysis of Cations*, Wiley, New York (1976), and are rounded off to one decimal place.
[b] λ is calculated for the species $[M(H_2O)_3(OH)]^{n+}$ and $[M(H_2O)_5(OH)]^{n+}$ (and, for aluminum, further hydrolyzed species) using γ values in Table 1 and equation (24) (see Appendix).

assumed throughout, then eight equilibria [equations (17)–(19)] can be written:

$$[M(H_2O)_4]^{n+} \rightleftharpoons H^+ + [M(OH)(H_2O)_3]^{(n-1)+} \tag{17}$$

$$[M(OH)(H_2O)_3]^{(n-1)+} \rightleftharpoons H^+ + [M(OH)_2(H_2O)_2]^{(n-2)+} \tag{18}$$

and so on until the final equilibrium:

$$[MO_3(OH)]^{(7-n)-} \rightleftharpoons H^+ + [MO_4]^{(8-n)-} \tag{19}$$

The various aquo, hydroxo, and oxo complexes involved in these equilibria can be regarded as a set with general formula $[M_aH_bO_c]^{(na+b-2c)+}$, n being the oxidation number of M. The actual species predominating under a given set of conditions depends upon the oxidation number n and also the chemical nature of M. For example, under mildly acidic conditions, if $n = 2$

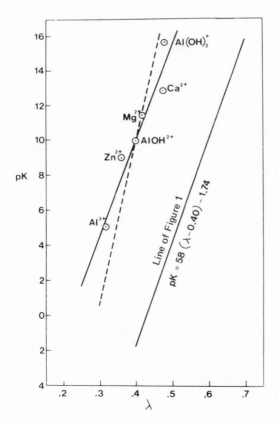

FIGURE 2. Plot of pK for the acid dissociation of hydrated cations denoted versus the optical basicity, λ, of the appropriately deprotonated $[M(H_2O)_4]^{n+}$ species. The plot of equation (8) is shown for comparison. (The pK–λ plot for hexaaquo ions is shown as the broken line.)

then the predominant species is $[M(H_2O)_4]^{2+}$, whereas if $n = 6$ it is $[MO_4]^{2-}$.

From the above discussion, it is quite justifiable to view the acidity of metal aquo and hydroxo complexes as no different from that of conventional oxyacids and proton-containing oxyanions. The acidity arises from the same mechanism, namely, the contracting of the electron cloud onto the oxygens by the atom M and by the hydrogens. It has already been shown (previous section) how this effect is related to the pK values of oxyacids by considering the optical basicities of their conjugate bases, and it follows that this relationship [i.e., equation (8)] might also be expected to apply to the acidity of aquo and hydroxoaquo complexes of metal ions.

pK values for such complexes are listed for a selection of metal ions in Table 3. Also listed is the λ value of the conjugate base in the equilibrium to which the pK value refers (except when the basicity-moderating parameter is not known). Two sets of λ values are presented, one assuming four coordination and the other six coordination. The values for four coordination are plotted against pK in Figure 2. It can be seen that, compared with the plot of equation (8) which applies to oxyacids (Figure 1), the trend is similar but that there is a displacement of approximately ten units of pK. A roughly similar displacement is also observed when the λ values of six-coordinate species are plotted (shown as a dashed line in Figure 2).

5. Concentrated Aqueous Solutions

The difference between the behavior of metal aquo complexes (and hydroxoaquo complexes) on the one hand and oxyacids (and proton-containing anions) on the other, which is illustrated so forcefully in Figure 2, prompts the question as to whether or not the pK values for metal ions in aqueous solution reflect the true acidity of the metal aquo complex. However, it must be remembered that the pK values are determined in dilute solutions where the aquo complex is extensively hydrated. Quite possibly this outer hydration sheath causes a more severe attenuation of acidity for aquo complexes than for oxyacids.

The determination of the "true" acidity of a metal aquo complex therefore demands reference to the conditions pertaining in concentrated solutions rather than in dilute solutions. As discussed earlier, this requires the abandonment of measurements that involve hydrogen ion concentration, and instead, acidity must be viewed in terms of the tendency of the aquo complex to donate protons (to some suitable acceptor).

For concentrated solutions of oxyacids, the tendency to donate protons is measured using weak organic bases:

$$HX + B \rightleftharpoons BH^+ + X^- \tag{20}$$

The concentration ratio c_{BH^+}/c_B is measured spectrophotometrically, and the protonating tendency of the acid HX is expressed as the *Hammett acidity function, H_0*:[11]

$$H_0 = pK^B - \log_{10} \frac{c_{BH^+}}{c_B} \qquad (21)$$

where pK^B is the dissociation of the conjugate acid of the indicator base (previously determined). (It should be noted that in dilute aqueous solution H_0 becomes identical with pH.)

This method of measuring acidity can be applied to concentrated aqueous solutions of metal salts.[2, 3] Compared with aqueous solutions of mineral acids, concentrated solutions of salts present a number of difficulties. Often the solution must be held at an elevated temperature, to maintain solubility, and it is therefore important that the salt should be chemically inert to avoid side reactions, e.g., nitration of the base. Also, it is necessary that the solutions should have good uv transparency for the necessary spectrophotometric analysis.

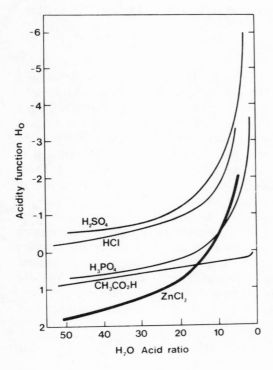

FIGURE 3. Increase in Hammett acidity function, H_0, for aqueous solutions of $ZnCl_2$, as the $H_2O : ZnCl_2$ ratio decreases (bold line). Trends for acetic, phosphoric, hydrochloric, and sulfuric acids are shown for comparison.

Of the various salt solutions considered, those of zinc chloride have so far been found to be best for these studies; it is possible to obtain solutions with a $ZnCl_2 : H_2O$ ratio of $1 : 4$ even at ambient temperature. When dilute, the Zn^{2+} ion imparts very feeble acidity to its solutions; since $pK = 8.96$ for the equilibrium

$$Zn^{2+} + H_2O \rightleftharpoons ZnOH^+ + H^+ \tag{22}$$

the acidity is less than for, say, acetic acid. Indeed, these dilute solutions have weaker acidity functions than do acetic acid solutions of similar concentration (Figure 3). However, with increasing concentration, H_0 actually becomes *negative*, and when the $ZnCl_2 : H_2O$ ratio is in the region of $1 : 10$, the trend, as can be seen in Figure 3, is similar to that of the strongest mineral acids such as sulfuric acid.

6. nmr Studies

So far, zinc chloride is the only salt for which Hammett acidity function trends have been directly measured. However, it is possible to obtain some idea of the acidity and indication of the H_0 values of solutions of certain other salts from proton nmr data.

It is expected that a covalently bound hydrogen atom would be more readily detachable (as a proton) if the negative charge, constituting the covalent bond, were distorted in such a way that there was less of it residing on the hydrogen atom. Thus it follows that there might be some relationship between protonating power and proton nmr chemical shift. Such a relationship has indeed been found for powerful mineral acids in aqueous solution, the H_0 values varying approximately linearly with the proton chemical shift δ_H, relative to pure water.[12] Figure 4 uses previously published data to show this for aqueous nitric acid and hydrochloric acid. Also included in Figure 4 is the point for $ZnCl_2 \cdot 6H_2O$,* and it can be seen that δ_H is somewhat too small; for the point to lie close to the line, the original shift should have been around -3.1 rather than -2.5.

There is a further indication that the Zn^{2+} shift is anomalous and that a "better" value would have been approximately -3.0. This will be referred to in the next section. It is unfortunate that this (apparent) anomaly exists, since it prevents us from using with confidence the straight line in Figure 4 for obtaining anything but very rough estimates of H_0 for

---

* The proton chemical shift has been experimentally determined[1] for $Zn(NO_3)_2 \cdot 6H_2O$ but not for $ZnCl_2 \cdot 6H_2O$; however, the (small) correction needed can be estimated from the behavior of solutions of nitrates and chlorides of other metal ions.

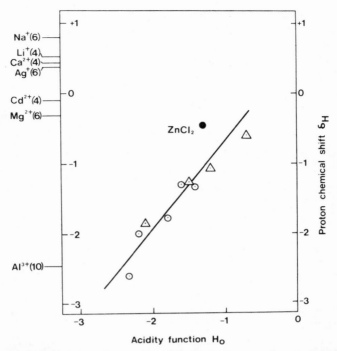

FIGURE 4. Plot of proton chemical shift δ_H (relative to pure water) versus Hammett acidity function H_0 for nitric acid ($\odot$) and hydrochloric acid ($\triangle$). [nmr data are from Reference 12 and H_0 data from C. H. Rochester, *Acidity Functions*, Academic, London (1970)]. Also plotted is the point for $ZnCl_2$. The proton chemical shifts for aqueous solutions of various metal nitrates (water : salt ratios in parentheses) are denoted on the left-hand ordinate. The latter data are taken from Reference 1, with the temperature equal to 25°C. It should be noted that they have been reduced by 2.0 to compensate for the change of solvent (see text).

aqueous salt solutions from their proton chemical shifts (relative to pure water). Shifts relative to tetramethylammonium ion have been previously obtained for concentrated solutions of Li^+, Na^+, Ag^+, Mg^{2+}, Ca^{2+}, Cd^{2+}, and Al^{3+},[1] and are converted to shifts relative to pure water by adding 2.0.† The resulting values of δ_H are shown in Figure 4, and it is apparent that the protonating power of most of these solutions is rather feeble, except for those of Al^{3+}, Mg^{2+}, and possibly Cd^{2+}. The solution of $Al(NO_3)_3 \cdot 10H_2O$ indeed appears to have an H_0 of perhaps -2 or -3, and it would be of interest to investigate the acidity of solutions with lower water content. In this context, it is worth recalling Angell *et al.*'s comment[1]

---

† -2.0 is the chemical shift of the tetramethylammonium ion relative to water.

that mixtures of $Al(NO_3)_3 \cdot 10H_2O$ and $AlCl_3 \cdot 10H_2O$ "can be used to dissolve noble metals at a much greater rate than can be achieved with boiling aqua regia." It is apparent that concentrated solutions of some metal salts provide a new range of solvents of widely varying acidity, and it is possible that these might find use for organic and inorganic synthesis and in the study of reaction mechanisms. Published work in this area is so far largely nonexistent.†

The acidic properties of salt hydrates are not confined just to the liquid state. The authors are indebted to Professor Y. Marcus (of the Hebrew University of Jerusalem, Israel) for pointing out previously published work concerning the interaction of solid $ZnSO_4 \cdot H_2O$, $MgSO_4 \cdot H_2O$, and other metal sulfate monohydrates with solid NaCl to produce HCl, even at low temperatures.[13] As far back as 1950, Walling[14] discovered that Hammett indicators absorbed on to the surface of certain metal sulfates underwent protonation. The importance of this surface acidity in heterogeneous catalysis has been pointed out.[15]

7. Predicted Acidity of Metal Aquo Complexes

Referring again to the Hammett acidity function of concentrated $ZnCl_2$ solutions, the rapid rise in protonating power of these solutions (Figure 3) shows how the true acidity of the $[Zn(H_2O)_4]^{2+}$ (or whatever ion is present) begins to be revealed as the outer sphere water molecules are removed. Thus the pK value of 8.96, obtained in dilute solution, is totally inappropriate for expressing the acidity of the aquo complex.

We have argued in this chapter that metal aquo complexes, neutral oxyacids, and oxyanions can all be regarded as members of one family. Thus, it is possible that the relationship between the ionization of an acid in dilute solution and the optical basicity of its conjugate base, i.e., equation (8), extends to metal aquo complexes, provided that the pK value refers to the aquo complex alone and not the *hydrated* aquo complex (which exists in dilute solutions). Hence, for aquo complexes of metal ions whose γ values are known, it is possible to calculate λ for the appropriate hydroxoaquo complex (the conjugate base) and to then use equation (8) for obtaining pK for the aquo complex. These values are given for some metal ions in Table 4.

It is known that mineral acids which dissociate readily in dilute aqueous solution are generally those that are strong protonating agents in concentrated solution (see Section 1). This connection between simple Arrhenius acidity and Hammett acidity is, however, only qualitative, and there

† But see also Reference 18.

TABLE 4. Predicted Acidities of Metal Hexaaquo Complexes

Metal ion	Optical basicity[a] of $[M(H_2O)_5(OH)]^{(n-1)+}$	Predicted pK[b]	Predicted H_0[c]
Li^+	0.45	1.3	-2
Mg^{2+}	0.41	-1.2	-6
Ca^{2+}	0.45	1.3	-2
Zn^{2+}	0.375	-3.2	-11
Al^{3+}	0.35	-4.6	-14

[a] Values from Table 3.
[b] Predicted pK is obtained using equation (8).
[c] Predicted H_0 is obtained using the straight line in Figure 6.

is no universal relationship between pK and H_0. Nevertheless, in view of the fairly successful correlation between proton nmr shift and acidity function, it is worthwhile to consider the predicted pK values of Table 4 together with the nmr data.

The nmr data available which allow the widest comparison of metal ions are for nitrates. The proton chemical shift (expressed as the constant, a, to avoid any slight temperature dependence) is plotted against the predicted pK values for the aquo complexes of Li^+, Mg^{2+}, Ca^{2+}, Zn^{2+}, and Al^{3+} (Figure 5). The points lie fairly close to a straight line, although a better fit for Zn^{2+} would have been obtained if the shift had been extended

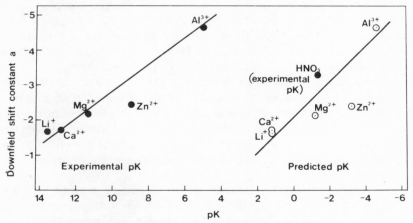

FIGURE 5. Proton magnetic shift constant, a, for concentrated aqueous solutions of metal nitrates (salt : water ratio of 1 : 10) versus (○) predicted pK values of the metal hexaaquo complex (see Table 4), and (●) pK values experimentally determined for aqueous solutions of the metal ions (see Table 3).

by approximately -0.5 (see previous section). It should be noted that the line runs close to the point for nitric acid for which the experimental pK value is plotted. Figure 5 includes the plot of chemical shift versus the experimental pK for the above metal ions in dilute solution, and it can be seen how these points lie well away from the experimental pK of nitric acid. The nmr data are therefore correlated very much better if the predicted pK values are taken to represent the acidity of the aquo complexes. The two sets of data in Figure 5 are separated by 10–12 units of pK, and this difference can be regarded as resulting from the enormous attenuating effect that outer sphere water molecules have on the powerful acidity of the metal aquo complexes.

In a recent publication[16] the arguments which led to the pK–optical-basicity relationship [equation (8)] have been extended to oxyacids (of formula H_xXO_4) in *concentrated* aqueous solution. It was envisaged that the stability of a protonated indicator base, BH^+ [and hence the ratio BH^+/B and the acidity function signaled by the base—see equation (21)] would be greater, the lower the optical basicity of the solvating cluster around the BH^+ ion. The solvating cluster, as the oxyacid transfers its proton to the base, B, has the formula $(H_{x-1}XO_4 \cdot nH_2O)^-$, and it has been found that the λ value of this species correlates particularly well with the acidity function of the solution for clusters having $n = 1$ or $n = 2$ (see Figure 6).

It is interesting to investigate what values of H_0 are predicted if this correlation is applied to metal aquo complexes, since the concept allows us to ignore completely the effect of the anions that are present in solution.

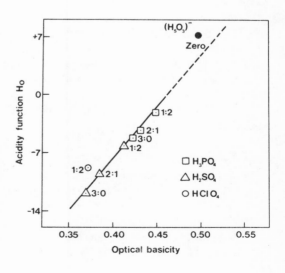

FIGURE 6. Plot of H_0 for oxyacid–water systems (acid : water mole ratio designated) versus the optical basicity of the deprotonated cluster solvating the Hammett indicator base; effective size of cluster is three molecules. (Reproduced from Reference 16 by kind permission of the Chemical Society, London.)

Anions almost certainly participate in solvating the BH^+ conjugate acid, and because of their negative charge they are expected to have an enormous effect in deprotonating BH^+. Thus it is anticipated that the values of H_0 predicted for the aquo complexes will be too negative by several units. If, after transference of a proton from the hexaaquo complex to the base, the resultant solvated species $[M(H_2O)_5(OH)]^{(n-1)+}$ is regarded as equivalent to a three-molecule cluster [i.e., writing it as $(H_7MO_4 \cdot 2H_2O)^{(n-1)+}$], then Figure 6 can be used for obtaining rough values of H_0. These values are in Table 4, and it is at once apparent that they represent very high acidities. For example, the value of $H_0 = -11$ for $[Zn(H_2O)_6]^{2+}$ is too negative by ten units of H_0, compared with the experimental value. However, it is again emphasized that the values in Table 4 represent acidities of metal hexaaquo ions in the absence of balancing anions. The difference between these values and experimental values can be thought of as arising from the severe attenuating effects of the anions. It is therefore expected that the acidity functions of concentrated solutions will depend not only upon the nature of the metal ion but also the anion. This effect is foreshadowed by further nmr results of Angell et al.[1]

Basicity-moderating parameters have not so far been assigned to transition metal ions, and thus it is not possible to predict the acidity of their aquo complexes. In view of the marked acid behavior of some of these ions even in dilute solution (for example, compare the experimental pK values of Fe^{3+} and Al^{3+} in Table 3), acidity function studies of their concentrated solutions should prove illuminating.

Appendix

When the optical basicity concept was first introduced,[4] it was applied mainly to glass systems where the ratio of basic oxide : acidic oxide often could be varied over a wide range. The symbol Λ was used for the optical basicity of a "bulk" medium, and this was regarded as representing the average basicity of the various oxyanion groups present in the medium. The basicity-moderating parameters were chosen so that

$$\Lambda = \frac{z_A r_A}{2\gamma_A} + \frac{z_B r_B}{2\gamma_B} + \cdots \tag{23}$$

where z_A, z_B, ... are the oxidation numbers of the various cations (e.g., $z_{Na} = 1$ and $z_B = 3$ in a Na_2O–B_2O_3 glass), r_A, r_B, ... their molar ratios with respect to the total number of oxygens present, and γ_A, γ_B, ... their basicity-moderating parameters.

Equation (23) can be rearranged[9] in a form which is convenient for obtaining the group basicities of individual oxyanion species. These optical basicities are symbolized λ to distinguish them from bulk Λ values. The rearranged form of equation (23) is

$$\lambda = 1 - \left[\frac{z_A r_A}{2}\left(1 - \frac{1}{\gamma_A}\right) + \frac{z_B r_B}{2}\left(1 - \frac{1}{\gamma_B}\right) + \cdots \right] \quad (24)$$

Equation (24) is used for obtaining the λ values in Tables 2, 3, and 4.

References

1. E. J. Sare, C. T. Moynihan, and C. A. Angell, *J. Phys. Chem.* **77**, 1869 (1973).
2. J. A. Duffy and M. D. Ingram, *Inorg. Chem.* **16**, 2988 (1977).
3. J. A. Duffy and M. D. Ingram, *Inorg. Chem.* **17**, 2798 (1978).
4. J. A. Duffy and M. D. Ingram, *J. Am. Chem. Soc.* **93**, 6448 (1971).
5. J. Wong and C. A. Angell, *Glass Structure by Spectroscopy*, Marcel Dekker, New York (1976), pp. 182–185.
6. J. A. Duffy and M. D. Ingram, *J. Chem. Soc. Chem. Comm.*, 635 (1973).
7. J. A. Duffy and M. D. Ingram, *J. Inorg. Nucl. Chem.* **37**, 1203 (1975).
8. J. A. Duffy and M. D. Ingram, *J. Inorg. Nucl. Chem.* **38**, 1831 (1976).
9. J. A. Duffy and M. D. Ingram, *J. Non-Cryst. Solids* **21**, 373 (1976).
10. J. W. Larson and L. G. Hepler, in *Solute–Solvent Interactions*, Vol. 1, Eds. J. F. Coetzee and C. D. Ritchie, Marcel Dekker, New York (1969), Chap. 4.
11. L. P. Hammett, *Physical Organic Chemistry*, 2nd ed., McGraw-Hill, New York (1970), pp. 263–313.
12. N. G. Zarakhani and M. I. Vinnik, *Russ. J. Phys. Chem.* **36**, 483 (1962).
13. L. Ben-Dor and R. Margalith, *Inorg. Chim. Acta* **1**, 49 (1968).
14. C. Walling, *J. Am. Chem. Soc.* **72**, 1164 (1950).
15. T. Takeshita, R. Ohnishi, T. Matsui, and K. Tanabe, *J. Phys. Chem.* **69**, 4077 (1965).
16. J. A. Duffy, *J. Chem. Soc. Faraday Trans. 1*, **75**, 1606 (1979).
17. J. H. Binks and J. A. Duffy, *J. Non-Cryst. Solids*, (1980) in press.
18. R. D. Dyer, R. M. Frono, M. D. Schiavelli, and M. D. Ingram, *J. Phys. Chem.* (1980) in press.

10

Molten Salts and Electrolyte Solutions

Some Aspects of Their Transport Properties with Respect to a Common Theory of Liquids

Joachim Richter

1. Introduction

In 1957 Fuoss and Onsager[1] wrote the following in a fundamental paper about the conductance of unassociated electrolytes:

> The problem of concentrated solutions cannot, in our opinion, be solved by any extension of the present theory (of liquids), which is based on a smoothed ionic distribution. The approach must start by an adequate theory for fused salts, which must then be followed by the theoretical treatment of the effect on the radial distribution function of adding uncharged (solvent) molecules.

Up to the present, we still do not have such a model to match the theory of Onsager and Fuoss, or even a common theory of liquids. Since water-free molten salt mixtures have simpler structures than concentrated aqueous electrolyte solutions, it would be reasonable first to develop a theory for molten salts and then to extend it to aqueous electrolyte solutions by approximation. In this chapter the transport properties of molten salt mixtures and concentrated aqueous electrolyte solutions will be discussed as far as the phenomenological data can contribute to such a theory.

In the last 15 years the transport properties of molten salts have been reviewed in several papers and books. In 1964 papers by Klemm[2] and

Joachim Richter • Lehrgebiet Physikalische Chemie der RWTH Aachen, Templergraben 59, D-5100 Aachen, Federal Republic of Germany.

Sundheim[3] appeared in the books edited by Blander and Sundheim, and in 1967 Janz and Reeves[4] wrote the paper "Molten-Salt Electrolytes— Transport Properties"[4] and this and the handbook by Janz,[5] presented the first comprehensive compilations of transport data in molten salts. Tomlinson[6] reported self-diffusion, conductivity, and equivalent conductance especially for one-component melts. It is interesting to note that in 1968 Tomlinson pointed out that the Nernst–Einstein relation has no phenomenological basis in the case of molten salts and, of course, of concentrated aqueous electrolyte solutions, because the Nernst–Einstein relation is a limiting law for very dilute solutions. Nevertheless one can still find discussions about this topic nowadays. Moynihan[7] reviewed the mass transport in fused salts in 1971, and in 1974 Copeland[8] discussed the transport properties of ionic liquids with respect to the models of ionic liquid transport. In the same year Bloom and Snook[9] provided a detailed statistical treatment of various molten salt models. In 1977 we reviewed the theory and the experimental possibilities, together with some results, for thermal diffusion in ionic melts.[10]

The transport properties of electrolyte solutions were reviewed by Newman[11] in 1967. Falkenhagen and co-workers[12] discussed the mass transport properties of ionized dilute electrolytes in terms of nonequilibrium statistical mechanics, and finally Miller[13] dealt with the application of irreversible thermodynamics to transport in electrolyte solutions in 1977.

As frequently mentioned, a theoretical treatment of concentrated aqueous solutions will be best founded on the view that these solutions are a special case of molten salts. Starting from this point of view, the statistical thermodynamics of molten salts and concentrated aqueous electrolytes have been reviewed by Braunstein[14] in 1971 and his conclusion still holds today: "No one simple model can be expected to cover the entire range of electrolyte solutions from dilute solution to fused salt, and an exact statistical-mechanical treatment remains in the future." Therefore we will discuss some theoretical aspects and experimental values of molten salts and of concentrated aqueous electrolyte solutions, with which we can calculate the phenomenological coefficients of the different species of the system. These coefficients are characteristic parameters for a given ionic liquid.

Transport phenomena in fluid systems can be expressed in terms of thermodynamics of irreversible processes or of nonequilibrium statistical mechanics. Experimental methods for the investigation of transport phenomena are in most cases based upon optical or electrochemical techniques. Research on transport phenomena in molten salt mixtures and aqueous electrolyte solutions has basically three aims: (1) the determination, for isothermal systems, of material data, such as the values of electric conductance, transport numbers, diffusion coefficients, viscosity, and, for nonisothermal systems, of transported entropies, etc.; (2) the description of the liquid phase by means of the phenomenological coefficients; and (3) the collection of pointers to the structure of the melt or the electrolyte solution and to the transport mechanism occurring in a microscopic volume of the system.

The measurement and significance of the material data are in principle no problem and the definitions of the transport data are obvious. The discussion of the structure of molten salts and concentrated electrolyte solutions, however, is essentially more complex. The elucidation of the transport mechanism cannot be performed solely by means of thermodynamics of irreversible processes because the measured transport data can only be interpreted on the basis of hypothetical structural models. Correlating these investigations with spectroscopic methods, one obtains, from the spectra, information about the bonds which are present in the liquid phase. Using this information we can define structural elements from which the melt or the electrolyte solution is built up. But, the microkinetic transport mechanism cannot be fully understood in this way. New force can be brought to bear on the discussion of microscopic transport mechanisms by computer simulation, mentioned below.

Another possibility for gaining insight into transport mechanisms, is through a comparison of a given transport property, e.g., diffusion or thermal diffusion, in a binary molten salt mixture, with the corresponding transport property in a highly concentrated aqueous electrolyte solution containing one of the components of the molten mixture as electrolyte.

2. Methods for Describing the Structure of Molten Salts and Electrolyte Solutions

In Figure 1 the entire composition range from pure water to the anhydrous molten salt is given in a mole fraction diagram, after Braunstein.[15] The central ordinate axis represents the mole fraction of water and of salt, respectively: in the lower part of this diagram is pure water ($x_{H_2O} = 1$ and $x_{salt} = 0$); in the upper part we have the anhydrous salt ($x_{H_2O} = 0$ and $x_{salt} = 1$). The logarithm of the mole fractions of water and salt is plotted using an arbitrary scale on the right and left sides, respectively, so that we obtain a two-dimensional geometrical illustration of the different concentration domains, for the aqueous solution, etc. These concentration ranges or areas overlap and have only qualitative meaning.

Area A characterizes the region of highly dilute aqueous electrolyte solutions, in which limiting laws hold. Area B is the region of concentrated aqueous electrolyte solutions, which is the topic of this chapter, together

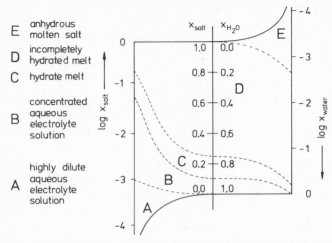

FIGURE 1. Concentration ranges for water + salt mixtures (after Braunstein[15]).

with the anhydrous melt (area E). Area C symbolizes the concentration range of the hydrate melts with complete hydration sheaths, and area D corresponds to the concentration range of incompletely hydrated melts with less than 4–6 mol of water per mole of salt. In the last two regions there are only very few measurements of the phenomenological transport processes to date, but new results which could support the development of a theory connecting the molten salts and the aqueous electrolyte solutions are imminent.

These concentration ranges can be considered as markers for the theories and structural models by means of which the transport mechanisms are discussed. In Table 1 the theories and models for the system water + salt mentioned above are presented with the corresponding concentration ranges of Figure 1 to which they are related.

Region A of the highly dilute aqueous electrolyte solutions is the region of validity of the Debye–Hückel equation, which extends up to $c = 0.001$ mol dm^{-3} for most aqueous electrolyte solutions. This corresponds to a value of $x_{salt} = 1.8 \times 10^{-5}$ in the mole fraction scale of Figure 1 with an assumed density $\rho \approx 1$ kg dm^{-1} of the solution. The Debye–Hückel equation gives a molecular interpretation of the thermodynamic properties of highly dilute electrolyte solutions. In this concentration range the concept of the ionic atmosphere can be defined. This ionic atmosphere is distorted by the relaxation and electrophoretic effects during ionic migration through the solution.[16] This molecular interpretation of the thermodynamic properties (concentration dependence of the activity

TABLE 1. Theories and Models of the Water + Salt System for the Discussion of the Transport Properties

Concentration range	Description of methods		
Anhydrous molten salt	↑ Hole models Liquid free-volume model	Computer simulation technique	↑
Incompletely hydrated melt	Lattice model Configurational model Quasilattice model		
Hydrate melt	Glass transition thermodynamics ↓		Friction coefficients
Concentrated aqueous electrolyte solution	Extended Debye–Hückel equation Series expansion of thermodynamic functions	↑ Computer simulation technique	
Highly dilute aqueous electrolyte solution	Debye–Hückel equation (relaxation and electrophoretic effect)	↓	↓

coefficients) is closely related to the transport-limiting laws of interdiffusion, conductance, etc.

In concentrated aqueous electrolyte solutions (area B) the ionic atmosphere cannot be defined precisely because the clear correlation between the central ion and the counterions in the hydration sheath is no longer possible on account of the stronger interactions in such solutions. Thus, the various extensions of the Debye–Hückel equation ignore the essential theoretical background. This is also the reason why a limiting law for molten salts analogous to the Debye–Hückel equation cannot be formulated:

1. The interactions between the ion constituents in a molten salt are stronger than in an electrolyte solution and are not shielded by any hydration sheath.
2. The concentration steps, in the mole fraction scale, in which measurements in molten salt mixtures are usually performed, are much wider than those, in the c scale, usually used in aqueous electrolyte solutions, so that the relevant limiting region of the Debye–Hückel law has never been investigated.

A further disadvantage of these extensions to the Debye–Hückel equation is the different description of the two components of the electrolyte solution which is expressed by the choice of the composition variables (concentration c_2, molality m_2 of the electrolyte). Thus a solvent which implies that one component is in excess needs to be assumed.

An extension to the thermodynamic description of concentrated binary electrolyte solutions is the series expansion of the activity coefficients in *mole fraction* units by Haase,[17] where the solvent and the electrolyte are both treated as components and which contains the description of the nonelectrolyte solution as a special case. These formulas are especially suitable for water + acid systems miscible over a wide concentration range.

In the concentration range of the hydrate melts (C), the incomplete hydration sheath melts (D), and the anhydrous melts (E) the different transport processes are discussed using different structure models in terms of statistical thermodynamics, e.g., the hole model, the liquid free-volume model, the Adam and Gibbs configurational-entropy theory, and the lattice model. In the region of the hydrate and incompletely hydrated melts especially, the quasilattice model and the glassy state concept are most commonly used. These model concepts have been treated in detail.[8, 9, 14] Currently, the temperature and pressure dependence of the transport coefficients at very high concentrations are being discussed in terms of glass transition thermodynamics after Angell[18] and the hydration and association equilibria in terms of the quasilattice model of Braunstein.[19] None of the cited models can cover the entire range of electrolyte solutions from pure water to anhydrous fused salt and all models only allow a qualitative interpretation of the transport phenomena.

For elucidating microscopic processes in ionic liquids the computer simulation technique provides a powerful new tool. Calculations are possible by the Monte Carlo method and by molecular dynamics.[20] The Monte Carlo method can provide only equilibrium properties of the melt or of the electrolyte solution. The molecular dynamic simulation can also yield simple transport coefficients which essentially depend on the colligative properties of the system. The computer simulation method can be applied to anhydrous molten salts on the one hand and to pure water and highly dilute aqueous electrolyte solutions on the other. In recent times, results have also been obtained for concentrated solutions (see Adams and Hills, Chapter 4). Presently, a close approach between theory and experiment is possible, e.g., for the self-diffusion coefficients of ionic liquids. However, calculations and computer-generated graphics make it clear that the calculated transport processes are not merely ionic migration of *single* particles, but complicated *cooperative effects* within small volume elements of the ionic liquid.

This last insight in particular discourages the development of any other molecular model or the improvement of an already existing one, because in most cases a single-ion transport process is assumed and the region of validity of these models is limited in concentration. Rather, we need quantities which are characteristic of the transport phenomena throughout the entire concentration range. These quantities are the friction coefficients of the different species of the system which are characteristic parameters for the entire composition range and which will probably offer possibilities for the analysis of transport phenomena, in close relation with the structure of the water + salt mixture.

3. Transport Phenomena in Molten Salts and Electrolyte Solutions

In the following, the fundamentally common, but also the essentially different properties of molten salts and aqueous electrolyte solutions are discussed. For this discussion we have to choose systems such that their respective transport coefficients would be comparable with each other. We

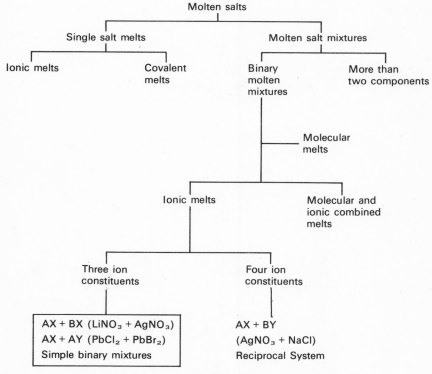

FIGURE 2. Molten Salt Tree similar to that given by Janz.[4]

therefore selected, among the variety of molten salts, such systems consisting of two components with a common ion constituent, e.g., $LiNO_3$ + $AgNO_3$, with the components $LiNO_3$ and $AgNO_3$, and the ion constituents Li, Ag, and NO_3 (Figure 2). We will call this sort of melt a "simple binary mixture." Such melts are comparable with aqueous solutions of binary electrolytes of the H_2O + $AgNO_3$ type, which also consist of three different sorts of particles, the components of which are identical with two of the components of the simple binary molten salt mixture.

Firstly we shall discuss the processes occurring in a simple binary melt. If a melt is at thermodynamic equilibrium, there exists no gradient of the intensive state variables in the melt. The thermodynamic equilibrium properties of the melt are described by the activity coefficients and the molar excess functions. In Figure 3a thermodynamic equilibrium is schematically represented using the example of a $LiNO_3$ + $AgNO_3$ melt. The dissociation of these salts has purposely not been considered in this figure in order to make clear that the thermodynamic properties generally refer to the components within the reference frame of local electroneutrality. The components are statistically distributed in space. In this equilibrium state the molecules and ions of the system, respectively, participate in Brownian motion.

In contrast to that situation, the ions exhibit a directional motion in an electric field; the cations migrate towards the cathode and the anions towards the anode, usually with different velocities. This ionic conduction mechanism is characterized by the conductivity, κ, of the system and the transport numbers, t_j, of the ion constituents. The transport number is defined by

$$t_j = \frac{\mathbf{I}_j}{\mathbf{I}} \tag{1}$$

$\mathbf{I}_j$ is the partial electric current density of ion constituent j, $\mathbf{I}$ the total current density, both along the same axis. If the partial electric current density is expressed by the diffusion density $\mathbf{J}_j$,

$$\mathbf{I}_j = z_j F \mathbf{J}_i = z_j F c_j (\mathbf{v}_j - \mathbf{v}_-) \tag{2}$$

where z_j is the charge number, F the Faraday constant, c_j the concentration, $\mathbf{v}_j$ the velocity of the ion constituent j, and $\mathbf{v}_-$ the velocity of the common anion constituent, we see that the transport numbers depend on the reference system, in this case on the mean velocity of the anion constituent. More generally, this means that all transport processes in simple binary mixtures are referred to the motion of the anions, implying that all transport coefficients depend on the reference system. If several transport coefficients are derived in different reference systems, they have to be converted to a common reference system before further calculation can

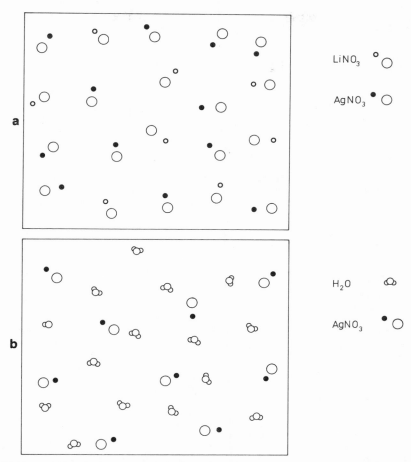

FIGURE 3. Schematic representation (a) of a simple binary molten salt mixture ($LiNO_3$ + $AgNO_3$) and (b) of an aqueous solution of a binary electrolyte (H_2O + $AgNO_3$) in the thermodynamic equilibrium state.

proceed. In Figure 4a the situation described above is graphically represented for the case of $LiNO_3$ + $AgNO_3$, which here is considered as completely dissociated. This contrasts with Figure 3a, since the electric transport is referred to the ions and not to the components. Here we can see that it is reasonable to consider the relative motion of the cations with respect to that of the anions and to choose the mean velocity of the anions as the reference velocity.

If a concentration gradient is applied in the simple binary molten salt mixture, interdiffusion occurs. A concentration balance is set up in which both components migrate respectively towards each other or, taking an

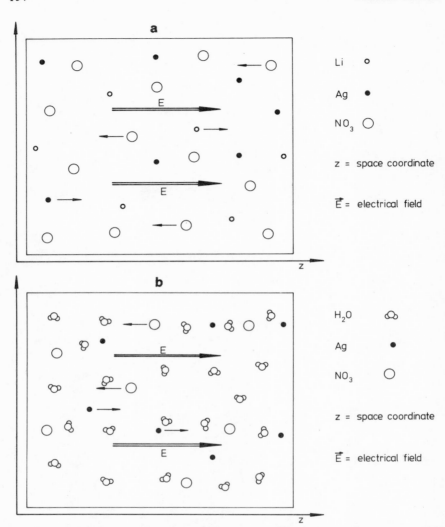

FIGURE 4. Electric transport (a) in a simple binary molten salt mixture (LiNO₃ + AgNO₃) and (b) in an aqueous solution of a binary electrolyte (H₂O + AgNO₃).

alternative view, in which the cation migrates relatively to the anion in the context of an electroneutrality condition (Figure 5a). Here it is especially clear that all transport processes have to be calculated relative to the motion of the common anion constituent. The diffusion coefficient D can be written in the usual form of Fick's first law

$$\mathbf{J}_2 = -D \text{ grad } c_2 \tag{3}$$

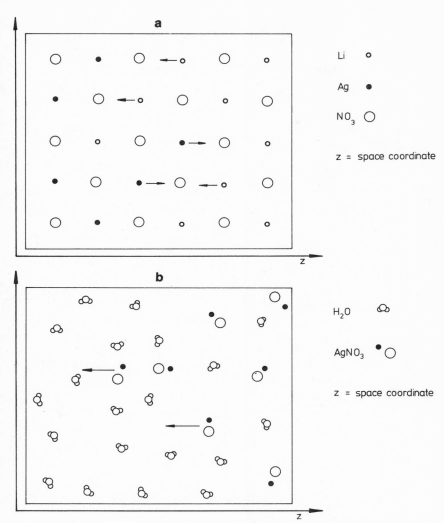

FIGURE 5. Interdiffusion (a) in a simple binary molten salt mixture (LiNO$_3$ + AgNO$_3$) and (b) in an aqueous electrolyte solution of a binary electrolyte (H$_2$O + AgNO$_3$).

but only for melts in which two ion constituents migrate independently of each other relative to a third one, which is the case in simple binary mixtures. The subscript 2 indicates the silver nitrate in the melt and the electrolyte in the solution below.

This example illustrates the analogy between the simple binary molten salt mixture and the aqueous solution of a binary electrolyte. In Figure 3 the LiNO$_3$ + AgNO$_3$ melt discussed above is schematically shown with the solution $H_2O + AgNO_3$ (Figure 3b) at thermodynamic equilibrium. In the figures concerning the aqueous electrolyte solutions (Figures 3b, 4b, and 5b) the representation of the hydration of the molecules and ions is deliberately omitted again in order to demonstrate the essential points of the processes discussed here. Melt and electrolyte solution both consist of three constituents, water, in the second system, being an electrically neutral constituent. Therefore, an important difference between the two systems arises: In aqueous electrolyte solutions the condition of electroneutrality concerns only the electrolyte. In our example we have

$$z_{Ag}v_{Ag} + z_{NO_3}v_{NO_3} = 0 \tag{4}$$

where v_j is the stoichiometric number of the ion constituent, j. In simple binary molten salt mixtures the reference particle is included in the condition of electroneutrality

$$z_{Li}v_{Li} + z_{NO_3}v_{NO_3} = 0 \tag{5a}$$

$$z_{Ag}v_{Ag} + z_{NO_3}v'_{NO_3} = 0 \tag{5b}$$

Some important differences result from the above in the mathematical formulation of the transport phenomena.

In an electric field the ions of an aqueous solution of a binary electrolyte migrate relative to the water (Figure 4b). The appropriate reference for such a solution is the mean velocity of the water molecules. In the case of diffusion in aqueous solutions of binary electrolytes (Figure 5b) the two differences to the simple binary molten mixtures are obvious: In aqueous solutions the condition of electroneutrality only concerns the electrolyte ($AgNO_3$), and the neutral component as a whole migrates relative to the water. But in both systems the transport data are given in the "Hittorf reference system"[21] and can be described using the same reference velocity.

4. Transport Experiments and Results

For the calculation of the phenomenological coefficients, which we will determine in the next section in order to describe the transport processes in aqueous solutions and melts, we need values of the activity coefficients, the equivalent conductances, the transport numbers, and the diffusion coefficients. The experimental methods for the determination of the activity coefficients and the conductivities are well known and are not considered further. The activity coefficients in the two examples $LiNO_3 + AgNO_3$ and $H_2O + AgNO_3$ were measured using a concentration cell without transference in the melt described by Richter and Sehm[22] and by vapor pressure determinations described by MacInnes and Brown[23] and by Pelzer.[24] The conductivities of $LiNO_3 + AgNO_3$ as a function of temperature and composition are given by Janz,[25] those of the system $H_2O + AgNO_3$ at 25°C by different authors.[26-32]

The transport numbers of the system $LiNO_3 + AgNO_3$ were investigated using a concentration cell with transference as a function of temperature and composition as described by Richter and Amkreutz.[33] For the aqueous system $H_2O + AgNO_3$ the transport numbers of the silver ions at 25°C were measured using different methods reported by several authors[27, 34-36] and were complemented by Roessler and Schneider[37] with Hittorf measurements up to 14 mol kg^{-1}. Only a few interdiffusion measurements in molten salts exist up to now. Optical methods, commonly used in aqueous electrolyte solutions, are still being developed for molten salts. Gravimetric and chronopotentiometric methods are not very accurate or relate only to a small concentration range. Measurements over the entire concentration range, as we need them here for the calculation of the phenomenological coefficients, only exist for $NaNO_3 + AgNO_3$ and $NaNO_3 + RbNO_3$ as described by Sjöblom[38] using a gravimetric method and for $NaNO_3 + AgNO_3$ and $LiNO_3 + AgNO_3$ after Richter[39] using the diaphragm cell method. The integral diffusion coefficients of the system $LiNO_3 + AgNO_3$ at 260°C are listed together with the activity coefficients, the equivalent conductance, and the transport numbers in Table 2.

The interdiffusion coefficients of the system $H_2O + AgNO_3$ at 25°C were measured by Miller and Albright[40] in 1972 using a Tiselius cell and

TABLE 2. Activity Coefficients f_2, Equivalent Conductivity Λ, Transport Numbers t_{Ag}, and Diffusion Coefficients D for the System $LiNO_3 + AgNO_3$ at 260°C. (The Transport Data Are Given in the Hittorf Reference System.)

x_{AgNO_3}	f_{AgNO_3}	$\dfrac{\Lambda \times 10^4}{S\ m^2\ mol^{-1}}$	t_{Ag}	$\dfrac{D \times 10^9}{m^2\ sec^{-1}}$
0.0	—	—	0.00	—
0.1	1.740	32.52	0.05	1.30
0.2	1.512	33.19	0.15	1.14
0.3	1.334	33.81	0.31	1.05
0.4	1.191	34.32	0.43	1.11
0.5	1.085	34.85	0.57	1.29
0.6	1.007	35.42	0.70	1.55
0.7	0.962	36.19	0.81	1.87
0.8	0.953	36.52	0.89	2.23
0.9	0.967	37.14	0.96	2.63
1.0	1.000	—	1.00	—

the Rayleigh method. The transport numbers, the equivalent conductance, the diffusion coefficients, and the derivative of the logarithms of the activity coefficients with respect to the molality have been tabulated by Miller and Pikal[41] for the system $H_2O + AgNO_3$ at 25°C in the concentration range $0 \leq m \leq 13.3$ mol kg^{-1}.

5. Phenomenological Coefficients

Knowing the thermodynamic and transport data mentioned above, we can now introduce the formalism of the phenomenological coefficients. In the following we will understand the term "phenomenological coefficients" as the general term for the linear "Onsager" coefficients, L_{ik},[13] the friction coefficients, R_{ik}, of the aqueous electrolyte solutions, and the friction coefficients, r_{ik}, of the molten salt mixtures. The introduction of these coefficients is useful for the discussion of the transport processes and the structural aspects of the ionic liquids, for these phenomenological coefficients are quantities which are characteristic especially of the transport processes throughout the entire composition range of the system water + salt.

We start with the dissipation function, ψ:

$$\psi = \sum_i \mathbf{J}_i \cdot \mathbf{X}_i > 0 \tag{6}$$

The $\mathbf{J}_i$ are the generalized fluxes, $\mathbf{X}_i$ the generalized forces of the thermodynamics of irreversible processes. Examples of the linear correlation of these fluxes and forces are Ohm's law for electric conductance and Fick's first law of diffusion. In aqueous electrolyte solutions material fluxes are usually linearly combined with generalized forces as "Onsager" coefficients[42]:

$$\mathbf{J}_i = \sum_k L_{ik} \mathbf{X}_k \tag{7}$$

L_{ik} depends on composition, temperature, and, what is important for our consideration here, on the reference system. For L_{ik}, the Onsager reciprocal relation (ORR)

$$L_{ik} = L_{ki} \tag{8}$$

is valid. The Onsager coefficients divided by the equivalent concentration c^* constitute the "interaction mobilities," L_{ik}/c^*, as interpreted by Miller and Pikal.[41] The inverse description

$$\mathbf{X}_i = \sum_k R_{ik} \mathbf{J}_k \tag{9}$$

leads, together with

$$\sum_k c_k^* R_{ik} = 0 \tag{10}$$

to the friction coefficients of the aqueous electrolyte solutions. There exists an interpretation of R_{ik},[41] too, but the physical meaning of L_{ik} and R_{ik} is not clear, especially relating to the maxima in the concentration slope of the Onsager coefficients in concentrated aqueous solutions.[43]

The definition of friction coefficients, r_{ik}, given by Klemm,[44] is a suitable form for molten salt mixtures, where no solvent exists (as in aqueous electrolyte solutions):

$$X_i = \sum_k r_{ik} \hat{x}_k (v_i - v_k) \qquad (11)$$

where $\hat{x}_k$ is the true mole fraction of the ion constituent, k. For r_{ik}, the ORR

$$r_{ik} = r_{ki} \qquad (12)$$

holds again.

The physical significance of the friction coefficients for molten salts is clearer than in aqueous solutions, because there is no solvent on which the r_{ik} could depend. The friction coefficients are a measure of the ionic interactions; they do not depend on the reference velocity because of the summation over all k values in equation (11), and they describe the interactions as a function of the composition of the mixture.

The Onsager coefficients result from transport data which are dependent upon the temperature and composition through the equations[43]

$$L_{++} = \frac{t_+^2 \kappa}{z_+^2 F^2} + \frac{v_+^2 D}{B} \qquad (13a)$$

$$L_{--} = \frac{t_-^2 \kappa}{z_-^2 F^2} + \frac{v_-^2 D}{B} \qquad (13b)$$

$$L_{+-} = \frac{t_+ t_- \kappa}{z_+ z_- F^2} + \frac{v_+ v_- D}{B} \qquad (13c)$$

The cations of the binary electrolyte are indicated by $+$, the anions by $-$. B is the thermodynamic part essentially determined by the activity coefficients:

$$B = \frac{vRT}{c} \left[1 + m \left(\frac{\partial \ln(\gamma/\gamma^+)}{\partial m} \right)_{T, P} \right] \qquad (14)$$

where v is the sum of the stoichiometric numbers, $v = v_+ + v_-$, R is the gas constant, T the temperature, P the pressure, m the molality, and γ the activity coefficient of the binary electrolyte, $\gamma^+ = 1$ kg mol^{-1}. The values of the friction coefficients, R_{ik} (i, $k = +$, $-$), of the aqueous electrolyte solutions were obtained by inverting the L_{ik} matrix. The R_{+-} values for the system $H_2O + AgNO_3$ at 25°C calculated by Miller and Pikal[41] are plotted in Figure 6 as a function of the molarity and of the mole fraction of the silver nitrate, respectively.

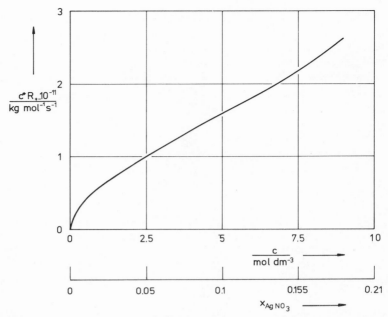

FIGURE 6. Caton–anion friction coefficients c^*R_{+-} vs. c and x_{AgNO_3} of the system $H_2O + AgNO_3$ at 25°C after Miller and Pikal[41] (c^* is the equivalent concentration, c the concentration, and x_{AgNO_3} the mole fraction of the electrolyte).

The friction coefficients of simple binary molten salt mixtures are given by[45]

$$r_{1+-} = \frac{2RT\Gamma}{D}\frac{t_{2+}}{x_2}\left(1 - \frac{t_{1+}}{x_1}\right) + \frac{2F^2}{\Lambda} \tag{15a}$$

$$r_{2+-} = \frac{2RT\Gamma}{D}\frac{t_{1+}}{x_1}\left(1 - \frac{t_{2+}}{x_2}\right) + \frac{2F^2}{\Lambda} \tag{15b}$$

$$r_{1+2+} = \frac{2RT\Gamma}{D}\frac{t_{1+}t_{2+}}{x_1 x_2} - \frac{2F^2}{\Lambda} \tag{15c}$$

$1+$ indicates the cation constituent of the first component, $2+$ that of the second one, and $-$ the common anion constituent. Γ describes the concentration dependence of the activity coefficients

$$\Gamma \equiv 1 + x_2\left(\frac{\partial \ln f_2}{\partial x_2}\right)_{T,P} \tag{16}$$

where f_2 is the activity coefficient of component 2.

In 1959 Laity[46] proposed the following:

A study of transport properties in three-ion mixtures will make it possible to see how the friction coefficients depend on composition. It was pointed out earlier, that three independent measurements (not counting the thermodynamic measurements which must also be performed) are necessary to determine the friction coefficients in such mixtures at a given composition. Unfortunately, this writer knows of no molten salt system for which all three measurements have been performed. The great need for experimental data in this area is obvious.

Up to recent times the calculation of friction coefficients failed because the values of the interdiffusion coefficients were lacking. We are now able to calculate the friction coefficients for the system $LiNO_3 + AgNO_3$ at 260°C with the data of Table 2. The results are given in Table 3 and plotted in Figure 7. (For the system $NaNO_3 + AgNO_3$ at 290°C see Reference 45.)

From these friction coefficients a simple binary molten salt mixture can be completely described in terms of thermodynamics for irreversible processes (except for self-diffusion). The r_{ik} shows a certain dependence on composition. The friction coefficients between the cations and anions are positive, those between the two cations are negative. The limiting values of the friction coefficients, $r_{ik}^{\bullet(1)}$, of the pure component 1 are finite. The same holds for the limiting values $r_{ik}^{\bullet(2)}$ of component 2. In this context the limiting values for pure silver nitrate are of interest, especially the value[45]

$$r_{2+-}^{\bullet(2)} = \frac{2F^2}{\lambda_{2+}^{\bullet(2)}} \tag{17}$$

which is a measure of the interactions between the silver and the nitrate ions in pure silver nitrate ($\lambda_{2+}^{\bullet(2)}$ is the ionic conductivity of pure silver nitrate).

TABLE 3. Friction Coefficients, r_{ik}, for the System $LiNO_3 + AgNO_3$ at 260°C

x_{AgNO_3}	$\dfrac{r_{1+-} \times 10^{-12}}{\text{kg mol}^{-1}\,\text{sec}^{-1}}$	$\dfrac{r_{2+-} \times 10^{-12}}{\text{kg mol}^{-1}\,\text{sec}^{-1}}$	$\dfrac{r_{1+2+} \times 10^{-12}}{\text{kg mol}^{-1}\,\text{sec}^{-1}}$
0.0	5.87	10.36	−4.45
0.1	5.56	8.91	−2.54
0.2	5.35	7.12	−1.08
0.3	5.59	5.34	−0.28
0.4	5.67	5.11	−0.86
0.5	6.01	4.85	−1.29
0.6	6.48	4.74	−1.61
0.7	6.97	4.72	−2.01
0.8	7.23	4.84	−2.50
0.9	7.47	4.91	−3.38
1.0	6.99	4.92	−4.17

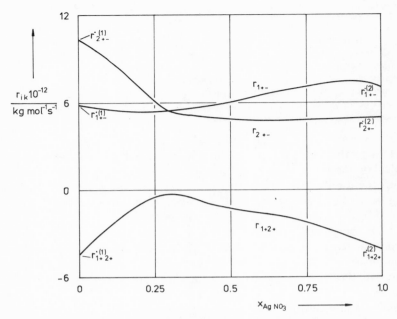

FIGURE 7. Friction coefficients r_{ik} of the system $LiNO_3 + AgNO_3$ at 260°C as a function of the mole fraction x_{AgNO_3} of the silver nitrate.

If we compare the two systems $LiNO_3 + AgNO_3$ and $H_2O + AgNO_3$, we qualitatively find that the friction coefficients resulting from equation (13c) for aqueous solutions of $AgNO_3$ tend to the limiting value (17) of the friction coefficient for pure silver nitrate as $x_{H_2O} \rightarrow 0$. At 25°C, this limiting value is only valid for silver nitrate in the hypothetical liquid state. For a quantitative calculation, the transport data together with these friction coefficients for the aqueous electrolyte solution have to be recalculated using the mean velocity of the nitrate ions as the reference velocity. Furthermore, measurements in these aqueous solutions have to be performed at temperatures at which the pure salt component in the molten mixture is liquid. In the future, friction coefficients obtained in this way will, perhaps, offer the possibility of relating transport phenomena to structure over the entire composition range of the system water + salt.

References

1. R. M. Fuoss and L. Onsager, *J. Phys. Chem.* **61**, 668 (1957).
2. A. Klemm, "Transport Properties of Molten Salts," in *Molten Salt Chemistry*, Ed. M. Blander, Interscience, New York (1964).

3. B. R. Sundheim, "Transport Properties of Liquid Electrolytes," in *Fused Salts*, Ed. B. R. Sundheim, McGraw-Hill, New York (1964).
4. G. J. Janz and R. D. Reeves, "Molten Salt Electrolytes—Transport Properties," in *Advances in Electrochemistry and Electrochemical Engineering*, Vol. 5, Ed. Ch. W. Tobias, Interscience, New York (1967).
5. G. J. Janz, *Molten Salts Handbook*, Academic, New York (1967).
6. J. W. Tomlinson, "Transport Properties of Molten Salts," *Rev. Pure Appl. Chem.* **18**, 187 (1968).
7. C. T. Moynihan, "Mass Transport in Fused Salts," in *Ionic Interactions*, Vol. I, Ed. S. Petrucci, Academic, New York (1971).
8. J. L. Copeland, *Transport Properties of Ionic Liquids*, Gordon and Breach, New York (1974).
9. H. Bloom and I. K. Snook, "Models for Molten Salts," in *Modern Aspects of Electrochemistry*, Vol. 9, Eds. B. E. Conway and J. O'M. Bockris, Plenum, New York (1974).
10. J. Richter, *Electrochim. Acta* **22**, 1035 (1977).
11. J. Newman, "Transport Processes in Electrolytic Solutions," in *Advances in Electrochemistry and Electrochemical Engineering*, Vol. 5, Ed. Ch. W. Tobias, Interscience, New York (1967).
12. H. Falkenhagen, W. Ebeling, and W. D. Kraeft, "Mass Transport Properties of Ionized Dilute Electrolytes," in *Ionic Interactions*, Vol. I, Ed. S. Petrucci, Academic, New York (1971).
13. D. G. Miller, *Faraday Discuss. Chem. Soc.* **64**, 295 (1977).
14. J. Braunstein, "Statistical Thermodynamics of Molten Salts and Concentrated Aqueous Electrolytes," in *Ionic Interactions*, Vol. I, Ed. S. Petrucci, Academic, New York (1971).
15. J. Braunstein, *Inorg. Chim. Acta Rev.* **2**, 19 (1968).
16. H. Falkenhagen, *Theorie der Elektrolyte*, Hirzel, Leipzig (1971).
17. R. Haase, *Angew. Chem.* **77**, 517 (1965).
18. C. A. Angell, *J. Phys. Chem.* **68**, 218, 1917 (1964); **69**, 2137 (1965); **70**, 2793 (1966); **81**, 232, 238 (1977).
19. J. Braunstein, *J. Phys. Chem.* **71**, 3402 (1967).
20. L. V. Woodcock, "Molecular Dynamics Calculations on Molten Ionic Salts," in *Advances in Molten Salt Chemistry*, Vol. 3, Eds. J. Braunstein, G. Mamantov, and G. P. Smith, Plenum, New York (1975).
21. R. Haase, *Thermodynamics of Irreversible Processes*, Addison-Wesley, Reading, Massachusetts (1969).
22. J. Richter and S. Sehm, *Z. Naturforsch.* **27a**, 141 (1972).
23. D. A. MacInnes and A. S. Brown, *Chem. Rev.* **18**, 335 (1936).
24. H. Pelzer, Thesis, RWTH Aachen, 1961.
25. G. J. Janz, U. Krebs, H. F. Siegenthaler, and R. P. T. Tomkins, *J. Phys. Chem. Ref. Data* **1** (3), 581–746 (1972). (*Molten Salts*, Vol. 3, *Nitrates, Nitrites, and Mixtures*).
26. I. D. McKenzie and R. M. Fuoss, *J. Phys. Chem.* **73**, 1501 (1969).
27. A. N. Campbell and K. P. Singh, *Can. J. Chem.* **37**, 1959 (1959).
28. A. N. Campbell and R. J. Friesen, *Can. J. Chem.* **37**, 1288 (1959).
29. T. Shedlovsky, *J. Am. Chem. Soc.* **54**, 1411 (1932).
30. G. D. Parfitt and A. L. Smith, *Trans. Faraday Soc.* **59**, 257 (1963).
31. C. H. Orr and H. E. Wirth, *J. Phys. Chem.* **63**, 1147 (1959).
32. A. N. Campbell and E. M. Kartzmark, *Can. J. Chem.* **28B**, 43 (1950).
33. J. Richter and E. Amkreutz, *Z. Naturforsch.* **27a**, 280 (1972).
34. R. Haase, G. Lehnert, and H. J. Jansen, *Z. Phys. Chem.* (Frankfurt am Main) **42**, 32 (1964).
35. D. A. MacInnes and I. A. Cowperthwaite, *Chem. Rev.* **11**, 210 (1932).

36. H. Strehlow and H.-M. Kroetz, *Z. Electrochem. Ber. Bunsenges. Physik. Chem.* **62**, 373 (1958).
37. N. Roessler and H. Schneider, *Ber. Bunsenges. Phys. Chem.* **74**, 1225 (1970).
38. D. Andréasson, A. Behn, and C.-A. Sjöblom, *Z. Naturforsch.* **25a**, 700 (1970).
39. J. Richter, *J. Chem. Eng. Data* **18**, 400 (1973); *Z. Naturforsch.* **28a**, 492 (1973).
40. J. G. Albright and D. G. Miller, *J. Phys. Chem.* **76**, 1853 (1972).
41. D. G. Miller and M. J. Pikal, *J. Solution Chem.* **1**, 111 (1972).
42. L. Onsager, *Phys. Rev.* **37**, 405 (1931); **38**, 2265 (1931); *Ann. N.Y. Acad. Sci.* **46**, 241 (1945).
43. R. Haase and J. Richter, *Z. Naturforsch.* **22a**, 1761 (1967).
44. A. Klemm, *Z. Naturforsch.* **8a**, 397 (1953); **17a**, 805 (1962).
45. J. Richter, *Ber. Bunsenges. Physik. Chem.* **78**, 972 (1974).
46. R. W. Laity, *J. Chem. Phys.* **30**, 682 (1959).

11

Water in Molten Salts: Industrial and Electrochemical Consequences

D. G. Lovering and R. M. Oblath

1. The Background to Molten Salt Technology

Now approaching its centenary, the Hall–Héroult electrolytic aluminum extraction cell[1,2] still reigns supreme as the largest industrial application of molten salts. Although rivals are now reappearing,[3] the most likely contender would merely involve a change from molten cryolite-alumina to a chloroaluminate, with a small reduction in temperature. Nevertheless, the rapid growth of molten salt technologies awaited the electrolytic separation of uranium from molten fluorides in the Manhattan project of the mid-1940s.

The use of molten cyanides, nitrates, and nitrites in heat treatment, cementation, cleaning, finishing, and brazing is now well established.[4] The benefits of high thermal capacity and good heat transfer have led to the use of molten salts in nuclear reactors.[5,6] Other applications in thermal and perhaps solar energy storage are envisaged. Molten salt batteries are currently receiving revived attention,[7] as are molten carbonate fuel cells.[8] New electrowinning and electrorefining processes using molten electrolytes are becoming increasingly attractive[9] with respect to commercial viability, especially since the pollution penalty of pyrometallurgical operations is absent. The direct electrodeposition of refractory metals[10] and of aluminum is receiving close attention.

D. G. Lovering and R. M. Oblath • R.M.C.S., Shrivenham, Swindon, England. *Present address* for R. M. Oblath: Goodyear International Tire Technical Center, Colmar-berg, Luxembourg.

Following the earlier pioneering efforts of Sundermeyer,[11] the excellent solvent properties of melts towards both inorganic[12] as well as organic materials are beginning to be exploited in engineering concepts. Not only may organic vapors be oxidized in high-yield, single-pass molten nitrate reactors,[13] but solid organic waste including refuse and motor tires may be pyrolyzed to give valuable, small-molecule organic reagents and fuels.[14, 15]

A penalty in molten salt technology has been the enhanced corrosion levels arising both from the elevated operating temperature and the ionicity of the system. The proliferation of studies of these problems[16] suggests that economic solutions will generally be forthcoming.

2. What Happens When the Molten Salt is Wet?

Water is surprisingly soluble in many molten salts.[17-19] In molten alkali metal nitrates,[20] the solubility follows Henry's law, at least at low concentrations. In chloride melts, the situation is less clear[21]: simple gas-like solubility at very low concentrations appears to be followed by hydrolysis at higher levels. In borates and silicates, at much higher temperatures, hydrolytic dissolution predominates.

Exactly why interactions between water and molten salts have received so little attention is something of a mystery, especially with the ever-increasing technological importance of these media. The effects of this ubiquitous solute in the industrial context can be quite deleterious, particularly with regard to corrosion and fuming. In slight mitigation, but never quantified, the rate of dissolution of atmospheric moisture into an industrial scale molten salt bath can be quite slow owing to the small surface area to volume ratio. Thus, thickening and other undesirable effects attributable to water uptake may, in a few cases, be unimportant during the working life of a bath or reactor.

The rates of corrosion of mild steel[22] and Armco iron[23] in molten LiCl–KCl fall rapidly to constant values after about five hours. The rates increase with temperature between 400 and 800°C. The oxidation kinetics are consistent with a parabolic rate law initially, indicating[24] growth of the oxide layer by diffusion of metal atoms through the layer. At long times, the kinetics exhibit an overall linear rate law which has been interpreted[24] as the ultimate achievement of constant film thickness. However, para-linear oxidation might be a more likely explanation, in which case "crack-back" of the growing oxide layer due to compressive stresses must be taken into consideration. The rate of corrosion of mild steel seems scarcely affected by the presence of traces of water in the melt, although the rate for Armco iron is significantly enhanced. Notwithstanding, the presence of residual traces of water in the original LiCl–KCl solvent has been seriously considered[24] as an initiating force prior to gross corrosion.

The consequences of a leak in the steam-generating plant of the homogeneous molten salt nuclear reactor, which utilized the $LiF-BeF_2$ eutectic melt, would obviously be serious.[25] Apart from physical effects such as explosive boiling, chemical reaction of the BeF_2 would yield hydrofluoric acid: both toxic and corrosion hazards would occur. The same might happen on a lesser scale, but perhaps more insidiously, if moist air leaked into the reactor. Metal fluoride corrosion products would either dissolve in the melt or be carried away in suspension by it. Design studies conclude[25] that, compared to the sodium-cooled fast reactor, such occurrences would be troublesome rather than catastrophic, since the acute hazards of fire and large-scale explosion are not realistic possibilities for the molten fluoride reactor.

More conventional steam boilers in generating plants are not at all free from "molten salt corrosion." Very highly concentrated solutions of sodium hydroxide or chloride can build up on boiler tubes[26] by condensation from the vapor phase. The problem is perhaps more familiar in boiler flues. Moreover, in the water cycle "wick boiling" can occur in crevices or regions of slack flow leading to concentration of salt deposits—a particular problem when the use of marginally saline cooling water seems otherwise viable.[27]

For good reasons, nickel and its alloys have always been attractive fabrication materials for use at high temperature and/or in corrosive environments. A number of recent studies have focused on the behavior of these metals in contact with molten salts. In wet, molten KNO_3 at 350°C, nickel, NI 90, and NIM 115 all become rapidly more noble at open circuit within 10–20 min of immersion in the melt, as an initial oxide phase is formed.[28] The potentials of specimens then become less anodic as a stable oxide film is asymptotically established during a subsequent 24-hr period. Whilst the limiting anodic reactions of all samples includes bulk dissolution, and often simultaneous oxygen evolution on the alloys, the cathodic limiting reaction is also one of corrosion, probably via chemical reaction with an active intermediate in the nitrate reduction scheme itself influenced by the presence of water, *vide infra*. In molten $NaNO_3-KNO_3$ and $NaNO_2-KNO_2$ within the temperature interval 230–310°C, the passivation potentials for nickel are 100 mV more anodic in the presence of water.[29] Furthermore, the exchange current density at this potential increases tenfold in the nitrate medium. Similar variation of the passivation potential of iron occurs in the nitrite melt,[30] although the exchange current density is apparently unaffected by the presence of water.

According to one report,[31] the presence of water in molten $NaNO_3-KNO_3$ at 300°C affects the nature of anodically formed oxide films on titanium. Application of $+30$ V for 20 min led to the formation of a black oxide film with poor dielectric properties and such a low leakage current that it was unsuitable for capacitor applications. Analysis of the film indicated a composition $TiO_{1.6}$; it was considered that the rutile structure contained excess interstitial titanium atoms. In contrast, a similar melt "dried" by the dubious procedure of "pre-electrolysis" at 1.6 V, produced very thin (~ 4000 Å) anodic oxide films on titanium, which appeared violet or green by interference of light; these had properties suitable for capacitor dielectric. In another study,[28] only a very thin yellow film was formed anodically on titanium at $+30$ V in undried KNO_3 at 350°C: a very thin blue film was similarly produced in an undried NH_4NO_3–$NaNO_3$ melt at 150°C. Thus, the role of water remains unclear in this context, although the melt and temperature appear to be important. A very recent and detailed study[32] of anodically grown oxide films on aluminum in nitrate melts shows that entirely different morphological features occur when using anhydrous systems as compared with melts containing up to ~ 0.2 m water. A fuller description of this work will be given elsewhere.

Studies of nickel and titanium alloys in contact with "wet" molten LiCl–KCl at 400°C are rather disturbing.[28] All the nickel alloys dissolve anodically at small applied potentials $\sim \frac{1}{2}$–1 V; however, titanium specimens are chemically attacked upon immersion into the molten electrolyte [cf. anhydrous LiCl–KCl (Reference 33)]. Furthermore, titanium is rapidly attacked *before* immersion when in contact with the vapor over a "wet" KCl–$ZnCl_2$ melt. Anodic polarization of nickel alloys in the latter melt produces black dendrites which subsequently react with the solvent. Again, the exact role of the water present in these cases is unknown, but might be guessed to be significant by analogy.[24]

An early report[34] suggests that the addition of water to a molten chloroaluminate bath assists the smooth, dendrite-free plating of aluminum onto mild steel. Either chemical polishing of high spots or preferential reduction, at active sites, of hydrogen ions, from the HCl generated in the melt, was considered to be responsible for the improved plating. However, in two more recent studies, the plating of aluminum onto brass[35] and steel[36] could be accomplished without difficulty from anhydrous chloroaluminate melts. In the latter case, good plating was contingent upon suitable specimen pretreatment such as an acid wipe or chemical or electrochemical polishing.

Apart perhaps from this last example, most of the documented effects of water as a contaminant in molten salt processes are detrimental to those processes. Acid gases frequently occur in the volatile products of hydrolytic reactions, thus contributing to fume losses as well as to the generation of highly corrosive pollutants. If hydrogen is also generated, then hydrogen embrittlement[37] may be a problem, not only in the case of metal extraction processes, but also for associated plant.

The Hall–Héroult cell well illustrates the problems of water contamination, which usually occurs by moisture uptake from the atmosphere. The most harmful component in the fume is undoubtedly HF, which is thought to be produced by direct reaction of water with fluoride ions.[38] The mist arising from smelting baths is, however, due to the release of hydrogen from the molten aluminum as it reacts with water.[39] The metallurgical, toxicological, environmental, and safety consequences of the release of HF and H_2 hardly need stressing.

In the TORCO segregation process for copper production, water is an essential component.[40] Copper ore is preheated to 1030–1120 K before transfer to the segregation chamber which contains NaCl, coal or coke, and silica. Water may be present, in the vapor, either from the wet ore or from the furnace fuel combustion reaction. Conversion reactions proposed[41]

$$2NaCl + H_2O + xSiO_2 \rightarrow Na_2O \cdot xSiO_2 + 2HCl$$

$$Cu_2O + 2HCl \rightarrow 2CuCl + H_2O$$

$$2CuCl + H_2 \rightarrow 2Cu + 2HCl$$

indicate that the water is recycled; these doubtless hide the true complexity of the process, since the carbon–steam reaction, to produce hydrogen, is a net consumer of water and itself requires continuous steam injection.[42] A recent review[19] has considered the effects of water specifically in molten salt extraction metallurgy.

3. The Electrochemistry of Water in Molten Salts

Combes will describe the chemistry of water in melts, sometimes using electrochemical analysis, in Chapter 15. However, it is worth noting here that the nitration of benzene in nitrate melts at 250°C will not proceed, even in the presence of the Lux–Flood acid, pyrosulfate, unless water is added.[43] In $Ca(NO_3)_2 \cdot 4H_2O + KNO_3$ (or $LiNO_3$) at $\sim 80°C$, the pyrosulfate apparently reacts to the bisulfate:

$$S_2O_7^{2-} + H_2O \rightarrow 2HSO_4^-$$

whereupon protons, arising from a self-dissociation, yield nitronium ions:

$$H^+ + NO_3^- \rightleftharpoons NO_2^+ + H_2O$$

(or $H_2NO_3^+$) which act as nitrating agents.[44] Nitration of organics may alternatively occur in the vapor phase above nitrate melts, where more exotic species might have transient existence.[45] Thus, the dependence of nitrobenzene yield upon water concentration in molten nitrates may not be related to the presence of nitrating agents such as $H_2NO_3^+$, etc. in solution; Zambonin's data[46] concerning the equilibrium concentrations of species in these melts would also support an alternative contention.

3.1. Nitrates and Nitrites

Two voltammetric waves can be observed[47] during the reduction of nitrate ions at platinum microelectrodes in molten, eutectic $NaNO_3$–KNO_3 at 250°C. The peak at -1.65 V vs. Ag/Ag(I), 0.07 M, signals the familiar onset of the irreversible reduction of nitrate ions from the bulk. The wave at less cathodic potentials, which can be removed by purging with a dry gas, is certainly due to the presence of water, as can be confirmed by a standard addition. The first detailed study of this "water wave" is due to Geckle,[48] who equilibrated water vapor with $NaNO_3$–KNO_3 melts at 350°C and examined these waves at the rotating disk electrode (r.d.e.). The waves can only be reproduced faithfully when the melts are "aged" and the electrodes arbitrarily "prebiased." Although the extremely sensitive mass spectrometer could not detect hydrogen or nitrogen in the supernatant gas even after prolonged electrolysis, direct electroactivity of the dissolved water was inferred on the basis of proportionality between concentration and wave height. This proportionality was later confirmed[49] at the r.d.e. and used as an analytical procedure. At water pressures up to 30 mm, this linearity is cited as evidence of a simple, gas-like dissolution process for water in the melt according to Henry's law. The variation of Henry's law constant with temperature may then be used to calculate the solvation energy for water in the melt. Notably, in $LiNO_3^+$-containing melts, even prolonged evacuation does not reduce the residual water level below 10^{-4} mol per mole of melt. The occurrence of the water wave is certainly not melt or temperature dependent; Hills and Power[50] confirmed this in molten eutectic $LiNO_3$–$NaNO_3$–KNO_3 at 150°C using stationary platinum microelectrodes. Proof of its origin was again evident when standard additions of $LiNO_3 \cdot 3H_2O$ were made. Unfortunately, the mechanism of its appearance was again ascribed to direct electron transfer to water molecules (or hydroxide ions resulting from hydrolysis).

The most extensive investigations of water (and oxide species) in nitrate melts have been made by the Italian school under Zambonin, himself a sometime co-worker of Jordan at Pennsylvania, the supervisor to Geckle. Massive electrolysis[51] in a $NaNO_3$–KNO_3 melt at ~ 230°C under a partial pressure of water vapor of 10 torr leads to the subsequent appearance of two anodic voltammetric waves at the r.d.e. These are assigned to the oxidation of hydroxide[52] and nitrite[53] ions:

$$4OH^- \rightarrow O_2 + 2H_2O + 4e^-$$

$$NO_2^- \rightarrow NO_2 + e^-$$

and can apparently be confirmed by the standard addition technique. A cathodic water wave is also evident. The latter was then considered to be due to the simultaneous reduction of water and nitrate:

$$NO_3^- + H_2O + 2e^- \rightarrow NO_2^- + 2OH^-$$

thus avoiding the objections to a mechanism involving direct water reduction. Whilst confirming that traces of water do not affect the general features of the nitrate voltammetric reduction wave, limited as it is by precipitation of sodium oxide, it was surmised that such traces would nonetheless prevent any buildup of superoxides and peroxides:

$$O^{2-} + H_2O \rightleftharpoons 2OH^-, \qquad\qquad K \sim 10^{18}$$

$$O_2^{2-} + H_2O \rightleftharpoons 2OH^- + \tfrac{1}{2}O_2, \qquad K = 2 \times 10^{10} \text{ mol}^{1/2} \text{ kg}^{-1/2}$$

$$2O_2^- + H_2O \rightleftharpoons 2OH^- + \tfrac{3}{2}O_2, \qquad K = 10^3 \text{ mol}^{1/2} \text{ kg}^{-1/2}$$

The kinetics of this last reaction were investigated[54] at the r.d.e. in molten equimolar $NaNO_3$–KNO_3 at 242°C; the reaction rate appears to be independent of hydroxide concentration. These results were rationalized[55] in terms of an initial dissociation of superoxide:

$$2O_2^- \rightleftharpoons O_2 + O_2^{2-}$$

$$O_2^{2-} + H_2O \rightleftharpoons [H_2O \cdot O_2^{2-}] \rightarrow OH^- + HO_2^-$$

$$2HO_2^- \rightarrow 2OH^- + O_2$$

followed by the formation of some rather unusual intermediates! After again confirming linearity between the diffusion current and water concentration at the water wave,[56] another mechanism has been proposed for this reduction process:

$$H_2O + e^- \rightarrow OH^- + H$$

$$H + NO_3^- \rightarrow OH^- + NO_2$$

and

$$H + H \rightarrow H_2$$

leading to two alternative overall mechanisms:

$$H_2O + NO_3^- + 2e^- \rightarrow 2OH^- + NO_2^-$$

or

$$H_2O + e^- \rightarrow OH^- + \tfrac{1}{2}H_2$$

with the suggestion that perhaps both schemes occur with the former predominating. Such proposals neglect the important evidence of mass spectroscopic studies and introduce the new possibility of the generation of NO_2. Unfortunately studies were not extended to another important permutation of experimental conditions—a wet melt which is also oxygen free!

Most recently, the potential of a (Pt or Au) H_2O/H_2, OH^- electrode in molten $NaNO_3$–KNO_3 at 230°C has been expressed[57] in terms of

$$E = K + \frac{RT}{F} \cdot \ln\left\{ \frac{[H_2O]}{[H_2][OH^-]} \right\}$$

although this is clearly at variance with the reaction stoichiometry for the overall reaction

$$2H_2O + 2e^- \rightarrow 2OH^- + H_2$$

for which standard electrode potential was calculated:

$$E^o_{H_2O/H_2, OH^-} = -1.96 \text{ V vs. Ag/Ag(I)},   0.07 \; M$$

This calculated electrode potential is more than a volt more cathodic to the observed onset of the water wave! A further study[58] of the (Pt) H_2O/H_2, OH^- electrode in the same melt but at 240°C on the r.d.e. followed the oxidation of hydroxide ions ($< 10^{-3} \; m$) at -0.25 V in the absence of nitrite and hydrogen. Nitrite was considered to have been removed by an NO_2 purge. The overall oxidation is then

$$4OH^- \rightarrow 2H_2O + O_2 + 4e^-$$

In the absence of hydroxide, no voltammetric wave attributable to hydrogen can be detected when the melt is "saturated" with moist hydrogen to a concentration of $\sim 10^{-5} \; m$. In the presence of hydroxide ($< 10^{-3} \; m$), however, the previously observed oxidation wave apparently gives way to an anodic component of a composite water wave at -0.75 V. The wave height of the anodic portion is proportional to hydroxide concentration, hence it was interpreted in terms of

$$H_2 + 2OH^- \rightarrow 2H_2O + 2e^-$$

For concentrations of hydroxide greater than $10^{-3} \; m$, a peaked response is evident; a current maximum occurs at -0.75 V, with a minimum at -0.5 V. The minimum current decreases with increasing concentration of hydroxide. A voltammetric wave also appears at -0.25 V and has been ascribed to the direct oxidation of hydroxide ions.[58] The mechanism for this process might involve the generation of hydroxyl radical intermediates:

$$2OH^- \rightarrow 2OH^\bullet + 2e^-$$
$$\updownarrow$$
$$H_2O + \tfrac{1}{2}O_2$$

It has been proposed[58] that in the anodic portion of the composite water wave occurring at -0.75 V, transport of hydrogen to the electrode surface becomes rate determining. Moreover, any hydroxide not oxidized in this step may then be available for direct oxidation in the wave at -0.25 V. According to this reasoning, the subsequent fall in current to the minimum at -0.5 V can be explained in terms of the consumption of hydrogen in a chemical reaction[59, 60] with the production of nitrite ions and catalyzed by hydroxyl radicals generated in the first step of the process occurring at -0.25 V:

$$OH^{\bullet} + H_2 \rightleftharpoons H_2O + H^{\bullet}$$

$$H^{\bullet} + NO_3^- \rightleftharpoons NO_2^- + OH^{\bullet}$$

$$\overline{H_2 + NO_3^- \rightleftharpoons H_2O + NO_2^-}$$

Alternatively,[32, 61, 98] nitrite ions themselves, arising either by spontaneous thermal decomposition of the melt[62, 63] or by the direct interaction of hydrogen gas with nitrate ions, may be responsible for the observed voltammetric response. Thus, the wave having a current maximum at -0.75 V might be due to the cooxidation of nitrite and hydroxide:

$$2OH^- + NO_2^- \rightarrow H_2O + NO_3^- + 2e^-$$

Furthermore, when the concentration of hydroxide exceeds 10^{-3} m the greater quantities of water arising both from this electrode reaction, as well as by chemical combination[64]

$$2OH^- \rightleftharpoons H_2O + O^{2-}, \qquad K = 10^{-18}$$

can lead to a decrease in nitrite concentration, at constant hydrogen concentration, according to the net equilibrium reaction above. The cooxidation, being responsive to nitrite concentration, can thus lead to the appearance of the observed current minimum[58] at -0.5 V, even though the hydroxide concentration is increasing. Confirmation of this hypothesis could readily be achieved by titrating the nitrate melt containing water and hydroxide with nitrite ions. Advantageously, this might be carried out in one of the lower-melting, $LiNO_3$-containing eutectic nitrate melts in which the thermal generation of nitrite is minimized.

In a reassessment of redox equilibria in melts, Jordan[65] reiterates that trivial electroreduction of water cannot be responsible for the voltammetric water wave. These assertions are based on the crucial and highly sensitive mass spectrometric results,[48] which failed to detect any hydrogen or oxygen molecules in the supernatant argon of a wet nitrate melt even after prolonged electrolysis. While accepting the plausibility of Zambonin's proposals,[56] Jordan points out that the voltammetric wave is not reversible and that this fact leads to the variation of $E_{1/2}$ with concentration of water. Further studies[66, 67] report voltammetric waves for water ($10^{-3} \rightarrow 10^{-2}$ m) at the r.d.e. in $NaNO_3$–KNO_3 at 250°C over a wide potential range [$+1.0$ V $\rightarrow -1.2$ V vs. Ag/Ag(I), 0.1 M]. The $E_{1/2}$ was at -1.0 V vs. reference electrode and the limiting currents of the water wave were again confirmed to be directly proportional to the bulk water concentration and to obey the Levich equation. A transient blue color occurring at the electrode[66, 67] was considered to be due to the generation of hydrated electrons[68]; no spectroscopic confirmation was obtained. According to these authors, the primary step in the electrolysis is the capture of an electron in a vacancy on the anion sublattice. This subsequently reacts with an hydrated sodium ion to yield an hydrated electron:

$$Na_{aq}^+ + e^- \rightarrow e_{aq}^- + Na^+$$

Thus, the complete scheme for the water wave is

$$NO_3^- + H_2O + 2e^- \rightarrow NO_2^- + 2OH^- \qquad \text{(overall)}$$

$$2H_2O + 2e^- \rightarrow 2(e_{aq}^-) \qquad \text{(transport controlling)}$$

$$NO_3^- + 2(e_{aq}^-) \rightleftharpoons NO_2^- + 2OH^-$$

$$NO_3^- + 2(e_{aq}^-) \rightleftharpoons NO_2^- + O^{2-} + 2H_2O$$

$$O^{2-} + 2H_2O \rightleftharpoons 2OH^-$$

in which hydrated electrons act as reducing agents. Further justification[66] for this surprising hypothesis was thought to be the similarity between the diffusion coefficient of the hydrated electron at 25°C (Reference 69) and that for "water" in the nitrate melt at 250°C. Unfortunately, however, even a crude extrapolation of the diffusion coefficient of the hydrated electron to 250°C using the Arrhenius equation suggests a difference of two orders of magnitude in the transport rates; using the Fulcher equation,* a difference of about four orders is apparent. Moreover, there appears to be some uncertainty about the potential at which the transient blue color appears—this was not recorded by Espinola.[66] When Swofford and Laitinen[47, 71] confirmed the earlier report by Hittorf[72] of a blue color in molten NaNO$_3$–KNO$_3$ this occurred at -2.8 V vs. Ag/Ag(I), 0.07 m, i.e., in the region where metallic sodium is deposited. In entirely separate studies,[28, 73, 74] the colored solutions observed at various metal cathodes in molten NaNO$_3$, $\sim$ 340°C, and molten KNO$_3$, $\sim$ 350°C, were thought to be due to nitrosyl complexes formed by the chemical reaction of active intermediates in the reduction of nitrate ions with the electrode substrates.

More recently, cyclic voltammetric studies of the water wave ($10^{-4} \rightarrow 10^{-2}$ m water) at *stationary* platinum microelectrodes in molten LiNO$_3$–KNO$_3$ at 145°C are described.[75] In this work, great care was taken to ensure that the melt was never overheated and that experiments were concluded as rapidly as possible; thus, thermal decomposition, or "aging," of the melt is minimized. Neither were arbitrary prebiasing procedures adopted to condition the electrodes. Furthermore, and importantly, cathodic potential excursions were terminated prior to the potential at which the

---

* See, for example, Reference 70.

onset of the reduction of nitrate ions occurs.[76] In this way, hopefully, all questionable and ill-defined procedures[47–51, 56, 65–67] are avoided. Evidently, the presence of both water and *nitrite* in the melt are essential for the appearance of the water wave, in agreement with a previous observation.[66, 67] However, in this present case[75] the nitrite had to be introduced into solution either by direct addition or by electrogeneration at more cathodic potentials, since there was no opportunity for its concentration to build up by other means.

According to these results,[75] the electrode reaction at the water wave is

$$NO_2^- + H_2O + 2e^- \rightarrow NO^- + 2OH^-$$

This has the stoichiometry required by the coulometric experiments[65] *and* involves the reduction of *two* tracer species possessing diffusion coefficients of an order such as to yield limiting currents of a magnitude[66, 67] similar to those observed *at* the operating temperature, *vide supra*. Furthermore, the presently formulated reduction is likely to be autocatalytic,[77] with the generation of 2 mol of nitrite per mole of reactant, by chemical reaction of the monohyponitrite intermediate with the melt:

$$NO^- + NO_3^- \rightleftharpoons 2NO_2^-$$

Hence, the build up of nitrite during wide potential excursions is readily explained. After 15 years there is still ample scope for further investigations of this reaction, especially perhaps with regard to a puzzling time dependence of the water wave[75] and the detailed mechanism. However, it has been suggested that the process may proceed via an adsorbed intermediate:[75]

$$Pt + NO_2^- + H_2O + e^- \rightarrow Pt(NO)_{ads} + 2OH^-$$

$$Pt(NO)_{ads} + e^- \rightarrow Pt + NO^-$$

Products generated by side reactions might provide valuable clues to the correct scheme.

Current–potential curves at platinum electrodes in molten $NaNO_3$-KNO_3 at 250°C, containing 10^{-4} *m* water, show a peak at -0.5 V vs. Ag/AgCl (sat), which has been attributed[78, 79] to the discharge of hydrogen ions. However, this is probably the water wave. It is not clear why the wave is absent on gold electrodes, although these authors claim that the formation of Au---O·OH at $+1.05$ V, close to the anodic potential sweep limit employed, is favored in the presence of water.

In molten $LiNO_3$-$NaNO_3$-KNO_3 at 130°C, the differential double-layer capacitance curves on mercury show a single minimum close to the point of zero charge.[80] When water is added up to ~ 0.1 mol H_2O/mol nitrate, the minimum capacity increases, but neither the shape nor position of the smooth parabolic curve is changed. At concentrations of 1.5 mol H_2O/mol nitrate the familiar "hump" characteristic of aqueous solutions begins to form, initially more as a plateau at the position of the minimum, and capacitance values increase significantly throughout the range. Similar results were reported much earlier by Randles and White.[81] At a concentration of 2.5 M ($T = 95°C$), the hump is completely developed. Double-layer capacitance–potential curves in molten $LiNO_3$–KNO_3 at 157–201°C and $NaNO_3$–KNO_3 at 225–250°C also show similar behavior,[82] although the minimum capacity of 19 μF cm^{-2} is slightly lower than the earlier value[81] of 20 μF cm^{-2}. This discrepancy can be explained[82] in terms of residual water in the melt, since a rise occurs in the minimum capacity as water is added to the molten nitrate; this is also in line with the Russian results.[80]

3.2. Chlorides

Traditionally, large voltammetric residual currents in halide melts, such as LiCl–KCl at 500°C,[83] are associated with impurity levels of dissolved water. This has been removed by pumping, gas purging (Cl_2 and/or HCl), or pre-electrolysis, often with the residual current itself being used to assess the purity of the melt—a topic extensively covered by S. H. White in Chapter 12.

The dissociation of alkali metal hydroxides in molten equimolar NaCl–KCl at 700–825°C may be monitored as a function of water vapor pressure using an oxide indicator electrode.[84] The equilibrium constant, $K = [O^{2-}] \cdot pH_2O/[OH^-]^2$ for this process:

$$2OH^- \rightleftharpoons H_2O + O^{2-}$$

may then be derived from the experimental electrode potential versus water vapor pressure response. A similar technique can also be used in molten NaCl–KCl.[85] If a hydrolytic equilibrium is written,

$$H_2O + 2Cl^- \rightleftharpoons 2HCl + O^{2-}$$

then an equilibrium constant, $K = [O^{2-}] \cdot (pHCl)^2/pH_2O$ may be determined from the variation in potential of the indicator electrode with water and HCl vapor pressures.

Apart from the recent work of Laitinen, considered below, the few other studies of wet chloride melts have been concerned mainly with chemical aspects including solubility or with residual currents. Clearly there is considerable scope for additional investigations of the electrochemistry of water, *per se* in this important group of molten salts.

3.3. Hydroxides

In anhydrous molten NaOH–KOH, between 450–500°C, chemical as well as electrochemical data indicate[86] that superoxide, peroxide, and oxide ions all have ranges of stability in solution. The introduction of water into a melt containing peroxide leads to the generation of superoxide,

$$3O_2^{2-} + 2H_2O \rightleftharpoons 2O_2^- + 4OH^-$$

which may further react with water to liberate oxygen:

$$2O_2^- + H_2O \rightleftharpoons O_2 + 2OH^-$$

On the other hand, the superoxide may, under suitable conditions, be electrochemically reduced to hydroxide:

$$O_2^- + 2H_2O + 3e^- \rightarrow 4OH^-$$

as discerned from analysis of appropriate voltammetric signals.[86] This co-reduction of water at -0.82 V versus a standard alkali metal reference electrode appears to limit cathodic potential excursions.[87]

The partial pressures of oxygen and water vapor also influence voltammetric responses at platinum, gold, silver, nickel, and iron electrodes in these melts.[88] Thus, the anodic dissolution of platinum is enhanced in the presence of water, whilst water is directly reduced at -1.8 V:

$$H_2O + e^- \rightarrow \tfrac{1}{2}H_2 + OH^-$$

The interdiffusion coefficient of water in molten, equimolar NaOH–KOH has been evaluated[89] by chronopotentiometry between 190 and 260°C. Water was introduced indirectly by the addition of silica; the water reduction potential is -0.88 V vs. Cu/Cu(I) reference electrode. Silica reacts with molten hydroxides according to[90]

$$SiO_2 + 4OH^- \rightleftharpoons SiO_4^{4-} + 2H_2O$$

The stoichiometry of this reaction can be verified by titrating the liberated water with sodium oxide. Confirming[90] the earlier report,[86] the coreduction wave for water and superoxide appears 1 V more anodic to the direct reduction of water upon addition of sodium oxide to the melt.

The peak current for the direct reduction of water in NaOH–KOH eutectic at 170°C at stationary platinum microelectrodes is directly proportional to the concentration of water.[91] Interestingly, the potential for the reduction of nitrate to nitrite,

$$NO_3^- + 2e^- \rightarrow NO_2^- + O^{2-}$$

shifts cathodically when water is added to an anhydrous melt. Furthermore, a second wave of more complex character appears and has been associated[91] with the coreduction of nitrate and water:

$$NO_3^- + H_2O + 2e^- \rightarrow NO_2^- + 2OH^-$$

As the concentration of water in the melt is raised, the first wave gradually disappears and the height of the second wave becomes dependent on the water content. Yet a further reduction wave at more cathodic potentials appears to be due to the reduction of nitrite ions and is independent of the water concentration. Although this work is reported by Claes and Glibert in more detail in Chapter 14, it is interesting to compare these deductions[91] with the most recent aspects of the water reduction wave in nitrate melts.[75] Of course, comparisons between different melt systems always need to be tempered with extreme caution.

3.4. Acetates

Water appears to be directly reduced at -1.8 V vs. Ag/Ag(I), 0.1 M, on platinum electrodes in a sodium–potassium acetate eutectic melt at 250°C, according to chronopotentiometric and voltammetric evidence.[92, 93]

4. How Water Affects Electrode Processes in Melts

It would be a reasonable surmise that water might influence electrode processes in molten salts. The above evidence as well as that contained in other chapters in this volume clearly indicate how some redox equilibria may be affected. Additionally, rather trivial changes may occur in solution species, and hence their reduction or oxidation behavior, simply because water may act as a ligand to form new complex ions. However, the effect of water on the kinetics and mechanisms of electrode processes in melts has received scant attention, partly perhaps because it raises conceptual difficulties about the possible role of water at the electrode surface: it is notable that Palanker and his colleagues[80] were unable to come to any definite conclusions about the significance of their results.

From measurements of the ac impedance of mercury electrodes in molten nitrates containing Ni(II) ions at 140°C at the half-wave reduction potential, it has been claimed[81] that the rate constant decreases by more than a half upon the addition of 0.5 mole fraction of water. Certainly, polarographic and galvanostatic results show[94] an increase in the rate constant for the discharge of Ni(II) ions in aqueous perchlorates when the supporting electrolyte concentration is increased from $0.4 \rightarrow 4$ M. However, this concentration is still a long way from the molten salt end of the spectrum. Moreover, aquonickelate(II) ions persist in aqueous melts such as $Ca(NO_3)_2 \cdot 4H_2O$, $Ca(NO_3)_2 \cdot 4D_2O$, $Ca(NO_3)_2 \cdot 4H_2O + 46$ mol % KNO_3 and $LiNO_3 \cdot 3H_2O$ at $\sim 50°C$ on the evidence[95] of polarographic and spectroscopic data. Indeed, even in such a complexing medium as $CaCl_2 \cdot 6H_2O$ at 35°C and 115°C, some aquonickelate(II) bonds remain, although the reduction kinetics on mercury electrodes, reflected in the polarographic behavior, is now quite different.[95] In the aqueous melt $Ca(NO_3)_2 \cdot 4H_2O + 56$ mol % KNO_3 at 50°C, Cd(II) ions are reduced at increasingly cathodic potentials as the system is titrated with additional water.[96] Although this would seem to indicate the formation of more highly water-substituted cadmium complexes in solution, and the corresponding large increases in limiting currents indicate radical changes in the melt density and viscosity, the constancy of the polarographic wave shape suggests that a common discharge mechanism for Cd(II) ions on mercury pertains right through from dilute aqueous solutions.[97] Most recently of all we have reported[96, 98] the effect of additions of small $(0 \rightarrow 3\ m)$ quantities of water on the polarographic reduction of Ni(II), Pb(II), and Tl(I) in molten, eutectic $LiNO_3$–KNO_3 at 145°C. At very low water concentrations $(< 4.5 \times 10^{-3}\ m)$ the Ni(II) polarogram is shifted up to 40 mV *anodic* and the wave changes from quasireversible to wholly reversible behavior. At higher water concentrations (up to 3.36 *m*), more familiar cathodic half-wave potential shifts occur. Within the experimental precision of linear sweep polarography (at a hanging mercury drop electrode) the rate constant for this reduction, after fully correcting for half-wave shifts, density, and diffusion coefficient variations, remains constant. This contrasts with the earlier investigation of this system,[81] although those authors did not make the necessary corrections. Nevertheless the dramatic variations occurring at the very lowest water contents signal a warning to workers who may not have adequately anhydrous molten solvents. The results of this study[96, 98] will be reported in greater detail elsewhere.[99]

A recent reappraisal[100] of some unexplained results obtained during an anodic chronopotentiometric study[101] of nitrite ions at oxidized platinum electrodes in molten $NaNO_3$–KNO_3 at 250°C and above, suggests that residual traces of moisture as well as nitrite might have been implicated. Need we emphasize, therefore, that many other electrochemical studies carried out in inadequately anhydrous conditions may benefit from similar reassessments, if not reinvestigations?!

The reduction of chromate[102] and of Cr(III)[103] in molten LiCl–KCl at 450–500°C both show marked changes in the presence of added water. A single chronopotentiometric reduction wave of -0.6 V vs. Pt/Pt(II), 1 *M*, is exhibited in the anhydrous melt for chromate ions.[102] Two reduction transitions at -0.6 V and -1.18 V appear when water is introduced, with a third, ill-defined wave evident between -1.25 V and -1.55 V due to the direct and irreversible reduction of water itself. The overall reduction scheme in the presence of water is

$$CrO_4^{2-} + 2H_2O + 3e^- \xrightarrow{\text{Li}^+} LiCrO_2(s) + 4OH^-$$

Detailed analysis of cyclic voltammetric waves at glassy carbon electrodes for the reduction of Cr(III) under varying vapor pressures of H_2O and HCl, indicate[102] that an initially reversible hydrolysis

$$CrCl_6^{3-} + H_2O \rightleftharpoons CrCl_5(OH)^{3-} + HCl$$

is followed by irreversible polymerization to form oxobridged, insoluble species. Furthermore, a rise in temperature accelerates the rate of this second process, although the simultaneous decline in the primary hydrolytic step may merely be a consequence of the lower solubility of water compared with HCl.

Aside from reassessments of earlier electrochemistry in "damp" melts, it would seem an opportune time to extend investigations into the effect of traces of water on electrode processes in molten salts.

ACKNOWLEDGMENTS

We thank the Science Research Council for a Research Grant. We are grateful to Dr. D. Inman and Dr. A. K. Turner for many helpful discussions and (A.K.T.) for much of the development of an alternative explanation of the composite wave in molten nitrates in the presence of hydrogen.

References

1. P. Héroult, Fr. Pat. 175711 (April 23, 1886).
2. C. M. Hall, U.K. Pats. 5669, 5670 (April 2, 1889).
3. G. A. Wolstenholme, *Chem. Ind.* (*London*) **9**, 383 (1975); also N. E. Richards in *Electrochemistry, the Past Thirty and the Next Thirty Years*, Eds. H. Bloom and F. Gutmann, Plenum, New York (1977).
4. M. Galopin and J. S. Daniel, *Electrodeposition Surf. Treat.* **3**, 1 (1975).
5. W. R. Grimes and S. Cantor, in *The Chemistry of Fusion Technology*, Ed. D. M. Gruen, Plenum, New York (1972).
6. G. Long, UKAEA, AERE Report M-1925 (1967).
7. D. A. J. Swinkels, in *Advances in Molten Salt Chemistry*, Vol. 1, Eds. J. Braunstein, G. Mamantov, and G. P. Smith, Plenum, New York (1971); see also Extended Abstracts, 154th Meeting Electrochemical Society, Pittsburgh (1978).
8. K. V. Kordesch, *J. Electrochem. Soc.* **125**, 77C (1978).
9. M. R. Edwards and D. G. Lovering, *Int. Metals Rev.*, 123 (1976).
10. D. Inman and D. E. Williams, in *Electrochemistry, the Past Thirty and the Next Thirty Years*, Eds. H. Bloom and F. Gutmann, Plenum, New York (1977).
11. W. Sundermeyer, *Angew. Chem.* (*Int. Ed.*) **4**, 222 (1965).

12. D. H. Kerridge, *M.T.P. Int. Rev. Sci. Inorg. Chem. Series One*, Vol. 2, Butterworths, New York (1972).
13. B. W. Hatt and W. D. Read, paper presented at a meeting of the Molten Salt Discussion Group, Southampton University (April 1978).
14. B. W. Hatt and M. J. Pitt, paper presented at a meeting of the Molten Salt Discussion Group, Southampton University (April 1978).
15. A. Bonomi, M. Lavaidy, and C. Gentaz, paper presented at EUCHEM Molten Salt Meeting, Noordwijkerhout, Holland (1976).
16. K. E. Johnson, in *Metal-Slag-Gas Reactions and Processes*, Eds. Z. A. Foroulis and W. W. Smeltzer, Electrochemical Society, New Jersey (1975), p. 581.
17. R. Battino and H. L. Clever, *Chem. Rev.* **66**, 395 (1966).
18. M. S. Hull and A. G. Turnbull, *J. Phys. Chem.* **74**, 1783 (1970).
19. E. R. Buckle and R. R. Finbow, *Int. Metals Rev.*, 197 (1976).
20. J. Braunstein, *Inorg. Chim. Acta. Rev.* **2**, 19 (1968).
21. W. J. Burkhard and J. D. Corbett, U.S. Atom Energy Comm. Report ISC-929 (1957); *J. Am. Chem. Soc.* **79**, 6361 (1957).
22. F. Colom and A. Bodalo, *Collect. Czech. Chem. Commun.* **36**, 674 (1971).
23. F. Colom and A. Bodalo, *Corros. Sci.* **12**, 731 (1972).
24. O. Kubaschewski and B. E. Hopkins, *Oxidation of Metals and Alloys*, Butterworths, London (1953).
25. S. Cantor and W. R. Grimes, *Nucl. Tech.* **22**, 120 (1974).
26. D. R. Holmes and G. M. W. Mann, *Corrosion* **21**, 370 (1965).
27. D. Lewis, D. G. Lovering, and B. Svensson, AB Atomenergi Report AE-RKK-381 (1969).
28. D. G. Lovering, Final Contract Report to NRDC/Rolls-Royce Ltd. (1970), and Extended Abstracts 23rd I.S.E. Meeting, Stockholm (1972), p. 103.
29. A. J. Arvia, R. C. V. Piatti, and J. J. Podesta, *Electrochim. Acta* **17**, 889 (1972).
30. A. J. Arvia, R. C. V. Piatti, and J. J. Podesta, *Electrochim. Acta* **16**, 1797 (1971); **17**, 901 (1972).
31. E. L. Krongauz, V. D. Kascheev, and V. B. Busse-Machukas, *Sov. Electrochem.* **8**, 1219 (1972).
32. A. K. Turner, Ph.D. thesis, London (1978).
33. *Encyclopaedia of Electrochemistry of the Elements*, Vol X, *Fused Salt Systems*, by J. A. Plambeck, Series Ed. A. J. Bard, Marcel Dekker, New York (1976).
34. R. C. Howie and D. W. MacMillan J. *Appl. Electrochem.* **2**, 217 (1972).
35. B. Nayak and M. M. Misra, *J. Appl. Electrochem.* **7**, 45 (1977).
36. B. Nayak and M. M. Misra, *J. Appl. Electrochem.* **9**, 699 (1979).
37. J. M. West, *Electrodeposition and Corrosion Processes*, VNR, London (1971).
38. K. Grjotheim, *Can. Metall. Q.* **11**, 585 (1972).
39. W. E. Haupin, *J. Electrochem. Soc.* **107**, 232 (1960).
40. J. K. Wright, *Miner. Sci. Eng.* **5**, 119 (1973).
41. M. I. Brittan and R. R. Liebenberg, *Trans. Inst. Min. Met.* **80**, C156, C262 (1971).
42. M. I. Brittan, *J.S.A. Inst. Min. Met.* **71**, 87 (1970).
43. F. R. Duke, in *Mechanisms of Inorganic Reactions*, Advances in Chemistry Series No. 49, American Chemical Society, Washington, D.C. (1965), p. 220.
44. R. F. Bartholomew and H. M. Garfinkel, *J. Inorg. Nucl. Chem.* **31**, 3655 (1969).
45. R. B. Temple and F. W. Thickett, *Aust. J. Chem.* **26**, 667 (1973).
46. P. G. Zambonin, *J. Electroanal. Chem.* **32**, App. 1 (1971).
47. H. S. Swofford and H. A. Laitinen, *J. Electrochem. Soc.* **110**, 814 (1963).
48. T. E. Geckle, M.S. thesis, Pennsylvania State University (1964).
49. M. Peleg, *J. Phys. Chem.* **71**, 4553 (1967).
50. G. J. Hills and P. D. Power, *J. Polarog. Soc.* **13**, 71 (1967).
51. P. G. Zambonin, *J. Electroanal. Chem.* **24**, 365 (1970).

52. M. Francini and S. Martini, *Electrochim. Acta* **13**, 851 (1968).
53. Yu. S. Lyalikov and R. M. Novik, *Uch. Zap. Kishinevsk. Gos. Univ.* **27**, 61 (1957).
54. P. G. Zambonin, F. Paniccia, and A. Bufo, *J. Phys. Chem.* **76**, 422 (1972).
55. P. G. Zambonin, *Anal. Chem.* **43**, 1571 (1971).
56. P. G. Zambonin, V. L. Cardetta, and G. Signorile, *J. Electroanal. Chem.* **28**, 237 (1970).
57. E. Desimoni, F. Paniccia, L. Sabbatini, and P. G. Zambonin, *J. Appl. Electrochem.* **6**, 445 (1976).
58. E. Desimoni, F. Palmisano, and P. G. Zambonin, *J. Electroanal. Chem.* **84**, 323 (1977).
59. E. Desimoni, F. Paniccia, and P. G. Zambonin, *J. Chem. Soc. Faraday Trans. 1* **69**, 2014 (1973).
60. E. Desimoni, F. Paniccia, and P. G. Zambonin, *J. Phys. Chem.* **81**, 1985 (1977).
61. A. K. Turner, private communication.
62. R. N. Kust and J. D. Burke, *Inorg. Nucl. Chem. Lett.* **6**, 333 (1970).
63. A. F. J. Goeting and J. A. A. Ketelaar, *Electrochim. Acta* **19**, 267 (1974).
64. J. Jordan, W. B. McCarty, and P. G. Zambonin in *Molten Salts, Characterization and Analysis*, Ed. G. Mamantov, Marcel Dekker, New York (1969).
65. J. Jordan, *J. Electroanal. Chem.* **29**, 127 (1971).
66. A. Espinola, Ph.D. thesis, Pennsylvania State University (1974).
67. A. Espinola and J. Jordan, Proceedings A.C.S. Meeting, San Francisco (August 1976).
68. D. C. Walker, *Q. Rev.* **21**, 79 (1967).
69. E. J. Hart and M. Ambar, *The Hydrated Electron*, Wiley-Interscience, New York (1970).
70. C. A. Angell and C. T. Moynihan, in *Molten Salts: Characterization and Analysis*, Ed. G. Mamantov, Marcel Dekker, New York (1969).
71. H. S. Swofford, Ph.D. thesis, Illinois University, Urbana (1972).
72. N. Hittorf, *Poggendorfs Ann. Phys.* **72**, 481 (1847).
73. H. E. Bartlett and K. E. Johnson, *Can. J. Chem.* **44**, 2119 (1966).
74. H. E. Bartlett and K. E. Johnson, *J. Electrochem. Soc.* **114**, 64 (1967).
75. D. G. Lovering, R. M. Oblath, and A. K. Turner, *J. Chem. Soc. Chem. Commun.*, 673 (1976).
76. G. J. Hills and K. E. Johnson, Proceedings of the 2nd International Conference on Polarography, Cambridge, England, 1959, Pergamon, London (1961), p. 974.
77. D. H. Kerridge and J. D. Burke, *J. Inorg. Nucl. Chem.* **38**, 1307 (1976).
78. R. Pineaux, *C. R. Acad. Soc. Paris Ser. C.* **267**, 1449 (1968).
79. R. Pineaux, *C. R. Acad. Sci. Paris Ser. C.* **268**, 788 (1969).
80. V. Sh. Palanker, A. M. Skundin, and V. S. Bagotskii, *Elektrokhimiya* **2**, 640 (1966).
81. J. E. B. Randles and W. White, *Z. Electrochemie* **59**, 666 (1955).
82. G. J. Hills and P. D. Power, *Trans. Faraday Soc.* **64**, 1629 (1968).
83. H. A. Laitinen, W. S. Ferguson, and R. A. Osteryoung, *J. Electrochem. Soc.* **104**, 516 (1957).
84. R. Combes, J. Vedel, and B. Tremillon, *C. R. Acad. Sci. Ser. C.* **273**, 1740 (1971); *Electrochim. Acta* **20**, 191 (1975); R. Lysy and R. Combes, *J. Electroanal. Chem.* **83**, 287 (1977).
85. R. Combes, J. Vedel, and B. Tremillon, *C. R. Acad. Sci. Ser. C.* **275**, 199 (1972).
86. J. Goret and B. Tremillon, *Bull. Soc. Chim. France*, 67 (1966).
87. J. Goret and B. Tremillon, *Electrochim. Acta* **12**, 1065 (1967).
88. H. J. Kruger, A. Rahmel, and W. Schwenk, *Electrochim. Acta* **13**, 625 (1968).
89. G. G. Bombi, S. Zecchin, and G. Schiavon, *J. Electroanal. Chem.* **50**, 261 (1974).
90. J. Goret and B. Tremillon, *Bull. Soc. Chim. France*, 2872 (1966).
91. P. Claes, paper presented at EUCHEM Molten Salt Meeting, Lysekil, Sweden (June 1978).
92. R. Marassi, V. Bartocci, and F. Pucciarelli, *Talanta* **19**, 203 (1972).
93. R. Marassi, V. Bartocci, F. Pucciarelli, and P. Cescon, *J. Electroanal. Chem.* **47**, 509 (1973).
94. J. Dandoy and L. Gierst, *J. Electroanal. Chem.* **2**, 116 (1961).

95. D. G. Lovering, *Collect. Czech. Chem. Commun.* **37**, 3697 (1972).
96. D. G. Lovering and R. M. Oblath, *J. Electrochem. Soc.* **127**, 1997 (1980).
97. D. G. Lovering, *J. Electroanal. Chem.* **50**, 91 (1974).
98. R. M. Oblath, Ph.D. thesis, CNAA (1978).
99. D. G. Lovering and R. M. Oblath, in preparation.
100. D. G. Lovering, private communication to D. Inman (1977).
101. D. Inman and J. Braunstein, *Chem. Commun.* 148 (1966).
102. I. Uchida and H. A. Laitinen, *J. Electrochem. Soc.* **123**, 829 (1976).
103. H. A. Laitinen, Y. Yamamura, and I. Uchida, *J. Electrochem. Soc.* **125**, 1450 (1978).

12

The Role of Water during the Purification of High-Temperature Ionic Solvents

S. H. White

1. Introduction

This paper is somewhat of a paradox because it is well known that when handling high-temperature solvents, water should be fastidiously avoided! However, it has been suggested recently that progress in understanding the behavior of concentrated electrolyte solutions will only be made through a consideration of dilute solutions of water in fused salts.[208, 210] It would appear that this interface is at last being bridged.

Generally, electrochemical studies involving aqueous solutions have required considerable attention to the preparation of highly purified solvent. The use of high-temperature molten inorganic liquids as solvents demands similar attention to such detail. Over the past 25 years, considerable expertise has been developed for the production of high-purity inorganic salts. Because the interest here concerns the interactions between water and molten solvent ions, as well as added solute ions, a major requirement for the high-temperature solvent is the initial absence of water and other impurity ionic species. These might, otherwise, moderate the behavior of water present at low concentrations and prevent the establishment of limiting behavior. Furthermore, because of the nature of the effects under investigation, high sensitivity of the instrumental techniques is often necessary, which can result in the ready observation of extraneous species

S. H. White • Imperial College, London, SW7 2AZ, England. Present address: EIC Corporation, 55 Chapel Street, Newton, Massachusetts 02158.

and interactions in the solvent itself. It is thus pertinent to deal with the principles of purification techniques, the role of water in these operations, and to set out generally applicable procedures which are known to produce the highest-quality materials and to elaborate methods which will characterize solvents in a reproducible way.

It has been a further objective of this chapter to bring together the knowledge so far acquired concerning the solubility and reactivity of water in a number of commonly employed molten solvent systems, together with data which will enable water contents to be monitored. Furthermore, it is clear that water is not always inert in such solvents, particularly as the temperature is increased, so that results relating to hydrolytic products have also been considered. It is hoped that this chapter will serve several purposes, among which are (1) the unification of procedures which are not always explicitly discussed in the literature but are often handed down from one Ph.D. student to another (with certain aberations on the way!), (2) to illustrate the principles on which commonly performed procedures are based so that perhaps further innovations may be achieved, and (3) to highlight the further data requirements for systems in common use and of technological importance, to encourage workers to look into some of the interesting problems associated with water as a solute, and which may lead to a better understanding of water in high-temperature milieux. That this may be fruitful can be judged by the subsequent work of Zambonin following the initial postulate by Jordan and Zambonin that the chemistry of oxide ions (derived from water!) was complicated by the hitherto unsuspected instability of oxide in molten nitrates.

2. Purity Requirements and Chemical Reactivity of Water

2.1. General

The requirements of purity can be easily defined for molten salt solvents based upon inorganic salts. The purity levels for systems based on organic salts are less well defined and will not be discussed further in this chapter; the papers by Ubbelohde,[1] Reinsborough,[2] Griffiths,[3, 4] and Kisza[5] may give useful entrées into this field. Generally speaking, freedom from moisture, oxide species, oxyanions such as sulfates, nitrates, phosphates, silicates, etc., metal ion impurities such as Fe^{3+}, Pb^{2+}, Zn^{2+}, etc., particulate matter such as dusts, carbon particles onto which metallic impurities often become absorbed, and organic residues arising from bulk salt preparative processing, is essential. Laboratory chemicals are generally available at purities between 95% and 98% with some being obtainable as higher-quality analytical grade materials 98%–99.5%. The demands of electronic and optical industries have required even high-quality materials, and a range of (high-cost) chemicals is now available under trade names such as Optran, Crystran, Patinal, etc. whose impurity levels may be less than 100 ppm. Thus careful selection of starting material (set against the cost of installing one's own purification systems!) can be made to ensure the use of high-quality solvent systems.

For studies involving interfaces and dilute solutions the degree of purification of the solvent required is high. For example, measurements involving millimolar quantities of solute dictate that impurity levels should be less by at least two orders of magnitude than the added solute. However, when the physical properties of solvents such as density, viscosity, or bulk thermodynamic quantities are to be measured it is unlikely

TABLE 1. Some Selected Solvent Components and the Range of Purity Available

Components		GPR, millimolal	AR, millimolal	Optran	Crystran
LiCl	H_2O	111			
	$\sum HM^a$	0.3			
	$\sum OA^b$	1			
KCl	H_2O		< 0.1		
	$\sum HM$		< 0.1		Available
	$\sum OA$		0.3		
NaCl			< 0.1		
			< 0.1		Available
			0.3		
$CaCl_2$	H_2O		Dihydrate		
	$\sum HM$		0.1		
	$\sum OA$		2		
LiF	H_2O	~ 100		< 100 ppm exc. S	Available
	$\sum HM$	7.0			
	$\sum OA$	4.0			
KF	H_2O	138	10.0	< 100 ppm	
	$\sum HM$	0.7	0.5		
	$\sum OA$	10.0	5.0		
NaF	H_2O		10.0	< 100 ppm	Available
	$\sum HM$		0.7		
	$\sum OA$		10.0		
$LiNO_3$	H_2O	Trihydrate	<0.6		
	$\sum HM$	1.0	low		
	$\sum OA$	2.0	4		
	Cl^-	5.0	0.28		
Li_2SO_4	H_2O		Monohydrate		
	$\sum HM$		0.3		
	$\sum OA$		0.3		

^{a} $\sum HM$ is the total heavy metal concentration.
^{b} $\sum OA$ is the total oxyanion impurity concentration.

TABLE 2. Solvents Containing Hygroscopic Components and Their Technological Interest

Solvent			Composition, mol %				Melting point, °C		
A	B	C	A	B	C	Ref.	Solvent B	Mixture	Use
LiCl	KCl	—	59	41	—	49	770	∼ 355	General purpose solvent
LiF	NaF	KF	46.5	11.5	42	49	995	454	Refractory metal plating and forming
MgCl$_2$	NaCl	KCl	50	30	20	49	801	396	Nuclear reprocessing
ZnCl$_2$	NaCl		33	67	—	172	801	475	Zinc electrowinning
LiF	LiCl	LiBr	22	31	47	173	610	430	High-temperature primary and secondary batteries
AlCl$_3$	KCl		67	33	—	49	770	124	Batteries
CaCl$_2$	NaCl		52	48	—	49	801	500	Sodium metal production

that millimolar levels of impurities will have any significant effect, and thus less stringent conditions of purity will apply. The attainment of such high levels of purification can often be made for many fused salt systems, by careful selection of the starting material (see Table 1), and processes such as desiccation, vacuum drying at elevated temperature,[6] displacement of impurities such as moisture by gas bubbling,[7] and metal ions by displacement by an active metal such as Grignard grade magnesium,[8] aluminium,[9] etc., or preelectrolysis.[10] Major problems of purification arise when the solvents consist of a salt or salts which are hygroscopic in nature, for example, any lithium, zinc, calcium, or magnesium salts which are commonly used as solvent components, because the melting point of the solvent mixture is reduced to a conveniently low level by their inclusion. Thus the presence of water of crystallization, the consequential solubility of this water in the molten salt, its reactivity with the salt at high temperature, and the subsequent chemistry of the hydrolysis products, form the central problem for the purification of melts based on these salts. Consequently, the chemistry and electrochemistry of water in, and its reaction products from, fused salt media must be understood when developing efficient methods of purifying salts and their molten mixtures. Table 2 shows some common solvent systems which employ some of the above-mentioned components and which have industrial as well as scientific interest.

2.2. Solubility and Reactivity of Water with Molten Inorganic Salts at High Temperatures

Since the number of salt combinations is large, the number of studies in this area is relatively small. The two groups of solvents which have received the most attention are the alkali metal nitrates and chlorides, and the results obtained will be used here to further the ends described at the end of the last paragraph. This information may also be used as a guide to systems where detailed knowledge is not available but care must be taken in making such extrapolations.

2.2.1. Chlorides and Fluorides

Some 20 years ago, Burkhard and Corbett[11] determined the equilibrium solubility of water in purified LiCl–KCl mixtures at 390 and 480°C, under partial pressures of water between 3 and 26 torr. The data are illustrated in Figure 1 and the results are summarized in Table 3. The solubility of water is related to its partial pressure by Henry's law, $\beta = k_H p$,

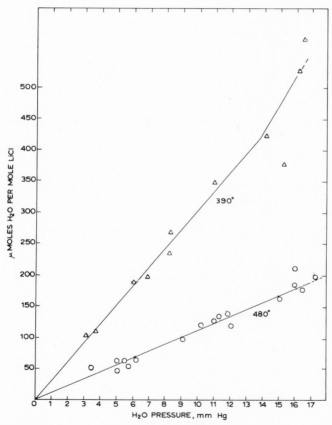

FIGURE 1. Plot of solubility of water versus partial pressure of water over 60 mol % LiCl, 40 mol % KCl at 390 and 480°C.[11]

TABLE 3. Water and Hydrogen Chloride Solubilities in Molten Chlorides

Melt composition, mol %	T, °C	Henrian constant, μmol mol^{-1} mm^{-1}	Range of Henry's law, mm	ΔH_{soln}, kcal mol^{-1}	Diffusion coefficient, $10^5 D_{\text{HX}}$ cm^2 sec^{-1}	Ref.
Water						
LiCl–KCl						
50/50	390	30.0[a]	14	−8.0	—	11
	480	14.0[a]	18		—	11
53/47	390	—	—			
	480	11.8[a]	18			11
60/40	390	30.5[a]	14	−11.0	3.1 ± 0.5[c]	15[f]
	480	11.3[a]	18		8.0 ± 0.5[c]	11
68.6/31.4	390	—	—		—	11
	480	10.8[a]	18			
Hydrogen chloride						
LiCl–KCl						
60/40	480	1.0[a]	90[d]	—	—	11
58.8/41.2	404	0.077[a]	760[e]	4.4	18–20	12
	462	0.096[a]			24–27	
	520	0.125[a]			20–21 $\Delta E_D = 4.4$	
	450	0.0938				
	ln $k_H = -2.232\, T^{-1} - 13.09$					
58.8/41.2	490	0.098[a]		3.3	24 ± 5	13
	570	0.113[a]			23 ± 5	
	675	0.142[a]				
	450	0.0871			$\Delta E_D = 2.9$	
	ln $k_H = -1676\, T^{-1} - 14.005$					
ZnCl$_2$–LiCl–KCl						
3.4 (1 M)	450	0.1218[g]	760	—	23–24	176
4.7		0.1375[g]			21–28	
8.2		0.1633[g]			10–10 ±20%	
11.9		0.2035[g]			9.2–10	
13.5		0.2187[g]			9–9	
17.4 (6.3 M)		0.2502[g]			7–8	
11.9/88.1	395	0.2145[g]			10.7 ± 2	
	500	0.1835[g]			8.9	

	T	k_H	$\ln k_H$					Ref.
NaCl–KCl								
75/25	800	0.0497[b]	$\ln k_H = -1464\,T^{-1} - 15.451$	+2.9	57.8	$\Delta E_D = +19.0$	$\ln D = -9574\,T^{-1} + 1.466_2$	14, 177[f]
50/50	800	0.107[b]	$\ln k_H = -1036\,T^{-1} - 17.013$	+2.1	46.5	$\Delta E_D = +19.8$	$\ln D = -9984\,T^{-1} + 1.631$	175, 14, 177[f]
	750	0.116						
25/75	800	0.144	$\ln k_H = -64.4\,T^{-1} - 15.692$	0.13	47.4	$\Delta F_{.} = +13.6$		14, 177[f]
LiCl	800	0.02695	$\ln k_H = -7562\,T^{-1} - 10.382$	+15.0	131	$\Delta E_D = -27.1$	$\ln D = +3570\,T^{-1} - 19.285$	14, 177[f]
NaCl	800	0.03279	$\ln k_H = -4224\,T^{-1} - 13.296$	+8.4	105	$\Delta E_D = -5.0$	$\ln D = 2536\,T^{-1} - 9.218$	177[f], 175, 174, 14
	930	0.0517						
	878	0.0474						
KCl	800	0.1471	$\ln k_H = -1576\,T^{-1} - 14.263$	+3.2	106	$\Delta E_D = +1.5$	$\ln D = -763\,T^{-1} - 6.14$	14, 175, 174, 174, 177[f]
	900	0.1620						
	900	0.1513						
	1000	0.1895						
RbCl	800	0.3079	$\ln k_H = +1187\,T^{-1} - 16.098$	−2.37	74.4	$\Delta E_D = +9.5$	$\ln D = -4762\,T^{-1} - 2.76_6$	14, 175, 174, 174, 177[f]
	830	0.3109						
	718	0.1329						
	942	0.2711						
CsCl	664	0.1211						174
	885	0.3776						

[a] Per mole LiCl.
[b] Per mole mixture.
[c] Diffusion coefficient based on water solubility.[11]
[d] Solubility determined over the range of HCl pressures.
[e] Determined at 1 atm HCl pressure.
[f] Diffusion coefficient data reference.
[g] Per mole LiCl.

in these dilute solutions. It was suggested that the changes in linearity between the solubility (β) and the water partial pressure (p) above 18 mm at 480°C and above 14 mm at 390°C (k_H becomes 140 at 480°C) are caused by the onset of hydrolysis of the lithium chloride:

$$2LiCl + H_2O \rightleftarrows LiOH + HCl_2^- + Li^+ \tag{1}$$

$$LiOH + 2LiCl \rightleftarrows Li_2O + HCl_2^- + Li^+ \tag{2}$$

This was substantiated by the observation that the HCl present in the water vapor in the apparatus was <2 μmol mol^{-1} (the limit of detection) at water pressures less than 16 mm, whereas the HCl content of the vapor rose to 200 μmol mol^{-1} when the water pressure reached 26 mm. The design of the apparatus prevented quantitative analysis of the hydrolytic reaction. These results suggest that, even at the relatively high temperatures involved, water has a significant equilibrium solubility in lithium-chloride-containing melts (and probably in other melts incorporating hygroscopic salt components). This behavior may arise from the highly polarizing (acidic) nature of the Li$^+$ ions and the polar nature of the water molecules themselves. Other polar molecules such as HCl, HBr, H$_2$S, etc., which may be hydrolysis products of the melts, might be expected to have measurable solubilities in solvents containing polarizing ions such as Zn^{2+}, Mg^{2+}, Li$^+$, etc.

The solubility of hydrogen chloride in 60 mol % LiCl–40 mol % KCl at 480°C was also reported by Burkhard and Corbett.[11] Their data for this solvent obeyed Henry's law up to 90 mm partial pressure of HCl with a k_H of 1 μmol (mol LiCl)$^{-1}$ mm^{-1}. Extrapolation to zero partial pressure of HCl, however, showed that the original melt contained 10–45 μmol OH$^-$ (mol LiCl)$^{-1}$, i.e., 0.8 mM OH$^-$ or less. Recently, other measurements of the solubility of HCl in LiCl–KCl have been reported[12, 13] at a pressure of 760 mm HCl. The Henrian constant k_H was approximately 0.1 μmol (mol LiCl)$^{-1}$ mm^{-1} at 480°C, an order of magnitude different from the earlier results. Using chronopotentiometry[12, 13] and linear sweep voltammetry,[12] diffusion coefficients, D, for H$^+$ ions were obtained (Table 3). These values show an order of magnitude difference from other ion diffusion coefficients at this temperature. However, on the basis of Burkhard and Corbett's data extrapolated to 1 atm HCl, the diffusion coefficient is reduced to a level consistent with those of other monovalent ions in molten salts, making speculation about unique transport modes for H$^+$ ions unnecessary (but see Reference 177). On the other hand, it is difficult to reconcile the solubility data reported by the different authors. unless one predicts that there is a limiting solubility of HCl in this melt well below 760 mm partial pressure of HCl. The range over which Henry's law is obeyed was ~ 90 mm according to Burkhard and Corbett, who gave no explanation for the lack of data above this level. If it is assumed that they were approaching the limits of solubility at these partial pressures, then the two sets of data may be reconcilable. Recent data by Ukshe and Devyatkin[14] for pure alkali halides show Henrian behavior up to 1 atm of HCl in the temperature range 700–900°C. Thus, a limiting solubility is rather surprising. Nevertheless the limit of solubility may be thought of in the sense of a solubility product for the two-phase system (gas–solution)

$$HCl(g) \rightleftarrows H^+(soln) + Cl^-(soln) \qquad (3)$$

$$K_{sp} = [H^+][Cl^-]$$

Since the chloride ion concentration is extremely high in the LiCl–KCl melt ($29\ M$) the common-ion effect may be responsible for the limitation of solubility. This is supported by the recent data of Minh and Welch,[12] who added zinc chloride to the eutectic and found an increase in the gas solubility. Zinc chloride will complex the chloride ions of the solvent reducing their activity and thus increasing the solubility of HCl. The presence of an equilibrium such as (3) might result in a coupling of the chemical reaction to the electrochemical process. (Such an effect was reported[177] for HCl in RbCl.) In this situation the use of the simple relationships employed to calculate D are not valid. A more detailed study of both the solubility relationship and the H^+ ion reduction mechanism is required in spite of the results of one of these authors.[12]

The solubility of HCl in the pure molten alkali metal halides[174, 175] shows a marked increase in the order $Li < Na < K < Rb \simeq Cs$ at a given temperature ($\sim 800°C$), indicating endothermic heats of solution except for RbCl. The diffusion coefficients reported[177] are large, $\sim 5 \times 10^{-5}$ to $20 \times 10^{-5}\ cm^2\ sec^{-1}$, consistent with the data reported for LiCl–KCl at 400–600°C, but somewhat inconsistent with diffusion coefficients for other ions at comparable temperatures. These systems warrant further investigation in view of the somewhat uncertain behavior of the solute HCl (Table 3).

The electrochemical reduction of water, per se, dissolved in LiCl–KCl eutectic has been investigated so that a convenient *in situ* method for detecting and determining the water content of this melt could be established.[15] Equilibration of the melt with water in the range of partial pressure used by Burkhard and Corbett enabled the concentration of water in the melt to be precisely defined. Cyclic voltammetry, at a platinum micro electrode, was then used to examine the cathodic reduction of water. The diffusion coefficient of water as a function of temperature and concentration was thus established. The cyclic voltammograms showed two reduction peaks, one at -1.35 V and the other at -2.50 V with respect to the standard Pt/Pt(II)(1 M) reference electrode. The latter peak was shown by standard additions to correspond to the reduction of hydroxide ions. Thus a product of the initial reduction of water must be OH^- ions. It was established from the dependence of i_p on the sweep rate and on the water concentration, that water is reduced in a fast-diffusion-controlled, one-electron transfer process, probably

$$H_2O + e \rightarrow OH^- + Pt(H) \tag{4}$$

followed by

$$OH^- + e \rightarrow O^{2-} + \tfrac{1}{2}H_2 \tag{5}$$

or possibly

$$Pt(H) + OH^- + e \rightarrow Pt + O^{2-} + H_2 \tag{6}$$

since at a clean Pt surface, alloying reactions involving reduced Li^+ ions at these cathodic potentials[16] (see Figure 12) can be expected in water-free melts. Using the reported data, the concentration of water can be expressed as a function of the observed voltammetric peak current and the relevant experimental conditions, thus

$$C_{H_2O} = 7.2331 \times 10^{-7} \frac{i_p T^{1/2}}{\omega^{1/2} A} \exp \frac{2629}{T} \tag{7}$$

for $350 < t°C < 500$, where i_p is the peak current in amps, T is in degrees Kelvin, ω is the sweep rate V sec^{-1}, A the microelectrode area in cm^2, and C is the concentration in mol cm^{-3}. The reversible removal and redissolution of water in the melt was not correlated with parallel changes in $i_p(H_2O)$ (cf. nitrates). The anodic oxidation of the dissolved water was not reported either. Takahashi[17] has calculated the dissociation constant for water in LiCl–KCl at 450°C, i.e., for the reaction at 450°C,

$$H_2O \rightleftarrows H^+ + OH^- \tag{8}$$

$$K_{(8)} = 8.0 \times 10^{-7} \quad \text{(ion fraction units)}$$

It can be concluded from this equilibrium constant that water exists in a largely undissociated form in systems containing low concentrations of water at equilibrium, for example, taking a typical mole fraction of $H_2O \sim 10^{-5}$, only 20% of the water molecules are dissociated.

The voltammetric data of Menendres showed no peaks for H^+ ion (or more likely HCl_2^-, *vide infra*,) reduction. The reduction of H^+ ions in HCl containing solutions occurs at potentials ~ -860 mV versus the standard Pt/Pt(II)(1 M) reference electrode.[18] However, from the data of Minh and Welch,[12] the peak current expected for hydrogen ions arising from 20%

dissociation of water is about 60 μA; this would not have been distinguishable from the residual currents reported by Melendres *et al.*[15] No assumptions about the nature of the solution species have been necessary in these experimental studies of water in this melt. However, it is likely that in this fused salt mixture, there will be considerable interaction of the water dipoles with the highly polarizing lithium cations, thus effectively increasing the stability of the water molecules and thus reducing still further the magnitude of $K_{(8)}$. However, at the same time, it might be expected that H^+ and OH^- ions will also interact with their ionic environments. Thus the energetics of reactions such as

$$Li^+ + nH_2O \rightarrow Li(H_2O)_n^+ \tag{9}$$

$$Li^+ + OH^- \rightarrow LiOH \tag{10}$$

$$H^+ + 2Cl^- \rightarrow HCl_2^- \tag{11}$$

will determine the overall stability of the system.

It would appear that, above a critical concentration of water (around 10 mM), HCl is lost from the system.[11] This is consistent with the solubility product concept for HCl in the melt (see above). From the data of Minh and Welch,[12] and Van Norman and Tivers[13] the solubility of HCl is $\sim 10^{-3}$ M. Thus the water concentration in equilibrium with this HCl content is (using $K_{(8)}$)

$$[H_2O] = \frac{[H^+]^2}{K_{(8)}} \quad \text{since } [H^+] = [OH^-] \tag{12}$$

$$[H_2O] = 4.0 \times 10^{-4} \text{ mole fraction} \simeq 12 \text{ mM}$$

which is in good agreement with the limits reported by Burkhard and Corbett.[11]

Under nonequilibrium conditions, e.g., if argon was being flushed through the melt containing entrapped water at concentrations greater than 5–10 mM, then considerable hydrolysis will occur as the HCl is removed from the melt. It should be clear from these results that water can exist in solution over a limited range of conditions. The normal situation, in which a hydrated salt is taken as a component of a solvent, will lead to considerable hydrolysis of the solvent unless extreme care is taken to remove this water prior to fusion. We shall return to this consideration later. Let us now assume that hydrolysis has been allowed to occur; what are the reaction products in the melt and how do they interact with their environment, i.e., the container, gas atmosphere, residual solutes, etc.?

The self-dissociation of water [equation (8)] suggests that as well as H^+ ions, hydroxide ions are likely to be formed. Lysy and Combes[19] (and see chapter 15 by Combes in this volume) have measured the equilibrium constant over a range of temperatures (642–742°C) for the reaction

$$2OH^- \rightleftarrows O^{2-} + H_2O \tag{13}$$

$K_{(13)} = 1.74 \times 10^{-3}$ (mole fraction scale extrapolated to 450°C)

which gives, for the dissociation of OH^-

$$OH^- \rightleftarrows O^{2-} + H^+ \tag{14}$$

$$K_{(14)} = K_{(13)} \times K_{(8)}$$

$$= 5.6 \times 10^{-9}   \text{(in mole fraction units)}$$

whence if $[H^+] \simeq 3.4 \times 10^{-5}$ mole fraction, then the equilibrium concentration of hydroxide is

$$[OH^-] = 4.8 \text{ mole fraction}$$

Thus, contrary to earlier assumptions that OH^- is a direct oxide ion source, it is seen that only very small concentrations of oxide can coexist under equilibrium conditions with hydroxide. Thus hydroxide ions will be a major hydrolysis product in these melts and their presence has been observed by a number of workers[15, 17, 20] using electrochemical techniques. Melendres[15] reported linear relationships between peak current density and concentration of hydroxide and the square root of scan rate. An approximate diffusion coefficient for OH^- can be calculated from the result at 390°C: $D = 3.8 \pm 0.8 \times 10^{-5}$ cm^2 sec^{-1}, and hence

$$C_{OH^-} = 1.309 \times 10^{-9} \frac{i_p T^{1/2}}{A\omega^{1/2}} \text{ mol cm}^{-3}   \text{at 390°C}$$

assuming a similar slope to that for $\log D_{H_2O} v(1/T)$:

$$D_{OH^-} = 1.0549 \times 10^{-1} \exp\left(-\frac{5257}{T}\right)$$

Therefore

$$C_{OH^-} = 6.5383 \times 10^{-7} \frac{i_p T^{1/2}}{A\omega^{1/2}} \exp \frac{2629}{T}   \text{for } 350 < t°C < 500 \tag{15}$$

Chloride melts prepared in a number of different ways are generally characterized with respect to their impurity levels by quoting a residual current density at -2 V vs. the standard Pt/Pt(II)(1 M) reference electrode at 450°C, typically[22–25] 100–300 μA cm^{-2}. Although the potential of -2.0 V is rather anodic of the reduction potential of hydroxide ions (-2.50 V) an estimate of the OH^- concentration can be made by assuming $\omega \simeq 10$ mV sec^{-1}, then the impurity concentration would be between 0.5 and 2.0 mM.

Burkhard and Corbett reported from their HCl solubility studies a residual OH$^-$ concentration of <0.8 mM. This may arise either from the incomplete reaction between lithium oxide and HCl during purification treatment or from inadequate handling procedures concommitant with the extreme hygroscopic nature of their lithium chloride. Takahashi *et al.*[17] have investigated the behavior of the half cell

$$Pt(s)\,|\,H_2(g)\,|\,OH^-(M)O^{2-}(M)LiCl\text{–}KCl\ \text{eutectic}$$

based on the redox reaction

$$2OH^- + 2e \rightleftarrows 2O^{2-} + H_2 \tag{16}$$

the electrode potential of which is given by

$$e = -2.278 - \frac{RT}{2F} \ln P_{H_2}[O^{2-}]^2/[OH^-]^2 \tag{17}$$

The $e°$ for this electrode is in reasonable agreement with the E_p value reported for OH$^-$ by Melendres of -2.50 V versus the standard Pt/Pt(II)(1 M) reference electrode. Although the equilibrium lies well to the left, another possible product of hydrolytic action is the oxide ion. The addition of oxide via Li$_2$O, CaO, etc. to melts held in noble metal containers under *inert* atmosphere gives rise to stable species which can be oxidized electrochemically at Pt[15], Au[21], and C (Reference 26) electrodes. In the case of the two former electrodes, at relatively high concentrations (1 → 100-mM solutions), the oxidation is probably

$$O^{2-} \rightarrow \tfrac{1}{2}O_2 + 2e \tag{18}$$

which occurs at -0.540 V versus the standard Pt/Pt(II)(1 M) reference electrode. The analysis of peak currents shows that this oxidation is reversible up to sweep rates ~ 1 V sec^{-1} and that two electrons are involved in the overall reaction. A diffusion coefficient for oxide ions of 2.4×10^{-6} cm^2 sec^{-1} at 400°C was reported by Takahashi in reasonable agreement with the results of Melendres. The concentration of oxide can be expressed as

$$C_{\text{oxide}} = 9.7314 \times 10^{-7} \frac{i_p T^{1/2}}{A\omega^{1/2}} \exp \frac{2540}{T} \tag{19}$$

for $350 < t°C < 500$.

Thus under conditions which must be carefully defined, water, hydroxide, oxide, and the proton can exist in melts containing lithium chloride; by analogy, these entities may also be expected to be present in solvents which contain other highly polarizing cations.[27, 28] These species are highly reactive and, as well as interacting with each other, may interact with their environment, as in the case of nitrates (vide ultra) to produce further oxycompounds. The simplest of these interactions is that between oxide and water:

$$H_2O + O^{2-} \rightleftarrows 2OH^- \tag{20}$$

i.e., $\qquad K_{(20)} = 1/K_{(13)}$

$\qquad K_{(20)} = 5.75 \times 10^2$ (mole fraction)   at 450°C

This reaction is rapid and is displaced well to the right. This has been confirmed experimentally by addition of Li_2O to the LiCl–KCl eutectic melt followed by equilibration with water vapor. The hydroxide peak was seen to increase markedly after equilibration.[15]

Oxide ions and hydroxide ions contained in chloride melts may also interact with silica-bearing glasses,[29] albeit slowly, to produce silicates and hydrogen chloride:

$$n(SiO_2) + O^{2-} \rightarrow (SiO_2)_n O^{2-} \tag{21}$$

$$nSiO_2 + OH^- + Cl^- \rightarrow (SiO_2)_n O^{2-} + HCl \uparrow \tag{22}$$

The latter reaction was studied by Roth and inadvertently by many other molten salt workers who, when their melts were insufficiently purified, observed severe etching of the Pyrex or silica containers! Considerable care is necessary to produce clean melts if siliceous materials are to be used as container materials.

The thermodynamic data for oxide species in Table 4 show that lithium forms a monoxide O^{2-} whereas sodium and potassium readily form peroxides O_2^{2-} and superoxides O_2^-. The solid state reactions

$$M_2O(s) + \tfrac{1}{2}O_2 \rightarrow M_2O_2(s) \tag{23}$$

$$M_2O(s) + \tfrac{3}{2}O_2 \rightarrow 2MO_2(s) \tag{24}$$

$$M_2O_2(s) + O_2 \rightarrow 2MO_2(s) \tag{25}$$

lie well to the right (Table 4), whilst the disproportionation reaction

$$\tfrac{3}{2}M_2O_2(s) \rightleftarrows M_2O(s) + MO_2(s) \tag{26}$$

is well to the left.[30]

Although these solid state reactions can act only as a guide to solution behavior, they do suggest that in the presence of oxygen, peroxide or superoxide may be the most stable species. Peroxide ions have been identified in chlorides[31] and together with superoxide have been detected by electroanalytical techniques in chlorides, [31, 40] hydroxides,[32] and nitrates.[51] The establishment of an oxygen electrode in the molten chloride[33] would clearly be valuable as an analytical aid *per se*, and as a means of measuring various oxide equilibria including the solubility of metal oxides. The results of attempts to establish such an electrode[21, 34, 35] are in direct contradiction with each other because of (1) the chemical interaction of O_2 with the oxide ion, (2) the interaction of oxygen atoms with the supposedly inert electrode substrate, i.e., the formation of PtO or C(O), which lead to either a different electrode function or overpotentials associated with the oxide discharge, and (3) the chemical interaction of oxide ions with the containment vessel.

Some of these difficulties can be resolved through considerations of the behavior of oxides already outlined. Takahashi[21] reported on the behavior of an oxygen/oxide electrode with an overall electrode reaction given by

$$\tfrac{1}{2}O_2 + 2e \rightleftarrows O^{2-} \tag{27}$$

the electrode potential of which, expressed with respect to the Cl_2 reference electrode, is

$$e = -1173 + 72 \log \frac{P_{O_2}^{1/2}}{N_{O^{2-}}} \qquad (mV) \tag{28}$$

On the other hand, for a similar electrode, Inman and Wrench[34] found that the electrode potential was given by

$$e = -1424 + 146 \log \frac{P_{O_2}^{1/2}}{N_{O^{2-}}} \qquad (mV) \tag{29}$$

when the measured cell voltage was plotted against $\log[O^{2-}]_T$. Such plots are not true Nernst plots, if the reaction between oxygen and oxide ions is

TABLE 4. Thermodynamic Data[26, 30] for Common Alkali Metal Oxides at 450°C

Compound	$-\Delta G°$, kcal mol^{-1}	Compound	$-\Delta G°$, kcal mol^{-1}
Li_2O	121	K_2O	63
Na_2O	76	K_2O_2	81
Na_2O_2	85	KO_2	47
NaO_2	39		

Reaction	Equation number	$^{450}\Delta G_R$, kcal	^{450}K
$M_2O + \tfrac{1}{2}O_2 \rightarrow M_2O_2$	(23)	-18	2.73×10^5
$M_2O + \tfrac{3}{2}O_2 \rightarrow 2MO_2$	(24)	-31	2.35×10^9
$M_2O_2 + O_2 \rightarrow 2MO_2$	(25)	-13	8.51×10^3
$\tfrac{3}{2}M_2O_2 \rightarrow M_2O + MO_2$	(26)	11.5	3.33×10^{-4}

significant. Now consider the reaction

$$O^{2-} + \tfrac{1}{2}O_2 \rightleftharpoons O_2^{2-} \tag{30}$$

$$K_{(23)} = [O_2^{2-}]/[O^{2-}]P_{O_2}^{1/2}$$

and

$$O_2^{2-} + 2e \rightleftharpoons 2O^{2-} \tag{31}$$

Since the partial pressure of oxygen is essentially unity, $P_{O_2}^{1/2} = 1$, then

$$K'_{(23)} = \frac{[O_2^{2-}]_c}{[O^{2-}]_f}$$

where $[O^{2-}]_f$ is the "free" oxide content and $[O_2^{2-}]_c$ is the complexed oxide. If $[O^{2-}]_T$ is the oxide solute, then

$$[O^{2-}]_T = [O^{2-}]_f + [O_2^{2-}]_c$$

leads to

$$e_{(31)} = e^{\circ}_{O_2^{2-}/O^{2-}} + \frac{RT}{2F} \ln \frac{[O_2^{2-}]_c}{[O^{2-}]_f^2}$$

$$= e^{\circ} + \frac{RT}{2F} \ln \frac{[O_2^{2-}]_c}{[O^{2-}]_f} - \frac{RT}{2F} \ln[O^{2-}]_f$$

Therefore

$$e_{(31)} = e^{\circ}_{O_2^{2-}/O^{2-}} + \frac{RT}{2F} \ln K'_{(23)} - \frac{RT}{2F} \ln[O^{2-}]_T - [O_2^{2-}]_c$$

Similarly, if one invokes the formation of superoxide,[36] the electrode reaction is given by

$$O_2 + e \rightleftharpoons O_2^- \tag{32}$$

then it can readily be shown that

$$e_{(32)} = e^{\circ}_{O_2/O_2} - \frac{RT}{F} \ln KI(24)^{1/2} - \frac{RT}{2F} \ln[O^{2-}]_T - \tfrac{1}{2}[O_2^-]_c$$

again demonstrating the nonvalidity of e vs. $\log[O^{2-}]_T$ plots when $[O_2^{2-}]_c$ is significant. Neither Takahashi[21] nor Inman and Wrench[34] have used the true concentration of oxide ions, although Takahashi has made a correction for loss of O^{2-} ion by reaction with glass. Unless the presence of glass prevents the formation of peroxide or superoxide,[37] it is difficult to see how the linear plots were obtained. (Both groups use noble metal containers for the melts but the use of Pyrex glass reference electrodes ensured contact between glass and the melts.) More work is certainly needed to resolve these problems. The possibility of operating a Pt/PtO/O^{2-} electrode seems somewhat unlikely since the solubility[33] of PtO is reported to be 3.3×10^{-3} M at 450°C (unless some kinetic factors related to the dissolution of PtO are involved).[38] The results of Weaver and Inman[39] are consistent with this view (the oxide concentration was $\sim 10^{-1}$ M). The characteristics of the peroxide and superoxide ions are not well established in these chloride melts (in contrast to nitrate melts). Mignonsin *et al.*[40] have briefly examined the reduction of superoxide using chronopotentiometry. They reported an irreversible reduction wave when sodium peroxide containing 5%–10% superoxide was added, which, on treating the solution with oxygen, doubled the value of $i\tau^{1/2}$ at a given current density. This could be interpreted as a doubling of O_2^- concentration consistent with the reaction [see (25) Table 4]

$$O_2 + O_2^{2-} \rightarrow 2O_2^- \tag{25}$$

They also investigated the interaction between Cl^- and oxygen in NaCl–KCl melt and found both electrochemical and spectroscopic evidence for reactions such as

$$Cl^- + O_2 \rightarrow \tfrac{1}{2}Cl_2 + O_2^- \tag{33}$$

and

$$2Cl^- + O_2 \rightarrow Cl_2 + O_2^{2-} \tag{34}$$

Fusion of chloride melts in oxidizing atmospheres should therefore be avoided.

Our understanding of the chemistry of simple oxygen containing species in molten chlorides is still at an early stage and considerable information relevant not only to melt purification but to potential technological applications of these systems remains to be obtained. The situation in the potentially useful fluoride melts is even less well understood than in chlorides.[41] Data is available for water in BeF_2 containing melts[42] and this leads to the conclusion that in such systems the water solubility is very small. Mamantov[41] discussed the behavior of water and hydrogen fluoride in bifluoride melts some ten years ago but no further progress has since been reported. For the more commonly employed fluorides, i.e., those based on LiF admixed with other alkali and alkaline earth fluorides, little or no data are available.[41, 43] The presence of oxide,[41, 47] hydroxide,[47] peroxide,[44, 46] superoxide[45, 46] will again be important and at the very least data for their *in situ* analysis would be of value.

To this end the electrochemical behavior of oxide and peroxide ions at gold electrodes in $LiF–BeF_2–ZrF_2$ (65.6 : 29.4 : 5.0 mol %) and $LiF–BeF_2–ThF_4$ (72 : 16 : 12) (melts of importance in the Molten Salt Reaction Experiment at Oak Ridge) saturated and unsaturated with oxide ions has been investigated recently.[209] The oxidation of oxide occurred in a two-electron transfer reaction with coupled chemical reactions, viz.,

$$O^{2-} \rightleftharpoons O + 2e$$

$$O + O \rightarrow O_2 \quad \text{and} \quad O + O^{2-} \rightarrow O_2^{2-} \tag{35}$$

The addition of peroxide led to the enhancement of the postwave suggesting that the chemical pathway indeed involves parallel processes. Because of the corrosive nature of these melts an iridium quasireference electrode was used and the value of E_p for the oxidation of oxide was quoted with respect to this $(+1.2 - 1.4$ V). No diffusion coefficients were reported. The instability of peroxide and superoxide ions in this melt was established and related to an absence of stabilizing countercations such as potassium. The affinity for oxide of the acidic cations present in the base melt would also enhance decomposition of the peroxideto oxide, viz.,

$$O_2^{2-} + M^{2+}(Be^{2+}) \rightleftharpoons MO_2(BeO) + \tfrac{1}{2}O_2 \qquad (36)$$

a heterogeneous process which certainly would require considerable time to reach equilibrium.

2.2.2. Oxyanionic Solvents

Little or no data on water solubility or oxide-containing species have been reported in the literature for oxyanion-containing melts other than nitrates. However, for these nitrate melts, the whole subject of oxide chemistry is rather controversial and beyond the scope of this chapter. The subsequent discussion will therefore only attempt to summarize the situation regarding molten nitrates in terms of the requirements for their purification.

The pure alkali metal nitrates are thermally stable only close to their melting points.[48] Therefore their binary and ternary mixtures, which have melting points that are considerably depressed, are commonly used.[49] Careful processing of these systems will ensure minimal decomposition to nitrite, viz.,

$$2NO_3^- \rightleftharpoons 2NO_2^- + O_2 \qquad (37)$$

The low-temperature requirement to prevent this thermal decomposition has an undesirable corollary with regard to residual water content, in that the solubility of water is considerably increased at the lower temperatures. In certain cases, e.g., melts containing Li^+ ions, irreversible reaction may also occur which leads to the formation of oxide-containing species.[50] Invariably, one of the final products of reactions such as these are hydroxide ions, e.g.,

$$2NO_3^- + H_2O \rightarrow 2OH^- + 2NO_2 + \tfrac{1}{2}O_2 \qquad (38)$$

These anions can interact further, albeit slowly, with oxygen to produce superoxide and water

$$2OH^- + \tfrac{3}{2}O_2 \rightleftharpoons 2O_2^- + H_2O \qquad (39)*$$

* Equation numbers (40)–(47) appear in Table 5.

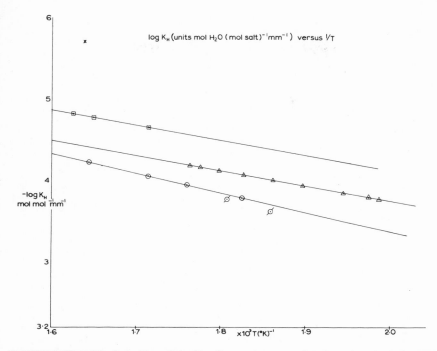

FIGURE 2. Plot of the logarithm of the Henrian constants as a function of the inverse of the temperature for different nitrate melts. ⊙, Lithium nitrate[50]; ∅, lithium nitrate[65]; ⊡, sodium nitrate[50]; ×, potassium nitrate[50]; △, sodium nitrate–potassium nitrate (50 mol %).[60]

Thus a major requirement for handling molten nitrates is the effective removal of water under inert conditions. Jordan and Zambonin[51, 52] proposed that in molten nitrates, oxide ions are rather unstable (in the absence of extraneous interactions with container material such as Pyrex). Subsequently, Zambonin has made an extensive study of oxide, peroxide superoxide, hydroxide, oxygen, and hydrogen[50–59] in sodium–potassium nitrate equimolar mixture in the temperature range 225–275°C. The influence of water[53, 60] on the stability of these species has also been studied. The major equilibria reported by Zambonin and other workers[61–63] are summarized in Table 5, which is taken in part from a recent review.[64]

The solubilities of water in several pure alkali metal nitrates[50, 65, 66, 67] and in the binary sodium nitrate–potassium nitrate,[60] lithium nitrate–potassium nitrate, and lithium nitrate–sodium nitrate[65] mixtures have been measured. These data are summarized in Table 6 and as a function of temperature in Figure 2. The solubilities expressed in terms

TABLE 5. Equilibrium Constants for Anionic Oxygen Reactions in Equimolar Sodium Nitrate–Potassium Nitrate

Equilibrium constant	Expression	Equation number	Units	Value at 502 K	Method	Ref.
$K_{(40)} = 1/K_{(25)}$	$\dfrac{[O_2][O_2^{2-}]}{[O_2^-]^2}$	(40)	—	3.5×10^{-6}	Voltammetric[a]	51
$K'_{(40)}$	$\dfrac{[O_2][O_2^{2-}]}{[O_2^-]^2}$		—	0.5×10^{-6}	Potentiometric[b]	55
	$\dfrac{[O_2][O_2^{2-}]}{[O_2^-]^2}$		—	6×10^{-6}	Manometric[c,d]	61
	$\dfrac{P_{O_2}[O_2^{2-}]}{[O_2^-]^2}$		$\dfrac{\text{atm}}{M}$	1.43	(See footnotes c and e)	61
	P_{O_2}		atm	1.2×10^2	Solid state reaction thermodynamics[f]	64
$K_{(41)}$	$\dfrac{[NO_2^-]^2[O_2]^2}{[O_2^{2-}][NO_3^-]^2}$	(41)	M	6.7×10^{-11}	Voltammetric	51
	$\dfrac{[NO_2^-]^2[O_2]^2}{[O_2^{2-}][NO_3^-]^2}$		M	11×10^{-11}	$K_{(41)} = K_{(44)}^2/K_{(40)}$[g]	64
$K_{(42)}$	$\dfrac{[NO_2^-][O_2^{2-}]}{[NO_3^-][O^{2-}]}$	(42)	—	3	$K_{(42)} = [K_{(41)}/K_{(43)}]^{1/2}$	51
	$\dfrac{[NO_2^-][O_2^{2-}]}{[NO_3^-][O^{2-}]}$	(42)	—	1.2	$K_{(42)} = K_{(44)}K_{(23)}$	64

$K_{(43)} = [K_{(26)}]^2$	$\dfrac{[O^{2-}]^2[O_2]^2}{[O_2^{2-}]^3}$	(43)	M	10^{-11}	Calculated for a hydroxide melt	51
	$\dfrac{[O^{2-}]^2[O_2]^2}{[O_2^{2-}]^3}$	(43)	—	1.6×10^{-12}	Solid state reaction thermodynamics[f]	64
$K_{(44)}$	$\dfrac{P_{O_2}^{1/2}[NO_2^-]}{[NO_3^-]}$	(44)	$atm^{1/2}$	1.3×10^{-5}	Chemical analysis NO_2^- ion[c,h]	63
$K_{(23)}$	$\dfrac{[O_2^{2-}]}{P_{O_2}^{1/2}[O^{2-}]}$	(23)	$atm^{-1/2}$	8×10^4	Solid state reaction thermodynamics[f]	64
$K_{(45)}$	$\dfrac{[OH^-]^2 P_{O_2}^{3/2}}{[O_2^-]^2[H_2O]}$	(45)	$M^{1/2}$	1.4×10^3	Potentiometry	55
$K_{(46)}$	$\dfrac{[O_2^{2-}][H_2O]}{[OH^-]^2[O_2]^{1/2}}$	(46)	$m^{1/2}$	3.6×10^{-10}	$K_{(46)} = K_{(45)}^{-1} K'_{(40)}$ [i]	55
$K_{(13)}$	$\dfrac{[OH^-]^2}{[O^{2-}][H_2O]}$	(13)	—	2.6×10^{17}	$K_{(46)}^{-1} K_{(23)}$	55
$K_{(47)}$	$[O^{2-}][NO_2^+]$	(47)	m^2	3.8×10^{-35}	Potentiometry	165

[a] Assumes $E^0_{O_2/O_2^-} = E_{1/2}$.
[b] Uses corrected $E^0_{O_2/O_2^-}$ [corrected value of $E^0_{O_2/O_2^-}$ based upon D_{O_2} and $D_{O_2^-}$].[55]
[c] Extrapolated to 502 K.
[d] Corrected to the solubility of oxygen.[166]
[e] Reported value[167] converted to oxygen pressure.
[f] Calculations based on data for sodium salts.[168,169]
[g] $K_{(40)}$ from Reference 61 used.
[h] Corrected enthalpy of 2 kcal mol^{-1} (Reference 64) used for extrapolation to 502 K.
[i] Zambonin gave $K'_{(40)} \times K_{(45)}^{-1} = 5 \times 10^{-7} \times 10^{-3} = 5 \times 10^{-11}$ mol$^{-1/2}$ kg$^{1/2}$! (an arithmetic error).[55]

TABLE 6. H_2O Solubility in Molten Nitrates/Nitrites

Solvent (ratios in mol %)	T, °C	Henry's law constant, μmol mol^{-1} mm^{-1}	Range of Henry's law, mm	ΔH_{soln}, kcal mol^{-1}	Diffusion coefficient, $10^5 D$ cm^2 sec^{-1}	Ref.
LiNO$_3$	265	232	30			65
	280	165			—	65
	275	161	30	−9.35		50
	335	59				50
NaNO$_3$	310	22	30	−8.15	—	50
	342	15	30			
KNO$_3$	337	2	30	—	—	50
	335	20 ± 7				67, 77
	360	15 ± 7				67, 77
	335	20 ± 2				66, 77
NaNO$_3$–KNO$_3$	227	176	20	−8.4	1.90^a	60
50 : 50	294	63	20			55
LiNO$_3$–KNO$_3$						
75 : 25	230	278	30		—	
	265	137	30			
50 : 50	230	162	30			65
	265	82	30			
25 : 75	230	99	30			
	265	54	30			
LiNO$_3$–NaNO$_3$	230	354	30		—	
75 : 25	265	165	30			
	280	118	30			
50 : 50	230	242	30			
	265	110	30			65
	280	81	30			
25 : 75	265	84	30			
	280	64	30			
NaNO$_2$	281	80	38^b		—	78
NaNO$_2$–KNO$_3$	143	610	20	−8.4	—	77
	201	167	20			
	260	59	20			

^{a} Diffusion coefficient at 229°C.
^{b} Solubility limit reached at ∼ 38 mm pressure H_2O.

of Henrian constants are highest in the lithium-containing melts; the presence of sodium ions also enhances the solubility of water compared with potassium ions. (Compare this with the behavior of halogens in molten chlorides.[68]) The relatively high solubilities and the small range over which Henry's law is obeyed suggest that hydrolytic reactions will

TABLE 7. Data Relating to Oxidic and Other Species in Nitrates
($NaNO_3$–KNO_3)

Species	Reference potential, V^a	Diffusion coefficient, $10^5 D$ cm^2 sec^{-1}	Temp., K	Method	Ref.
NO_2^-	$E_{NO_2/NO_2^-}^{\tau/4} \simeq -0.44$	5.2^b	571	Chrono-potentiometry	74
	$E^{1/2} \simeq -0.44$	0.525	502	r.d.e.	70
		2.48 ± 0.7	523	Chrono-potentiometry	170
H_2O	$E^{1/2} = -1.19$	1.9	502	r.d.e.	60
O_2^-	$E_{O_2/O_2^-}^{1/2} = -0.74$	0.475	502	r.d.e.	51
O_2^{2-}	$E_{O_2/O_2^{2-}}^{1/2} = -1.28$	0.31	502	r.d.e.	51
OH^-	$E_{O_2/H_2O/OH^-}^{\circ} = -0.495$	—	502	Potentiometry	55
O_2	$E_{O_2/H_2O/OH^-}^{\circ} - 0.65$	31^c	520^d	r.d.e.	57
Cl^-	$E^p \simeq -0.27$	0.152	423	LSV^e Hg	147
	$E^{1/2} = -0.212$	0.691	518	PP^f Hg	171

^{a} Versus 0.07 m Ag$^+$/Ag reference electrode.
^{b} The large discrepancy here may arise from the inadvertent presence of oxidic species which in the chronopotentiometric method consume current leading to τ being enhanced, hence resulting in D being too large (cf. r.d.e. technique).
^{c} This based on K_H 4.8×10^{-6} mol kg^{-1} atm^{-1} and using wet oxygen, i.e., reaction (48).
^{d} Measured over the temperature range 525–575 K.
^{e} Linear sweep voltammetry.
^{f} Pulse polarography.

occur if the water content exceeds about 10–15 mM (cf. the chlorides). Peleg[50] showed that in $LiNO_3$ approximately 2 mM of water were irretrievably lost to the melt at temperatures close to the melting point (252°C).

The electrochemical detection of water in nitrates is an area of considerable controversy (see Chapter 11 by Lovering and Oblath). The so-called "water wave" reported by Swofford and Laitinen[69] at -1.65 vs. 0.07 M Ag$^+$/Ag reference electrode was used independently by Peleg[50] and Zambonin[60] to correlate the water content of their melts (from solubility measurements) with the limiting current density for the reduction of water at a rotating platinum disk electrode (r.d.e.).[70] The diffusion coefficient obtained by Zambonin[60] is given in Table 7.* The temperature dependence of diffusion of water in nitrates has not been studied, so that an accurate assessment of the water content by the use of this technique can only be made close to the melting point in the equimolar $NaNO_3$–KNO_3 mixture. The temperature coefficient of diffusion of water for this melt, together with

* *Editors' note:* Some reservations concerning its precise value have been expressed verbally to one of us.

those melts which contain lithium ions, would be necessary to standardize and control the purification procedures for molten nitrates.*

The formation and presence of nitrite ions in these melts can be readily detected by polarography,[71] sweep voltammetry,[71] or chronopoten-tiometry at stationary[73, 74, 170] or rotating platinum electrodes,[70, 72] and some of the corresponding data are collected in Table 7. It is important to recognize that although much of the nitrite can arise from thermal degra-dation it may be readily generated by electroreduction of the nitrate ions.[75] This has important implications in the application of preelectrolysis methods to molten nitrates.[76, 130]† A comparison[72] of the nitrite con-tent of variously prepared equimolar $NaNO_3$-KNO_3 batches analyzed by chemical analysis and by observing the nitrite oxidation wave on a r.d.e. at 200°C showed good agreement. Further, it was shown that fusion of the recrystallized salts at the eutectic temperature led to an approximately tenfold increase in nitrite content (dependent on temperature, time, and drying gas). This enhanced nitrite concentration could be reduced to $\sim 10^{-5}$ m (the original level in the pure unmelted salts) by flushing the melt with NO_2 gas followed by argon. Alternatively, fusion under pure oxygen has also been claimed to reduce the nitrite content of the melt[71, 85] [cf. expression (44) and $K_{(44)}$ in Table 5].

The presence of oxy- species such as O_2^-, O_2^{2-} might only be expected if contamination, or hydrolysis of the melt had occurred in the absence of an excess of water. Significant water content would, through the following reactions, ensure that these species are converted to hydroxide (Table 5):

$$O^{2-} + H_2O \rightleftarrows 2OH^- \tag{13}$$

$$O_2^{2-} + H_2O \rightleftarrows 2OH^- + \tfrac{1}{2}O_2 \tag{46}$$

$$2O_2^- + H_2O \rightleftarrows 2OH^- + \tfrac{3}{2}O_2 \tag{45}$$

Thus contamination by hydroxide ions of nitrate solvents can arise via these reactions. Zambonin[53] has investigated the anodic oxidation of hydroxide ions. He proposes the reaction

$$OH^- \rightleftarrows \tfrac{1}{2}H_2O + \tfrac{1}{4}O_2 + e \tag{48}$$

which he also reported to be irreversible. However, a linear plot of i vs. $[OH^-]_{molal}$ is given and can be used to determine the OH^- ion content of the equimolar $NaNO_3$-KNO_3 melt. Reaction (48) was also studied potentiometrically.[55] It was reported to be reversible at concentrations greater than 10^{-3} m. The r.d.e. experiment which was carried out using

---

* Editors' note: Always assuming the voltammetric wave proves to be diffusion controlled.[146]
† Editors' note: See also Reference 146.

potential scan rates of 3 V/min (which is approaching steady state conditions) and in melts containing $\sim 10^{-4}$ m OH$^-$ are thus consistent, but the rate-determining step remains to be identified. More work is clearly required on this redox system, the results of which might help in understanding similar processes at the slightly higher temperatures encountered in halide melts.

Hull and Turnbull[77] have determined the solubility of water in a mixture of potassium nitrate (55.5 wt %) and sodium nitrite (44.5 wt %) over the temperature range 143-279°C using nitrogen equilibrated with known partial pressures of water. Water injection tests showed that water could be reversibly added and removed from the melt. Henry's law was obeyed up to 24 mm water partial pressure. The Henrian constant was related to temperature by

$$\log K_H = \frac{6040}{T} + 3.4917 \ln T - 31.4638 \qquad (49)$$

The dissolution of water into this melt is strongly exothermic. This has also been found to be the case for nitrates and chlorides and was correlated with strong cation–water molecule interactions. Kozlowski and Bartholomew[78] determined the solubility of water in sodium nitrite at two partial pressures (7 and 19.5 mm) with similar results to those found in sodium nitrate. They suggested on this basis that the dissolution process involved interstices within the melt structure, as suggested earlier by Frame *et al.*[66] No electrochemical studies have been made of water in nitrites although their decomposition potentials have been discussed.[79, 132] Water solubilities in sulfates and carbonates have not been reported but a recent preliminary result for a Li$_2$SO$_4$-containing melt has been obtained.[80]

Thus as a rough guide to the possible behavior and influence of water in other oxyanionic melts, the data so far obtained for nitrates and nitrites at the lower temperatures can be used. It must be emphasized that specific data should be obtained for individual melts. There is clearly an extensive field for study not only characterizing water from the point of view of purification of such melts but to investigate the physical and chemical interactions in these diverse systems.

3. Principles of Purification Methods

3.1. Introduction

The preceding discussion should leave no doubt that carefully worked out purification procedures are necessary to avoid complications which may arise from the presence of water and its derivatives. The primary objective of any purification procedure should be to remove as much moisture as possible under the mildest possible conditions. The complexity of subsequent operations will be determined by the success or failure of this initial step because generally they will be designed to remove residual water and its decomposition products as well as trace metal ions.

Methods of melt purification could be classified according to the above division but because other impurities also arise which are independent of water, the procedures are perhaps best classified as follows: (a) physical methods which include desiccation, vacuum drying, recrystallization, distillation, sublimation, and zone refining, and (b) chemical methods which include acid–base reaction, halogenation, chemical displacement, reduction by hydrogen, and electrochemical reduction and oxidation.

The principles necessary for the application of the first group will be dealt with only in certain specific cases since most of the relevant information is readily available in various text books.[81, 82] On the other hand, the principles underlying the second group of methods are not so accessible, so that a more detailed discussion will be necessary in connection with the more important systems, the practice for which will be described in the final section.

3.2. Physical Methods

Salts such as LiCl, $CaCl_2$, $MgCl_2$, $ZnCl_2$, $LiNO_3$, Li_2SO_4, LiF, and KF form one or more hydrates which often will possess low water vapor pressures. Indeed, $CaCl_2$ is a well-known desiccant. Because the water vapor pressure of the laboratory atmosphere may range between 6 and 16 mm, the handling of these materials in such atmospheres will lead to deliquescence (observe dry LiCl in air for a few minutes!). Thus, initial handling of these materials should be carried out in either a drybox* or a dry bag, e.g., when weighing out the salt, etc., to minimize the amount of water that it will be necessary to remove in the subsequent operations.

Desiccation and vacuum drying are the most effective methods of water removal; sublimation (aluminum chloride[9, 84]) and recrystallization ($LiNO_3 \cdot 3H_2O$)[85] will be valuable methods for upgrading salts which are insufficiently pure in commercial form, e.g., aluminum chloride.[86, 87]

The interaction between bulk water and salts may be divided into two classes: (i) those that involve the reversible combination of salt with water, and (ii) those that involve irreversible chemical reaction of the water with the salt. This broad definition is important both as regards the solvent components which are being dried and to the desiccant chosen. Many of

* For example, see Reference 83.

TABLE 8. Some Desiccants and the Equilibrium Water Vapor per Liter of Air at 30.5°C

Desiccant	Residual water,[a] μg	Desiccant	Residual water,[a] μg
P_2O_5	0.02	$CaCl_2$ fused, anhydrous	400
BaO	0.7	H_2SO_4 concentrate	3
$Mg(ClO_4)_2$ anhydrous	2	95% H_2SO_4	300
CaO	2	$CaCl_2 \cdot H_2O$	150
Silica gel	30		

[a] 1000 μg water in 1 liter of gas at 30°C exerts 1 mm partial pressure of water.

the common salts of interest for solvent preparation fall into the first group. Therefore, under carefully controlled conditions at low temperatures, water may be completely removed by simple drying procedures. To ensure that there is effective water removal, the water vapor pressure of the hydrated desiccant should be very much lower than that of the salt to be dried. It can be seen from Table 8 that the desiccant must be carefully chosen. Probably the best desiccants are regenerative molecular sieves dried according to precise procedures or consumable phosphorus pentoxide, calcium oxide, or barium oxide. (Note the possible complications which could arise when using these latter materials as oxide precursors!) The removal of water by the desiccant phosphorous pentoxide, which involves a chemical reaction with the water, would appear to be the most effective (but expensive) method of drying hygroscopic salts. The use of reduced pressure techniques also depends upon the magnitude of the vapor pressure of the salt hydrates. The vapor pressure of the container system must be lowered sufficiently, so that effectively the salt to be dried will effluoresce. Continued pumping and incorporation of a suitable cold trap ensures that the water vapor is removed from the system.[6, 88] The advantage of this procedure, and hence its frequent use, is that it can be operated sequentially in a closed system established to carry out other purification operations. However, the purification cell often contains the salt tightly packed with a low surface area/volume ratio, i.e., the area exposed to the vacuum (see Figures 7, 11) is so low that a long pumping time will be necessary before maximum water removal is achieved. Thus in any scale up of material quantities there will be required a considerable change in operational conditions.

Sublimation procedures can be used where high-vapor-pressure salt systems with small liquidus range are involved. The most common system is that involving the aluminum halides. $AlCl_3$, for example, is a component of the acid–base solvent system $MCl–AlCl_3$ available until the recent development of the Alcoa process was poor in quality, containing much iron and carbonaceous matter.[9, 89] Sublimation has been utilized as a rapid and convenient method of improving this material prior to final chemical processing.

Recrystallization from aqueous or nonaqueous solutions may be used to upgrade many salts possessing moderate to low solubility. The impurities to be removed must not form solid solutions with the crystals of the salt to be purified. The application of the common ion effect enables moderately soluble materials to be recrystallized, for example, CsCl can be purified from cold aqueous solutions saturated with HCl.[90] Lithium nitrate trihydrate whose anhydrous product forms a major component of low-melting nitrate solvents, presents a problem of purification which is best solved by recrystallization as the demihydrate.[91, 92] Lithium nitrate or its trihydrate can contain high levels of chloride ion (1–10 mM) and investigations of double-layer capacitance, absorption, and complexation can only be meaningful if this initial chloride ion concentration is reduced by at least two orders of magnitude. Figure 3 shows a cyclic voltammogram at 150°C for a stationary mercury drop in $LiNO_3$–KNO_3–$NaNO_3$ which has been inadequately purified. The peak represents the oxidation of Cl^- ions to soluble Hg_2Cl_2. By standard addition methods the purity of nonrecry-

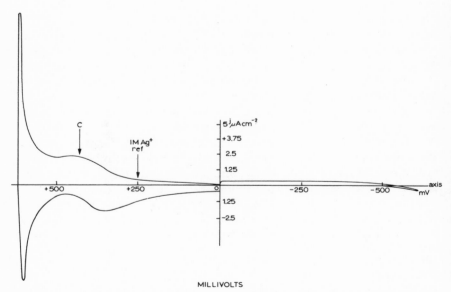

MILLIVOLTS

FIGURE 3. Voltammogram at hanging mercury drop in $LiNO_3$–$NaNO_3$–KNO_3 eutectic, 150°C.[92] C, chloride oxidation peak; sweep rate, 100 mV sec^{-1}.

stallized $LiNO_3$ containing melts and those containing recrystallized $LiNO_3$ was 2 and 0.04 mM, respectively.[92]*

The technique of zone refining[93] has rarely been applied to the purification of inorganic salts[94] for solvent purposes, although manufacturing companies make use of this method as well as single-crystal techniques to prepare relatively large amounts of high-purity material. This method is worthy of further consideration.

3.3. Chemical Methods

Assuming that contamination by water has been minimized, the final removal of impurities must be made in the molten state so that, inevitably, some further hydrolysis may occur. If the containers are made of Pyrex, further contamination by leaching of oxides from the borosilicate network may also occur.[29] The final stages of purification require methods which will deal with millimolar quantities of impurity or less. Thus chemical and electrochemical reactions are favored to achieve this, although zone refining might be more effectively used for the purification of single salts when the success of this powerful procedure for electronic grade materials such as silicon, germanium, etc. is considered.

3.3.1. Gaseous Reagents for Purification

In general, these methods are concerned with the removal of final hydrolytic residues. Gaseous reagents offer the following advantages:

(i) They are readily introduced into and removed from a closed system.

(ii) They can be easily purified to extremely low levels of oxygen, water, nitrogen, etc. contamination.

(iii) The reactions in which they are involved are essentially homogeneous (compare the use of cementation reactions with Mg, Al, etc.)

Unfortunately data for the corresponding equilibria involving such processes are scarce and confined essentially to chlorides.

It has been shown that hydroxide and oxide are the major products of hydrolysis; thus reactions such as

$$O^{2-} + 2HCl \rightleftarrows H_2O + 2Cl^- \tag{50}$$

$$OH^- + HCl \rightleftarrows H_2O + Cl^- \tag{51}$$

must be carried out to remove these contaminants. Free energy data for the pure compounds[95] can be used to show that such reactions are thermody-

---

* *Editors' Note:* These analyses are for batch-selected materials. Also, see Reference 211, where batch material was employed.

namically favorable, even if the alkali or alkaline earth oxide reacts further with metallic ion impurities to form partially soluble metal oxides.

Consider the reaction

$$MO(soln) + 2HCl(g) \rightleftarrows MCl_2(soln) + H_2O(g) \tag{52}$$

$$K_{(52)} = \frac{a_{MCl_2} P_{H_2O}}{a_{MO} P_{HCl}^2}$$

Since the concentrations of the solution species are in the range where $\gamma \to 1$

$$\exp\left[-\frac{\Delta G_{(45)}^\circ}{RT} \right] = K = \frac{N_{MCl_2}}{N_{MO}} \frac{P_{H_2O}}{P_{HCl}^2} \tag{53}$$

Specifically, iron is found at levels of 0.5 mM in LiCl. $\Delta G_{(52)}^\circ = -27$ kcal at 450°C,[95, 96] then $K_{(52)} = 1.75 \times 10^8$. Thus if the ratio $P_{HCl}^2/P_{H_2O} = 10^6$, i.e., $P_{HCl} = 1$ atm, $P_{H_2O} = 10^{-6}$ atm, which might be typical of the HCl gas stream, then

$$\frac{N_{MO}}{N_{MCl_2}} = \frac{10^6}{1.7 \times 10^8} \simeq 5.8 \times 10^{-3}$$

i.e., the FeO content has been reduced by 170 : 1.

The reaction of HCl with alkali metal oxides has been investigated and the equilibrium constant measured. The values of K for the reaction

$$O^{2-} + 2HCl \rightleftarrows H_2O + 2Cl^- \tag{50}*$$

in LiCl–KCl[19, 20] and NaCl–KCl[97–99] is given as a function of temperature in Table 9. The influence of solvent cations is not markedly significant.

Oxide impurities as well as metal ion impurities may be removed from molten halides by the use of hydrogen–hydrogen halide gas mixtures. This procedure has been employed for the large-scale processing of molten fluorides used in the Molten Salt Reaction Experiment at Oak Ridge.[41, 100] The highly reducing conditions result in the reduction and removal of many metal halides and oxides dissolved in the molten halide, viz.,

$$MO(soln) + HX(soln) \to MX_2(soln) + H_2O\uparrow(soln) \tag{56}$$

$$MX_n(soln) + \frac{n}{2} H_2(g) \to M(g) + nHX(soln) \tag{57}$$

The reaction requires the activity of HX in solution to be small, which effectively increases the number of metal halide systems which can be reduced, i.e., it tends to make $\Delta \bar{G}$ for reaction (57) more negative. The

* Equation numbers (51) and (52) do not exist. For (53)–(55) see Table 9.

solubility[68] of HX where HX is HCl or HF follows Henry's law for LiCl–KCl (Reference 11) and BeF_2–LiF,[101] respectively, suggesting that γ is a constant over the range of partial pressures used and if $\gamma \rightarrow 1$ for these dilute solutions then the activity of HX in these melts is indeed very low. The fact that metal ions such as Fe(II), Ni(II), Co(II) can be reduced in fluorides[41] supports this qualitative analysis. Unfortunately there are little or no explicit data for the hydrogen–hydrogen fluoride electrode except in LiF–BeF_2 (Reference 102) and cryolite.[103] Data for the chloride melts are also sparse, the only definitive study being that of Laitinen and Plambeck,[104, 105] who report an E°_{MF} for the reaction

$$HCl(g) + e \rightleftarrows H_2O(g) + Cl^-(soln) \qquad (58)$$

to be -0.800 V versus the standard Pt/Pt(II) (unit mole fraction) reference electrode. These data may be combined with those of Burkhard and Corbett[11] to estimate a value of E°_M for the reaction

$$HCl(soln) + e \rightleftarrows H_2(g)(1\ atm) + Cl^-(soln) \qquad (59)$$

which is pertinent to the reduction reactions (57) in chlorides. It is found that E°_M is approximately -1.48 V versus the standard Pt/Pt(II) (1 M) reference electrode. Hence, the strongly reducing nature of the hydrogen/H^+ couple in such media is explained. Furthermore, metal ions such as Fe(II), Cr(II), etc. should be reducible. Such cations are certainly removed in fluorides.[41]

It was mentioned during the earlier discussion of oxide-containing species in halide melts that oxygen can react with chloride to produce peroxide and superoxide.[40] Thermodynamically the reverse reaction of Cl_2 with oxide-containing species is more favorable and thus halogenation has been considered as a means of purifying melts,[106, 107] e.g.,

$$AO + X_2 \rightarrow AX_2 + \tfrac{1}{2}O_2 \qquad (60)$$

$$2AOH + X_2 \rightarrow H_2O + 2AX + \tfrac{1}{2}O_2 \qquad (61)$$

$$MO + X_2 \rightarrow MX_2 + \tfrac{1}{2}O_2 \qquad (62)$$

where A is an alkali or alkaline-earth metal and M is a heavy metal.

Equilibria for the different metal oxide–chlorine systems have been reported[108] and common impurities such as FeO, PbO should be eliminated by this treatment assuming their chlorides are volatile. Maricle and Hume[106] have claimed that direct treatment of salts such as NaCl–KCl and LiCl–KCl with chlorine produces melts free of hydroxide as demonstrated by their polarograms before and after treatment (Figures 4 and 5). The author[109] has used this treatment for large batches of LiCl–KCl and found that the Pyrex container remained unetched, a sign that hydrolysis products have largely been removed. Water was seen to be

TABLE 9. Oxide Equilibria in Molten Chlorides

Equilibrium constant	Expression	Equation number	Units	Value	T, °C	Method	Ref.
LiCl-KCl							
$K_{(13)}$	$\dfrac{[x_{O^{2-}}]P_{H_2O}}{[x_{OH^-}]^2}$	(13)	(mole fraction)$^{-1}$ atm	1.74×10^{-3}	450	Potentiometry extrapolation from $742 > t > 642$ using solid electrolyte probe	19
				$\log K_{(13)} = 7.86 - 7.68 \times 10^3 T^{-1}$			
$K_{(50)}$	$\dfrac{[x_{O^{2-}}]P_{HCl}^2}{[x_{Cl^-}]^2 P_{H_2O}}$	(50)	(mole fraction) atm	2.63×10^{-12}	450	Potentiometry, extrapolation from $742 > t > 642$, using solid electrolyte probe	19
				$\log K_{(50)} = 2.29 - 10.03 \times 10^3 T^{-1}$			
$K_{(14)}$	$\dfrac{[x_i(O^{2-})][x_i(H^+)]}{[x_i(OH^-)]}$	(14)	Ion fraction	5×10^{-12}	450	$2k_H[K_{(13)}K_{(50)}]^{1/2a}$	19
	$\dfrac{[x_i(O^{2-})][x_i(H^+)]}{[x_i(OH^-)]}$		Ion fraction	1.25×10^{-11}	450	Potentiometry	17
$K_{(54)}$	$\dfrac{[x_i(H^+)^2][x_i(O^{2-})]}{[x_i H_2O]}$	(54)	Ion fraction	8.3×10^{-7}	450	Potentiometry	17

					Temp.	Method / formula	Ref.
$K_{(55)}$	$\dfrac{[P_{HCl}][x_{O^{2-}}]}{[x_{OH^-}][x_{Cl^-}]}$	(55)	—	6.8×10^{-8}	450	$[K_{(13)} K_{(50)}]^{1/2}$	99
				$\log K_{(55)} = 5.075 - 8.85_3\ 10^3 T^{-1}$			
NaCl–KCl							
$K_{(13)}$	$\dfrac{[x_{O^{2-}}]P_{H_2O}}{[x_{OH^-}]^2}$	(13)	(mole fraction)$^{-1}$ atm	6.07×10^{-2}	750	Potentiometry	99
				$\log K_{(13)} = +6.75 - 8.15\ 10^3 T^{-1}$			
$K_{(50)}$	$\dfrac{[x_{O^{2-}}]P_{HCl}^2}{[x_{Cl^-}]^2 P_{H_2O}}$	(50)	(mole fraction) atm	1.4×10^{-14}	750	Potentiometry	
				$\log K_{(50)} = +40.2 - 55.3 \times 10^3 T^{-1}$			
$K_{(14)}$	$\dfrac{[\bar{x}_{O^{2-}}][\bar{x}_{H^+}]}{[x_{OH^-}]}$	(14)	Ion fraction	5.2×10^{-12}	750	$2k_H[K_{(13)} K_{(50)}]^{1/2b}$	
$K_{(55)}$	$\dfrac{[P_{HCl}][x_{O^{2-}}]}{[x_{OH^-}][x_{Cl^-}]}$	(55)	—	2.9×10^{-8}	750	Potentiometry $[K_{(13)} K_{(50)}]^{1/2}$	
				$\log K_{(55)} = 23.47_5 - 31.72_5 \times 10^3 T^{-1}$			

[a] k_H obtained from data of Reference 13 = 3.7×10^{-5} mole fraction.
[b] k_H obtained from data of Reference 14 = 5×10^{-5} mole fraction.

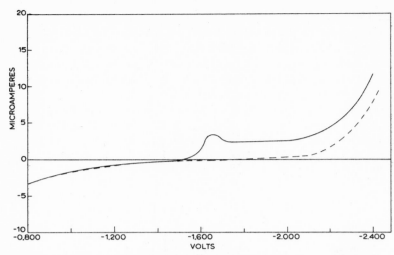

FIGURE 4. Residual currents in NaCl–KCl (50 mol %) before (————) and after (– – – –) removal of "hydroxide ion" impurities.[106] Volts versus Pt/Pt(II) reference electrode.

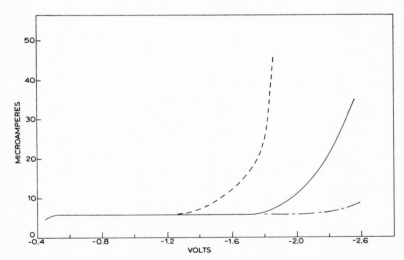

FIGURE 5. The residual currents during the purification of LiCl–KCl eutectic.[106] – – – – –, before; ————, after 20 min Cl_2 and 20 min argon; — – – —, after 40 min Cl_2 and 20 min argon. Volts versus Pt/Pt(II) reference electrode.

evolved during the treatment. Unfortunately the polarography carried out by Maricle and Hume did not detect the presence of oxide ions but the increasing anodic current suggests that such species may have still been present. Little or no experimental data are available concerning reactions (60)–(62). These equilibria might well be studied using the methods developed by Tremillon and Combes.

Oxyanionic melts such as sulfates, nitrates, and carbonates may be treated with their conjugate acid gas to ensure that the thermal decomposition (catalyzed by impurity?) which might lead to oxide is reversed:

$$CO_3^{2-} \rightleftarrows CO_2 + O^{2-} \tag{63}$$

$$SO_4^{2-} \rightleftarrows SO_3 + O^{2-} \tag{64}$$

$$2NO_3^- \rightleftarrows N_2O_5 + O^{2-} \tag{65}$$

In the case of sulfates, pyrosulfate ions can be used as an alternative sulfur trioxide source,

$$S_2O_7^{2-} \rightleftarrows SO_3 + SO_4^{2-} \tag{66}$$

to improve the quality of such melts.

3.3.2. Other Chemical Reagents for Purification

Ammonium[110] and sodium bifluorides[111, 112] have been used as precursors for the HF needed to remove oxidic impurities from molten fluorides. Ammonium chloride[113] has been used in a similar way to generate HCl for similar purposes in chlorides. In the former case the use of bifluorides must be considered a convenient method for providing HF, bearing in mind the requirements for handling hydrogen fluoride gas.

It has been suggested that the removal of oxide can be achieved by precipitation, e.g., ZrO_2 in ZrF_4 containing melts[114, 209] and Al_2O_3 in aluminum halide melts, or simply by introducing small quantities of Cr^{3+}, Al^{3+} into the bulk halide melt to remove oxide. A similar approach, that of introducing Ag^+ ions or Hg_2^{2+} ions, might be considered suitable for the removal of Cl^- ions from $LiNO_3$-based melts, but as we shall see, there are limitations. The former method depends upon the oxides of Zr, Al, and Cr having extremely low solubility products in the appropriate solvents.[99, 115, 117] Although precise data are not available, K_{sp} may be expected to be $\sim 10^{-20}$ (contrast this with data for PtO earlier[33]) for these oxides; thus

$$M_nO_m \rightleftarrows nM^{m+} + mO^{n-} \tag{67}$$

$$K_{sp} = [M^{m+}]^n[O^{n-}]^m$$

Consider the following two situations:

(i) M^{m+} is a cation of the solvent, e.g., Al^{3+} in $AlCl_3$ or Zr^{4+} in ZrF_4-based melts. The concentration of Zr^{4+} or Al^{3+} is high (0.05 → 0.66 mole fraction). Hence in the case of $AlCl_3$-based melts

$$[O^{2-}] = \left[\frac{K_{sp}}{(0.66)^2}\right]^{3/2} \simeq [K_{sp}]^{3/2}$$

for which the value lies in the region of 10^{-7} to 10^{-6}. Thus oxide ion may successfully be removed in such melts by using this technique.

(ii) M^{m+} is added to diminish the oxide concentration. In this case the addition of up to 10^{-3} mole fraction of M^{m+} should be possible without seriously altering the nature of the solvent then

$$[O^{2-}] = \left[\frac{K_s}{(10^{-3})^2}\right]^{3/2}$$

for which the value lies in the region of 10^{-4} to 10^{-5}. Because of the limitation on the value of M^{m+}, the concentration of oxide cannot be reduced as effectively as in case (i). An addition of M^{m+} for $m = 2$, or 1, would be most effective. However, the values of K_{sp} tend to be much higher for these lower oxidation state ions.

The removal of chloride ion impurities by precipitation or common ion effect from lithium-nitrate-based solvents illustrates a further difficulty with this approach. This arises when the anion acts as a powerful ligand, producing complex ions, which results in an *increase* in solubility. Thus, the lower level to which the impurity may be reduced will not be as low as predicted theoretically on the basis of the K_{sp} alone.[117] This is certainly true if Ag^+ or Hg_2^{2+} ions are used to remove Cl^- ions from nitrates.[118, 119, 130] The formation of oxy- complexes[120] has recently been reported in molten chlorides and, certainly in fluorides, one might expect the formation of soluble aluminates, for example, which will ultimately lead to an increase in the oxide ion solubility. Unless the precipitates are removed by filtration they will of course buffer the solutions with oxide ions, which is undesirable.

The removal of other oxyanionic contaminants from halide melts has received very little attention. It can be particularly serious in fluoride melts where silicate and silicofluoride ions are present in considerable quantities.[121] Sulfate, nitrate, and phosphate ions may also be present in these and other halide melts at levels ranging from 0.1–10 mM (Table 1). Small additions of calcium carbide will reduce these and other impurity metal ions,[122] viz.,

$$4CaC_2 + SO_4^{2-} \rightarrow S^{2-} + 4CaO + 8C \tag{68}$$

$$CaC_2 + MO \rightarrow CaO + M + 2C \tag{69}$$

Subsequent treatment of the melt with HCl/Cl_2 gas will remove oxide and sulfide and the carbon may be removed by filtration. Synthetic calcium carbide[123] is preferred because of its low oxide content.

Heavy metal ions can be removed by displacement reactions using either a metal corresponding to the solvent cation, e.g., Al in $AlCl_3$-based melts,[9] or the use of a base metal such as Mg, whose ions will not generally interfere with the studies for which the melt is to be used. This latter procedure has been exploited by Laitinen *et al.*[22, 124] and is used commercially by Anderson Physics Laboratories, Inc. However, in the strictest sense the melt is still contaminated and this can be important, say, in battery studies where Li or (Li)Al electrodes are to be studied.[15, 125] The efficiency of this displacement–cementation process depends upon a clean, high surface area, metal surface being available (free from oxide) for the displacement reactions. The displacing metal and displaced metal are easily removed by filtration after reaction. Roe[22, 124] reported that prior to the use of magnesium to displace impurities in LiCl–KCl eutectic, polarographic residual currents at -2.0 V versus the standard Pt/Pt(II) (1 M) reference electrode were between 400–700 $\mu A\ cm^{-2}$. The addition of magnesium reduced these currents to 75–300 $\mu A\ cm^{-2}$, corresponding to impurities in the concentration range 0.4–1.5 mM. In aluminum-chloride-based melts, a renewable surface of aluminum has been formed by electrolysis between two Al electrodes. Under the conditions used, Al dendrites grow out from the cathode and become detached if the applied current is reversed in polarity. Thus by periodically reversing the current impurity metal ions may be removed not only by primary deposition on the cathode but also by concomitant displacement reactions on the clean (high surface area) detached dendrite surfaces.[126]

3.3.3. Electrochemical Methods of Purification

Many of the impurities found in the salts used to produce inorganic solvents are electroactive. It is logical therefore to consider electrolytic methods for their removal. Apart from the relative simplicity of introducing two inert electrodes into the melt to remove impurities, the method offers precise control and a means to monitor the progress of purification. In practice the degree of purification can be as good as that using, say, Mg metal addition for the removal of metal ions, but it is uncertain whether oxidic impurities are so effectively removed to low levels (< 2 mM). In recent years there has been a tendency to combine gaseous and electrolytic procedures in an attempt to achieve a higher degree of purity, particularly for halide melts.[115, 127, 128, 129] The application to oxyanionic systems of the electrolytic method[76, 130] is somewhat more limited because of the electroactivity of their anions and a lack of knowledge about their electrolytic behavior on various electrode substrates.

Preelectrolysis, by which the technique is known, has been used extensively to purify aqueous and nonaqueous solutions for research. It has also been applied in industrial processing where, for instance, a plating bath is "run in," before productive plating is commenced. These applications and molten salt preelectrolysis depend on the principle that each and every redox system has a characteristic electrode potential expressed against some arbitrary reference electrode in the solvent considered. (Tabulated data are available for a number of solvents.[96, 105, 131]) The anodic and cathodic limiting potentials are defined for each solvent system by the least cathodic and anodic potential of electroactivity for the solvent cations–anions.[132] Within this definitive voltage span there is a range of electroinactivity for the pure solvent, i.e., for an inert electrode, the electrode will be perfectly polarizable. By the selection of such electrodes, redox systems whose potentials lie within the voltage span of the solvent can undergo oxidation or reduction reactions. Thus metal ion impurities can be removed by electrodeposition, whereas oxidic impurities might be removed by electrooxidation to, say, O_2. It is important to recognize that the material of which the electrode is constructed may be significant because of the following:

(i) Chemical interactions with, for example, a reduction product of a component of the solvent which will lead to a decrease in the range of solvent inactivity, e.g., the use of aluminum[133] or vitreous carbon[134] in the lithium-containing melts can produce a considerable lowering of the Li metal activity by alloying or intercalation reactions, which in turn leads to the production of lithium at much more anodic potentials than on an inert cathode.

(ii) The electrode reaction may be polarized such that the range of inactivity is extended compared to that expected for reversible behavior, e.g., the discharge of oxide ions on various carbonaceous material in halide melts shows considerable anodic overvoltage.[26, 135]

(iii) The electrode material itself may be electroactive before the solvent limit is reached, for example, platinum[136, 137] and gold[137] in molten chlorides will dissolve at potentials close to those for the oxidation of Cl^- ions.

(iv) The formation of films on the electrode may alter the properties expected, say, in (iii), e.g., the presence of oxide films on platinum can prevent dissolution and may even promote chlorine discharge (electrocatalysis[136, 138, 139]).

Careful choice and full knowledge of the behavior of the electrode material is therefore needed for each solvent system to optimize the conditions for purification. Preelectrolysis aims to remove low-level impurities, a consequence of which is that for reasonable time scales to be employed, the current flowing must be maximized to achieve significant purification (Faraday's laws) in spite of the low current *densities* which result from the low concentrations concerned. Optimization can be achieved by effectively increasing the current *densities* involved; this may be achieved by enhancing mass transfer by rotating or vibrating the electrode or by gas sparging of the malt. Alternatively the current may be increased by having electrodes of high surface area. Usually a combination of methods will be used. In this context, the use of reduced pressure (vacuum) to expand the volume of gaseous products of electrooxidation reactions, e.g., for oxide ions, has been used for many years in our laboratories both for chlorides and fluorides and was recently reported in some detail independently by Townsend.[140]

Consider an idealized current voltage curve (Figure 6) for a solvent whose cathodic limit is

$$M^+ + e \rightarrow M \tag{70}$$

and whose anodic limit is

$$X^- \rightarrow \tfrac{1}{2}X_2 + e \tag{71}$$

Two situations may arise during purification:

(i) The case where impurities exist which are both anodically and cathodically electroactive. Such impurities can be shown as limiting current densities $j_L^A(1)$, $j_L^C(1)$, $j_L^C(2)$ (see Figure 6). Removal of these impurities can be achieved simultaneously in one of two ways. A constant current density, say, 0.1–0.01j (corresponding to a 10–100-fold reduction of impurities) is applied. When the impurity limiting current density falls below the imposed current density, the voltage will increase rapidly, thus detecting completion of the process. Alternatively, a fixed voltage compatible with the potential span between the impurities to be removed, i.e., greater than DE but less than AC in Figure 6, can be applied. An exponentially decaying current will be observed, and the preelectrolysis terminated at a suitably acceptable current level.

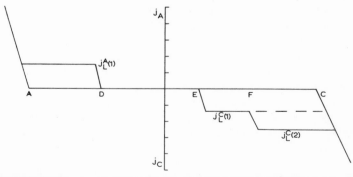

FIGURE 6. Idealized current density–voltage curves for two cathodically active impurities and one anodically active impurity.

(ii) The situation may arise in which there is only one major cathodic or anodic impurity or that one (anodic–cathodic) impurity is in much greater excess than (cathodic–anodic) other impurities. In these cases consideration must be given to the counterelectrode reaction either initially or after removal of the minor component. If the major impurity is to be cathodically removed then the anodic reaction may conveniently be the discharge of the anionic component of the solvent; the products of this are invariably gaseous, which by reducing the cell pressure can readily be removed as well as producing a beneficial stirring action. Such a situation may arise after treating a melt with, say, $HCl–Cl_2$ to remove the oxidic impurities. Preelectrolysis is often used following this treatment to remove the final residual metal ions. Obviously it is important to ensure that the chlorine gas evolved at the anode does not backreact with cathode or the deposited impurities. The use of a separate anode compartment is therefore desirable. On the other hand, if a major impurity is to be anodically oxidized to remove it from the melt then the countercathode reaction may be more difficult to accommodate. Alkali or alkaline-earth metals will generally be the cathodic product in halide melts and these are readily soluble in the melt, possibly resulting in cyclic processes. This problem can be overcome either by employing higher operating temperatures and lower pressures or by using an electrode which interacts with the alkali metal and reduces its activity by many orders of magnitude. Bismuth is a convenient liquid electrode for this purpose for low-temperature halide melts since alkali metals will readily dissolve in it[141, 142] to produce solutions in which the activities of alkali metals are very low.[141] When the amounts of charge to be consumed are small, graphite or vitreous carbon are also suitable and can be readily fabricated into cathodes. The cathodes should be contained in a separate compartment which is provided with electrical contact to the bulk melt through a side arm or fritted disk. In the purification of molten cryolite liquid silver[143] has been used as a low-activity sink for aluminum while anodically removing oxide species from the melt.

Melts based upon oxyanions lead to problems of a different kind. The oxyanions themselves are cathodically active, being reduced to lower oxidation state anions which are then contaminants of the melt. The problem of removing anodically active impurities in oxyanionic melts is overcome by providing for the cathodic process, a reducible cation in a compartment physically separated, but in electrical contact with, the solvent melt. A solution, say, of $AgNO_3$ in a molten nitrate solvent contained in a Pyrex tube and separated by a sintered disk from the bulk melt might be used as catholyte along with the appropriate noble metal. This situation might be best treated in a three-electrode potentiostatic system for maximum control.

For the large-scale preparation of refractory metal fluoride plating baths[140] utilizing oxide solutes, the cathodic element to be plated subsequently from the bath is usually the solute cation. This approach rests on the premise that the cations do not interact with oxide ions as they certainly would if the solvent were a chloride. This procedure is apparently quite successful for fluorides.

The removal of the minor component will concomitantly lead to an increase in cell voltage, which will result in the discharge of the solvent cations or anions to sustain the applied constant current density. At constant voltage a significant current will flow if and only if the applied cell voltage is sufficiently large to encompass the solvent decomposition potentials. Judicious selection of the cell voltage ensures that the cell current is that due to the impurities only. After the removal of the minor impurity (as indicated by the very low cell current) it would be necessary to introduce the auxiliary cathode assembly (as discussed previously) to complete the removal of the major impurity.

The control of constant current electrolysis is probably best made by measuring the total charge passed (now that electronic integrators are readily available), which may also be used in the case of constant voltage electrolysis as an alternative to the decay of current.

If the melt is stirred, then an equation developed by Meites[144] can be applied to calculate the time necessary to reduce the initial concentration to a given level. Thus for reversible processes under well-stirred conditions,

$$i = i(0)\exp - S_0 t$$

where the initial current $i(0) = S_0 nFC_0^b V$ and $S_0 = DA/V\delta$ sec^{-1}, D is the diffusion coefficient (cm^2 sec^{-1}), A is the electrode area (cm^2), V is the melt volume (cm^3), and δ is the diffusion layer thickness (cm).

Example: Let the residual metal ion content of a melt be 2.5 mM. In the apparatus described later (Figure 11), V is typically 200 cm^3 and the electrode areas are ~ 15 cm^2. If the pressure is reduced, then gassing can occur and stirring results through the anode reaction, and $\delta \simeq 10^{-2}$ cm and $D_{ion} \simeq 10^{-5}$ cm^2 sec^{-1} for M^{2+} ions $\sim 500°C$. Hence with $S_0 = 3.75 \times 10^{-5}$ sec^{-1} and $i(0) = 7.2$ mA, the time for the current to reach 1.8 mA is 5 hr. In practice a typical electrolysis time to achieve such a current is 4 hr. This current corresponds to approximately 75% reduction in the concentration of metal ion impurities. To achieve 99% removal a time of 20 hr is required, assuming δ can be maintained at $\sim 10^{-2}$ cm by effective stirring. These calculations suggest that preelectrolysis prior to HCl–Cl$_2$ treatment should be particularly effective because the presence of substantial oxide concentrations ensures adequate stirring by the anodic gases during electrolysis. For 90% removal of metal ions (in the above system) one can calculate that the residual current would be ~ 130 μA cm^{-2}. Values reported in the literature range from 150 to 300 μA cm^{-2}, suggesting that rather less than 90% removal of metallic impurities is achieved and that the residual concentrations of impurities are

0.2–0.5 mM. Boisde *et al.*[145] have measured the residual oxygen concentration by neutron activation analysis and found that this was 20 ppm after HCl treatment.

4. Specific Methods of Purification for Some Commonly Employed Solvents

In this section a brief description of the apparatus required and the method of proceeding will be given for $LiNO_3$–KNO_3–$NaNO_3$, LiCl–KCl, LiF–NaF–KF eutectic and cryolite, and $AlCl_3$–MCl mixtures. These procedures will illustrate the variety of techniques that are available and which should be applicable to many related solvent systems. Table 10 gives references to purification techniques for other less common solvents. Perhaps one can appeal here for workers to establish the purity of their melts by a commonly accepted technique such as voltammetry, using as far as possible a suitable reference electrode preferably based on or easily related to silver–silver(I) in the appropriate solvent. This may go some way towards removing the doubts about purity often expressed when conflicting results are reported for a particular system. This is especially important when, say, spectroscopic measurements are made on a system which has been or may be studied electrochemically.[76, 146]

4.1. The Eutectic Mixture Lithium Nitrate–Sodium Nitrate–Potassium Nitrate (30 mol % $LiNO_3$, 17 mol % $NaNO_3$, 53 mol % KNO_3, mp ~ 120°C)

4.1.2. General Comments

The major problems with the preparation of this solvent are the hygroscopic nature of the lithium nitrate, its high impurity content with respect to chloride and sulfate, and the high water solubility in the solvent at the low fusion temperature. Lithium nitrate is generally available as the laboratory grade trihydrate and more expensively as an anhydrous material from Alfa Ventron.* One approach to this problem is to use the trihydrate and recrystallize the $LiNO_3 \cdot 0.5H_2O$ from hot $LiNO_3$ solutions. This procedure repeated twice reduced the chloride content from about 5–10 mM to 0.04–0.07 mM as shown by sweep voltammetry at a hanging mercury drop. A diffusion coefficient for chloride ions of 1.51×10^{-6} cm^2 sec^{-1} was obtained at 156°C (after suitable correction for residual chlorides).[147] Melts which are fused in the absence of fine particles of silica are

---

* *Editors' Note:* However, the latter does not involve the simultaneous purchase of large amounts of water which must then be removed in time-consuming, costly processes.

pale yellow in color. This is thought to be due to the presence of peroxide possibly arising from the hydrolytic action of residual water on nitrate. The addition of 1 g of silica powder to the mixture[37] promotes decomposition of the peroxide and colorless, transparent melts are obtained. Preelectrolysis is also reported to have a similar effect.[76]

4.1.3. Procedure*, [147]

Lithium nitrate is twice recrystallized from a hot (60–70°C) filtered, concentrated solution to yield (60%) $LiNO_3 \cdot 0.5H_2O$. About 50 g of this hydrate is spread on a large Petri dish over excess P_2O_5 in a vacuum desiccator. The sample is left under a good vacuum for about one week. The required proportion of this dried sample is combined with recrystallized and P_2O_5-dried samples of sodium nitrate and potassium nitrate, 1 g of fine silica powder, and thoroughly mixed. This mixture is placed in tube P of the apparatus shown in Figure 7. The purification cell is evacuated by a rotary pump through K with the receiving crucible L and caps G and Q in place. The cathode B and anode A are replaced by stoppers. The system is pumped through a N_2 cold trap until the pressure is less than 5–10 μ (readily achieved by a two-stage rotary pump), whereupon the cell is heated to about 40–50°C and the pressure allowed to fall to the initial lowest level. The temperature is raised in two further successive heating and pumping steps to about 140°C after which the bulb D is broken by lowering rod C. The melt, which is forced through the No. 4 sintered Pyrex disk E by purified argon and a reduced pressure below the disk, is collected in crucible L. The purification apparatus is returned to room temperature under vacuum. The melt is removed via an argon-purged plastic bag, into the drybox. A more elaborate removal system has also been described.[148] Subsequent storage of the melt in a small empty desiccator within the drybox will minimize contamination. It is well to remember that dehydrated lithium salts are amongst the best-known desiccants.

$LiNO_3$–KNO_3, $LiNO_3$–$NaNO_3$ are purified by similar procedures. The equimolar $NaNO_3$–KNO_3, mixtures of nitrites and nitrate–nitrite mixtures should also be subjected to a recrystallization, vacuum-melting procedure ensuring that the final temperature is close to the melting point and the residence time at this temperature is minimized in the nitrate systems to control the level of nitrite present. Typical voltammograms for some purified melts are shown[147] in Figures 8 and 9. The apparatus could readily be used to introduce a nitrogen dioxide purge to reduce nitrite

---

* In all procedures it is assumed that glassware–silicaware have been thoroughly cleaned. 50/50 (by volume) HNO_3/H_2SO_4 is recommended for this.

TABLE 10. Some Additional References to Solvents Commonly Used

Solvent	Composition component, A, B mol %	mp, °C	Comments	Refs.
(a) Simple ionic melts				
LiCl	—	610	Requires HCl treatment	6
KCl	—	770	Highly volatile	178, 90, 6
NaCl	—	801	Highly volatile	6, 178
CsCl	—	646		178
MgCl$_2$	—	708	Purified bromides also	181
BaCl$_2$	—	962		181
PbCl$_2$	—	501		181
CaCl$_2$	—	773	CaCO$_3 \rightarrow$ Ca oxalate + Cl$_2$	180, 181
LiF	—	845		179
NaF	—	993		110, 111, 121
KF	—	852		110
MgF$_2$	—	1261	HF treatment	110
CaF$_2$	—	1423	Highest melting	110
SrF$_2$	—	1473	salts, stable over	110
BaF$_2$	—	1355	several 100°C	110
(b) Simple ionic mixtures				
NaCl–KCl	50	658		106, 152
KCl–CsCl	40	616		90, 129
KCl–MgCl$_2$	67.5	435	Useful since MgO tends to precipitate out	183
MgCl$_2$–NaCl–KCl	50, 30	396		183
LiCl–NaCl–KCl	43, 33	357		76
PbCl$_2$–KCl	52	406		182
LiF–KF	50	492		184
KHF$_2$		512		185, 186
LiBr–KBr	60	348	Use of HBr and Mg	149

(c) Compounds "covalentlike" and their mixtures

Compound	Composition	Temp.	Comments	Ref.
$AlCl_3$	—	192[a]		187, 192
AlF_3	—	1272(S)		188
$ZnCl_2$	—	318	Polymeric	189
$SnCl_2$	—	246		190
$KCl-AlCl_3$	33 and various	128		191
$NaCl-AlCl_3$	Various	—		194, 195
$AlCl_3-NaCl-KCl$	66, 20	89		9
$KCl-ZnCl_2$	51	228		156
$KCl-SnCl_2$	Various	180 upward		196
$LiF \cdot AlF_3$	75	792	Problem with oxides	197
$NaF \cdot AlF_3$	75	1023	Problem with O^{2-}, Si	143, 198, 199

(d) Oxyanionic melts and their mixtures

Compound	Composition	Temp.	Comments	Ref.
$NaNO_3$	—	310	Thermal storage applications	200
KNO_3	—	337		200
$CsNO_3$	—	417		200
$KCNS$	—	177	Rather unstable	202
$LiClO_4$	—	247		201, 203
$NaClO_4$	—	468		
$KClO_4$	—	580		201
$LiClO_4-KClO_4$	72.5	207		
$LiClO_4-NaClO_4$	76	208		
$Li_2SO_4-Na_2SO_4-K_2SO_4$	78.8, 13.5	512	Convenient low melting, stable sulfate mixture	204
$Li_2CO_3-Na_2CO_3-K_2CO_3$	43.5, 31.5	397	Fuel cells, large-scale usage	150, 205
$KCN-NaCN$	40	~500	Important for precious metal plating; inert atmosphere	206, 207

[a] Determined under pressure.

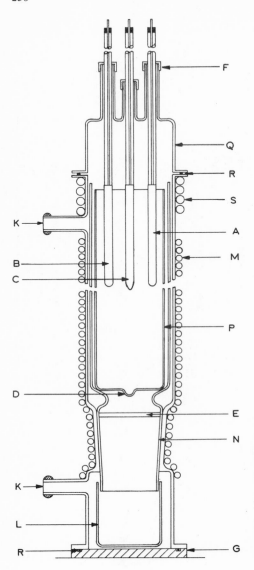

FIGURE 7. Apparatus for the purification of molten nitrates and low-melting halide systems (< 500°C). A, Anode; B, cathode; C, Pyrex breaker rod; D, breakable glass dimple; E, glass sinter (No. 4 porosity); F, quickfit screw thread joint and cap SQ13; G, brass plate; K, S29 ball to vacuum system; L, Pyrex receiver; M, heating element; N, B45 or B55 of inner filter tube; P, inner Pyrex tube; Q, Pyrex "top hat" on F.G. 50 or 75 flange; R, "O" rings.

contamination.[72] As far as the author knows this has not been done in a routine way to purify nitrates (but see Reference 74).

Eutectic mixtures of carbonates and sulfates formed by the alkali metal salts could be treated in similar procedural sequence in the apparatus shown in Figure 7, with suitably modified tube P. The use of SO_3 and CO_2 to reduce the oxidic content of the melts could easily be achieved. Some procedures similar to this are described in earlier works.[150, 151]

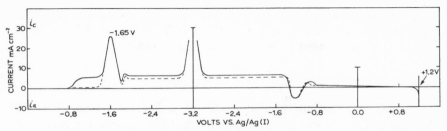

FIGURE 8. Polarogram for molten $NaNO_3$–KNO_3.[69] $E^{1/2} \approx -0.8$ V. ——————, water wave; – – – – –, after bubbling nitrogen.

4.2. Lithium Chloride–Potassium Chloride (0.59 mol % LiCl, 0.41 mol % KCl; mp ~355°C)

4.2.1. General

This mixture has probably received the most attention with respect to the variety of studies made with it as a solvent and to its purification.[7, 23, 106, 115, 124, 127, 149] The hygroscopic nature and relatively poor quality of commercial LiCl thus require substantial efforts to avoid the deleterious effects of water, the traces of residual metal ions and carbonaceous matter in this and the potassium chloride. Treatments involving vacuum drying,[6] chlorine bubbling,[106] hydrogen chloride bubbling,[7, 124] cementation,[23, 124] preelectrolysis,[115, 127] and filtration[127] have been used singly or in one or more combinations. The procedure to be described will

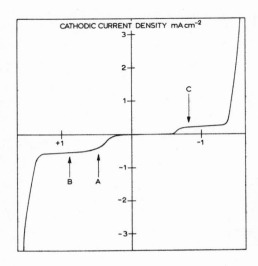

FIGURE 9. Current–voltage curve for $LiNO_3$–$NaNO_3$–KNO_3 eutectic at 150°C. 1-cm-long Pt needle microelectrode.[71] A, Cl^- oxidation; B, NO_2^- oxidation; C, water reduction.

involve all of these excepting cementation. Cementation and preelectrolysis are essentially alternatives to each other. Preelectrolysis is favored by the author and is advocated in the following description.

4.2.3. Procedure

Apparatus[26] similar to that described in Figures 7 and 10 can be employed, but the upper temperature limit is 500°C. Since the procedure to be described can be used to purify melts based on $MgCl_2$, $CaCl_2$, LiBr, etc., the apparatus in Figure 11 has been developed and will be described because of its versatility.

LiCl should be of anhydrous grade and predried under vacuum over P_2O_5 before combination with a similarly dried potassium chloride sample in the appropriate proportions. The handling of these salts is conveniently made in either a dry bag or drybox. The apparatus is prepared using either

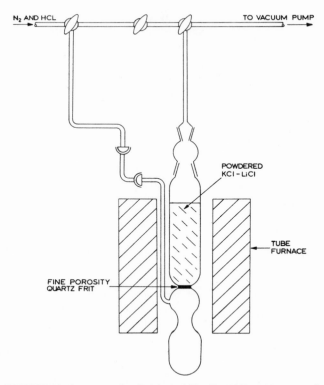

FIGURE 10. Apparatus for drying and filtering LiCl–KCl eutectic.[7]

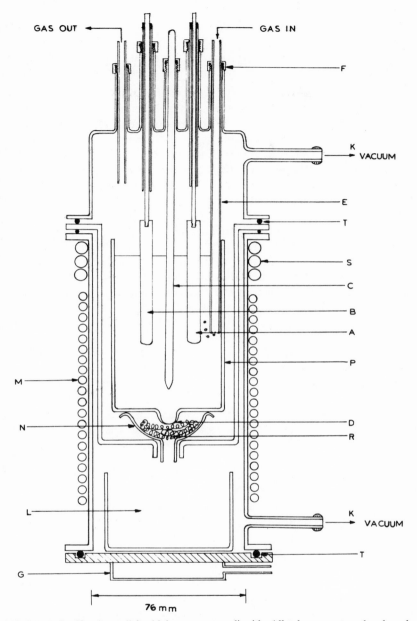

GAS OUT

GAS IN

F

K VACUUM

E

T

S

C

B

A

P

M

D

N

R

L

K VACUUM

T

G

76 mm

FIGURE 11. Purification cell for high-temperature liquids. All tubes except on head made of transparent/translucent silica. Gas in/out controlled through Rotaflow taps not shown. Horizontal scale, 1 : 1; vertical scale, 1 : 3. A, Graphite anode; B, tungsten cathode; C, silica breaker rod; D, breakable dimple in silica tube; E, silica tube with frit to disperse HCl or Cl_2; F, SQ13 capped tube; G, water-cooled base plate; K, S29 ball joint; L, silica receiver; M, heat element of Nichrome; N, silica cup containing silica filter pad; P, inner silica tube containing salts; R, silica wool pad; S, water cooling; T, viton "O" rings.

a Pyrex frit or silica wool as the filtering medium (at temperatures above 500°C, silica frits can be used).

Tube P is loaded with a batch of the mixture (in the apparatus shown 300–500 g can be prepared) and the main cell closed with the cell head—less the preelectrolysis electrodes A and B—and melt receiver. The cell is evacuated with a two-stage rotary pump until the pressure reaches ~ 5–10 μ when an oil diffusion pump is switched in to reduce the pressure further to less than 0.5 μ. This initial pumping takes approximately 24 hr or more and is essential to reduce the initial water content to a minimum before heating commences. The temperature is raised in two stages to 300°C with continuous pumping. Hydrogen chloride gas, dried with molecular sieve AW 500 and activated charcoal, is introduced into the system, which has been brought to 1 atm with purified argon. The gas is flushed over the surface of the solid whilst the temperature is raised to 400–430°C, and then the gas is dispersed through the melt through a frit; after about 2–3 hr the hydrogen chloride is replaced by argon followed by dried chlorine. This is allowed to flow for 1–2 hr and is finally displaced, again by argon. The preelectrolysis electrodes A and B are introduced quickly under high argon flow and the 2.7-V, constant-voltage supply connected. The anode is lowered into the melt followed by the cathode (to ensure cathodic protection of the tungsten) and the pressure in the cell immediately reduced to aid stirring by expanding the evolving anodic gases. For electrodes 15–20 cm^2, the initial current will be ~ 10–20 mA, rapidly falling to 2–3 mA in 4 hr, corresponding to a residual current of 100–200 μA cm^{-2}. The bulb D is broken by use of the rod C, and, with a suitable argon gas pressure above the frit and reduced pressure below the frit, the melt is filtered through into the receiver L. The apparatus is cooled under high vacuum; the billet of melt is removed and stored in the dry box, as described for the nitrate melts.

The use of prolonged, initial preelectrolysis before gaseous treatment will enable the residual current to be reduced further. Egan and Heus[153] report much lower residual currents (0.5 μA cm^{-2}) after 72 hr electrolysis, but their cell voltage was only 1.5 V. Hence their residual current relates only to this limited range (many of the trace impurities to be removed require 2–2.5 V), which was necessary with vitreous carbon (and/or graphite) electrodes, in order to avoid the intercalation reaction of lithium. Since iron, chromium, etc. impurities often require removal, it is desirable to use the more noble tungsten cathode[115] rather than stainless steel preferred in earlier work.[26]

Cyclic voltammetry has been employed to confirm the high degree of purity obtained by this procedure. The purity is comparable with that described by Laitinen et al.[22, 124] using a method now adopted commercially. The voltammetric measurements also emphasize that further im-

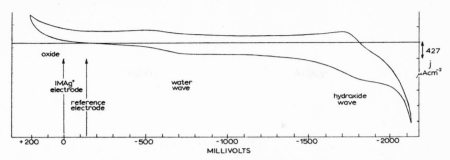

FIGURE 12. Voltammetric curve for LiCl–KCl after purification showing the presence of remaining traces of water and hydroxide visible when using tungsten as microelectrode 100-mV sec^{-1} sweep rate. Current density scale 427 μA cm^{-2} per vertical division.[152]

provements in quality should be possible on the basis of polarographic results (Figures 12 and 13).[152]

Many halide melts such as MgCl$_2$–KCl,[154, 155] CaCl$_2$–NaCl, NaCl–KCl,[152] (Figure 14) ZnCl$_2$–KCl,[156] and mixtures of bromides[115, 149, 157] and iodides[158] are readily purified in this type of apparatus using similar techniques. The major parameter which would need to be varied with melt composition is the preelectrolysis voltage, but with a tungsten cathode 2.5 V or slightly more will generally prove satisfactory (excluding Zn-containing melts).

4.3. Lithium Fluoride–Sodium Fluoride–Potassium Fluoride (FLiNaK)[159] (46.5 mol % LiF, 11.5 mol % NaF, 42 mol % KF; mp 454°C)

4.3.1. General

This mixture represents the lowest melting composition of the alkali metal fluorides. It has been used as a model system in developing electroanalytical techniques for the molten salt reactor experiment as well as a medium for metal deposition from solutions of refractory metal fluorides. Purification presents difficulties because of the corrosive nature of the impure alkali metal fluorides and the more hygroscopic nature of its components. Recently Optran grade LiF, NaF, KF have become available (at a price!) with typical impurity levels <100 ppm. These may offer ideal materials for solvent preparation which aims only to achieve dryness and freedom from hydrolysis products. Alternatively, commercially available laboratory chemicals form the starting material and a more elaborate purification procedure is then required. The treatment of the FLiNaK

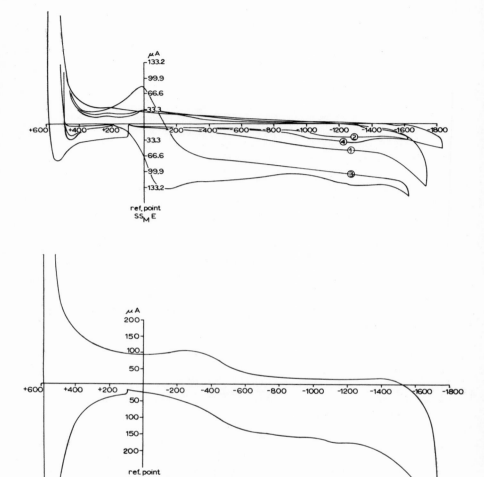

FIGURE 13. Bottom: Voltammogram for LiCl–KCl eutectic at 425°C after vacuum fusion but prior to preelectrolysis.[152] Pt microelectrode; sweep rate, 100 mV sec^{-1}; area, 0.145 cm^2. Top: Voltammograms taken during purification of LiCl–KCl eutectic.[152] (1) After electrolysis between W cathode and graphite anode 24 hr. (2) After HCl treatment for 2 hr following preelectrolysis. (3) In the presence of HCl. (4) After argon purging and short evacuation; sweep rate, 100 mV sec^{-1}; area, 0.145 cm^2. SS$_M$E = standard Ag/Ag(I) (1 M) reference electrode.

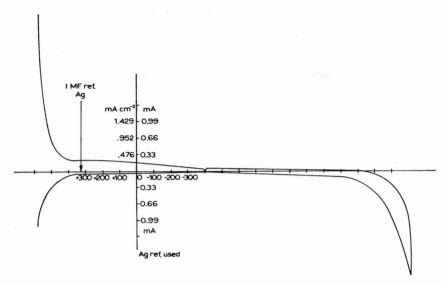

FIGURE 14. Voltammogram at Pt microelectrode for purified (HCl/Cl$_2$) NaCl–KCl equimolar mixture at 720°C.[152] Sweep rate, 100 mV sec^{-1}.

would best involve hydrogen–hydrogen fluoride gas mixtures, used in a similar manner to that for chloride melts, although the problems of handling these gases will limit their use to specialized establishments.[110, 159, 160, 161] The alternative of using a chemical precursor of HF is more universally applicable and the procedure based on this approach with laboratory grade chemicals will now be described.[162]

4.3.2. Procedure

The individual salts should be dried separately before being combined in the appropriate proportions to be used in the purification procedure.

About 15 wt % of ammonium bifluoride is combined with the FLiNaK mixture and placed in a graphite crucible. The mixture is melted and transferred to a Pt or Ni crucible. This is then evacuated within a nickel or Inconel tube (Figure 15) and followed by melting and sparging with hydrogen at 750°C for 48 hr. During this time NH$_3$ and HF will be removed together with moisture produced from the reaction between oxidic species and HF. These products should be collected in a cold trap (Teflon). The H$_2$ also serves to prevent corrosion of the nickel parts of the cell. Preelectrolysis is then carried out between graphite electrodes at a constant current density of 5 mA cm^{-2}, with a total current of 100 mA. This preelectrolysis might be better carried out with a tungsten cathode

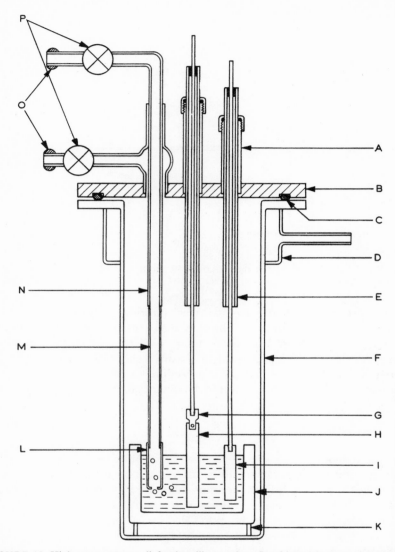

FIGURE 15. High-temperature cell for handling molten fluoride solvents. A, SQ13 screw capped tube; B, brass head; C, "O" ring neoprene or viton; D, water-cooling jacket; E, Pyrex support tube; F, Inconel or nickel cell body; G, stainless connector block; H, graphite/tungsten cathode; I, graphite anode; J, graphite/vitreous carbon crucible; K, alumina plinth; L, molybdenum or graphite bubbler; M, steel tube; N, Pyrex gas inlet tube; O, S13 ball joints; P, vacuum taps.

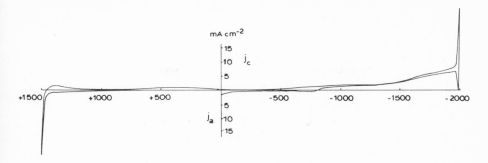

FIGURE 16. Voltammogram for FLiNaK after H_2/HF treatment.[163] Pt microelectrode; sweep rate, 100 mV sec^{-1}; reference electrode is

$$Ni/NiF_2(S)\ FLiNaK \parallel$$
$$\underset{\text{membrane}}{\overset{LaF_3}{}}$$

and about 3 V or more applied between the electrodes. The residual current of 5 mA cm^{-2} is high and the reported residual chronopotentiometric $i\tau^{1/2}$ product $(i\tau^{1/2})$ corresponds to ~ 5 mM level of impurities. Clayton *et al.*[163] have reported residual currents of less than 1 mA cm^{-2} for FLiNaK produced via HF/H_2 treatment (Figure 16). The use of graphite cathodes in preelectrolysis is unsatisfactory because of the alkali metal intercalation reaction[134] and subsequent lowering of the cell voltage at constant current. The impurity levels in these melts are considerably higher than in the corresponding chloride melts. More selective preelectrolysis would achieve a more favorable result.

4.4. Cryolite (Na_3AlF_6; MPt 1023°C)

4.4.1. General

This solvent, which is clearly important to the aluminum industry, has recently been of interest in other metallurgical applications.[121] The hand-picked, natural Greenland cryolite is contaminated with both silica, alumina, and iron, which may be acceptable for studies of the Hall–Héroult cell, but generally for electrochemical–spectroscopic work a higher level of purity is necessary. Thonstad[143] has reported that the quality of cryolite can be improved by preelectrolysis. The following procedure is an adaptation of this method.[121]

4.4.2. Procedure

The cryolite is weighed into a graphite or molybdenum crucible into which has been melted a 50-g billet of silver to act as cathode. A graphite or vitreous carbon rod forms the anode. The cell is evacuated at room temperature to 20 μ and slowly heated to about 500°C; the pressure is then raised to 1 atm of argon (Figure 15). The cell temperature is now raised to 1050°C and the anode lowered into the melt, with a voltage of 1.6 V applied. Initially, the current is 100 mA cm^{-2} and it slowly falls to about 5 mA cm^{-2} over a 24-hr period. The cell is cooled and the cryolite stored under argon or in a drybox. A voltammogram for a batch of melt produced in this way is shown in Figure 17. Silver–copper alloy cathodes have also been used in a similar way to purify LiF–YF$_3$ melts[164] and current–voltage

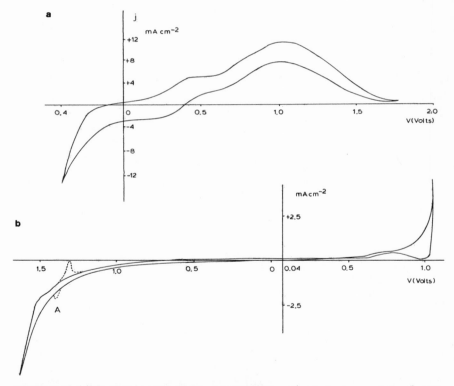

FIGURE 17. (a) Purified cryolite: Sweep rate, 100 mV sec^{-1}; electrode area, 1.5 cm^2; WE, graphite; CE, graphite; RE, nickel wire; vitreous carbon crucible. (b) Purified sodium fluoride: sweep rate, 200 mV sec^{-1}; electrode area, 0.12 cm^2; WE, platinum; CE, vitreous carbon; RE, Ni/Ni(II) (0.1 m)

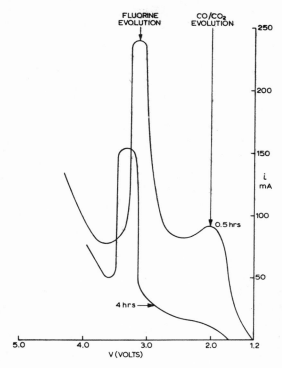

FLUORINE
EVOLUTION

CO/CO_2
EVOLUTION

4 hrs →

0.5 hrs

i
mA

V (VOLTS)

FIGURE 18. Current–voltage curve for $3LiF–YF_3$ showing gradual removal of oxide at a graphite anode, using Ag–Cu cathode.[164] After short electrolysis 0.5 hr, after longer electrolysis 4 hr.

characteristics such as those in Figure 18 used to moniter the preelectrolytic treatment. Sodium fluoride can also be produced in highly purified form using techniques similar to those described[121] as well as by using NaF–NaHF mixture.[111] Voltammetric data for the former are shown in Figure 17.

4.5. Aluminum Halide–Alkali Metal Halide Mixtures (Typically 67 mol % $AlCl_3$, 33 mol % KCl; mp 128°C)

4.5.1. General

The binary and ternary mixtures of aluminum halides with alkali metal halides provide a variety of low-melting reaction media in which the Lewis acid–base character can readily be varied. The ease of hydrolysis of the aluminium halides and the presence of iron compounds require elaborate purification procedures for these solvents. The high-purity aluminum chloride now available as a byproduct of the new Alcoa process should, however, make their preparation somewhat easier.

4.5.2. Procedure

A typical approach[86] for the 2AlCl$_3$·KCl mixture will be described using AlCl$_3$ which had previously been sublimed (for example, 9, 84, 126). The aluminum chloride and dried A.R. potassium chloride are mixed together in a dry box in the molar ratio 2 : 1. This mixture is subsequently transferred to an apparatus similar to that shown in Figure 7, suitably modified[86] for work with AlCl$_3$. After evacuation and backpurging with high-purity argon at room temperature to about 100 torr to remove volatiles, a slight positive pressure of argon is maintained in the cell whilst the mixture is melted. A dark solution is usually observed at this stage due to impurity iron, carbonaceous matter, and oxide precipitates. Two aluminum electrodes are introduced and electrolysis at a constant current density of 1.5 mA cm^{-2} is maintained for at least 48 hr or until the melt is colorless (other workers[87, 193] report that up to ten days are required when the current density is 0.5 mA cm^{-2}). Considerable solid matter is produced at the electrodes during this process. Some loss of AlCl$_3$ may occur in the apparatus in spite of efforts to maintain a constant temperature throughout the cell. A sealed cell would be desirable but the need to filter the melt dictates a demountable system. The melt is finally filtered into a Pyrex receiver in a manner similar to that described for LiCl–KCl, etc.

The cell is cooled and the salt transferred to a dry box for subsequent operations. It is beneficial to *operate* with these melts under dry box conditions, i.e., a furnace must be built into the dry box assembly. In this way highly purified melts can be handled over considerable periods of time. Residual currents of the order 0.95 mA cm^{-2} at a sweep rate of 100 mV sec^{-1} were found using a vitreous carbon electrode over the range 0–1.2 V versus an Al reference electrode, i.e., comparable with those reported by others.[192, 193]

References

1. J. J. Duruz, H. J. Michels, and A. R. Ubbelohde, *Proc. R. Soc.* London, **332A**, 281 (1971); *J. Chem. Soc. B*, 1505 (1971).
2. V. C. Reinsborough, *Rev. Pure Appl. Chem.* **18**, 281 (1968).
3. G. P. Smith, C. H. Liu, T. R. Griffith, *J. Am. Chem. Soc.* **86**, 4796 (1964).
4. T. R. Griffiths, *J. Chem. Eng. Data* **8**, 568 (1963).
5. A. Kisza and M. Grzeszezuk, *J. Electroanal. Chem.* **91**, 115 (1978), and references therein.
6. H. J. Gardner, C. T. Brown, and G. J. Janz, *J. Phys. Chem.* **60**, 1458 (1956).
7. H. A. Laitinen and W. S. Ferguson, *J. Electrochem. Soc.* **104**, 516 (1957).
8. D. L. Hill, J. Perano, and R. A. Osteryoung, *J. Electrochem. Soc.* **107**, 698 (1960).
9. U. Anders and J. A. Plambeck, *Can. J. Chem.* **47**, 3055 (1967).
10. S. Senderoff and A. Brenner, *J. Electrochem. Soc.* **101**, 16 (1954).
11. W. J. Burkhard and J. D. Corbett, *J. Am. Chem. Soc.* **79**, 6361 (1957).

12. N. Q. Minh and B. J. Welch, *Austr. J. Chem.* **28**, 965, 2579 (1975); see also N. Q. Minh and B. J. Welch, *J. Electroanal.* **92**, 179 (1978).
13. J. Van Norman and R. Tivers, *J. Electrochem. Soc.* **118**, 258 (1971).
14. E. A. Ukshe and V. N. Devyatkin, *Russ. J. Phys. Chem.* **39**, 1222, 1641 (1965) (Eng. Trans.).
15. C. A. Melendres, J. P. Ackerman, and R. K. Steunenberg, *Proceedings of the International Symposium on Molten Salts*, Ed. J. Braunstein *et al.*, Electrochemical Society, New York (1976), p. 575.
16. S. H. White, unpublished work (1978).
17. Y. Kanzaki and M. Takahashi, *Electroanal. Chem.* **58**, 349 (1975).
18. S. H. White, unpublished work (1979).
19. R. Lysy and R. Combes, *J. Electroanal. Chem.* **83**, 287 (1977).
20. J. B. Raynor, *Ber. Bunsenges. Phys. Chem.* **67**, 360 (1963).
21. Y. Kanzaki and M. Takahashi, *J. Electroanal. Chem.* **58**, 339 (1975).
22. H. A. Laitinen, R. P. Tischer, D. K. Roe, *J. Electrochem. Soc.* **107**, 546 (1960).
23. R. Baboian, D. L. Hill, and R. A. Bailey, *Can. J. Chem.* **43**, 197 (1965).
24. R. D. Caton and H. Freund, *Anal. Chem.* **36**, 15 (1964).
25. S. H. White, D. Inman, and B. Jones, *Trans. Faraday Soc.* **64**, 2841 (1968).
26. N. S. Wrench, Ph.D. thesis, London University (1967).
27. A. Komura, H. Imanago, and N. Watanabe, *Denki Kagaku* **40**, 306, 762 (1972).
28. S. I. Rempel, *Dokl. Akad. Nauk SSSR* **74**, 331 (1950).
29. E. Briner and P. Roth, *Helv. Chim. Acta* **31**, 1352 (1948).
30. *Metal Reference Book*, Ed. C. J. Smithells, Butterworth, Washington, D.C. (1976), p. 206.
31. K. H. Stern, *J. Phys. Chem.* **66**, 1311 (1962).
32. J. Goret and B. Tremillon, *Bull. Soc. Chim. Fr.*, 67 (1966).
33. H. A. Laitinen and B. B. Bhatia, *J. Electrochem. Soc.* **107**, 705 (1960).
34. N. S. Wrench and D. Inman, *J. Electroanal. Chem.* **17**, 319 (1968).
35. D. Inman, in *Electromotive Force Measurements in High Temperature Systems*. Ed. C. B. Alcock, Institution of Mining and Metallurgy, London (1968), p. 163.
36. P. G. Zambonin, *J. Electroanal. Chem.* **24**, App. 25, (1970).
37. G. C. Barker and R. L. Faircloth, Atomic Energy Research Establishment, C/R2032, Harwell, England (1956).
38. D. Inman, *Electrochim. Acta* **10**, 11 (1965).
39. D. Inman and M. J. Weaver, *J. Electroanal. Chem.* **51**, 45 (1974).
40. E. P. Mignonsin, L. Martinot, and G. Duyckaerts, *Inorg. Nucl. Chem. Lett.* **3**, 511 (1967).
41. G. Mamantov, in *Molten Salts, Characterization and Analysis*, Ed. G. Mamantov, Marcel Dekker, New York (1969), p. 529.
42. A. L. Mathews and C. F. Baes Jr., *Inorg. Chem.* **7**, 373 (1968).
43. S. Pizzini, R. Morlotti, and E. Romer, *Nucl. Sci. Abstr.* **20**, 43 (1966).
44. S. Pizzini and R. Morlotti, *Electrochim. Acta* **10**, 1033 (1965).
45. F. L. Whiting, G. Mamantov, and J. P. Young, *J. Am. Chem. Soc.* **91**, 6531 (1969).
46. F. L. Whiting, G. Mamantov, and J. P. Young, *J. Inorg. Nucl. Chem.* **34**, 2475 (1972).
47. F. L. Whiting, G. Mamantov, G. M. Begun, and J. P. Young, *Inorg. Chim. Acta.* **5**, 260 (1971).
48. P. W. M. Jacobs and B. D. Bond, *J. Chem. Soc. A*, 1265 (1966).
49. G. J. Janz, *Molten Salts Handbook*, Academic, New York (1967).
50. M. Peleg, *J. Phys. Chem.* **71**, 4553 (1967).
51. P. G. Zambonin and J. Jordan, *J. Am. Chem. Soc.* **89**, 6365 (1967); **91**, 2225 (1969).
52. P. G. Zambonin and J. Jordan, *Anal. Lett.* **1**, 1 (1967).
53. P. G. Zambonin, *Anal. Chem.* **43**, 1571 (1971).
54. P. G. Zambonin and A. Cavaggioni, *J. Am. Chem. Soc.* **93**, 2854 (1971).
55. P. G. Zambonin, *J. Electroanal. Chem.* **33**, 243 (1971).

56. E. Desimoni, F. Paniccia, L. Sabbatini, and P. G. Zambonin, *Proceedings of the International Symposium on Molten Salts*, Ed. J. Braustein *et al.*, Electrochemical Society, New York (1976), p. 584.
57. E. Desimoni, F. Paniccia, and P. G. Zambonin, *J. Electroanal Chem.* **38**, 373 (1972).
58. E. Desimoni, F. Paniccia, and P. G. Zambonin, *J. Chem. Soc. Faraday Trans. 1* **69**, 2014 (1973).
59. P. Paniccia and P. G. Zambonin, *J. Chem. Soc. Faraday Trans. 1* **68**, 2083 (1972).
60. P. G. Zambonin, V. L. Cardetta, and G. Signorile, *J. Electroanal. Chem.* **28**, 237 (1970).
61. J. Schlegel and D. Priore, *J. Phys. Chem.* **76**, 2841 (1972).
62. J. Goret and B. Tremillon, *Bull. Soc. Chim. Fr.*, 97 (1966).
63. R. N. Kust and J. D. Burke, *Inorg. Nucl. Chem. Lett.* **6**, 333 (1970).
64. J. D. Burke and D. H. Kerridge, *Electrochim Acta* **19**, 251 (1974).
65. G. Bertozzi, *Z. Naturforsch* **22a**, 1748 (1967).
66. J. P. Frame, E. Rhodes, and A. R. Ubbelohde, *Trans. Faraday Soc.* **57**, 1075 (1961).
67. H. Haug and L. F. Albright, *Ind. Eng. Chem. Process Des. Develop.* **4**, 241 (1965).
68. S. N. Flengas and A. Block-Bolten, *Advances in Molten Salt Chemistry*, Vol 2, Ed. J. Braunstein, G. Mamantov, and G. P. Smith, Plenum, New York (1973), p. 55.
69. H. S. Swofford and H. A. Laitinen, *J. Electrochem. Soc.* **110**, 814 (1963).
70. P. G. Zambonin, *Anal. Chem.* **41**, 868 (1969).
71. G. J. Hills and P. D. Power, *J. Polar. Soc.* **13**, 71 (1967).
72. A. F. J. Goeting and J. A. A. Ketelaar, *Electrochim. Acta* **19**, 267 (1974).
73. D. Inman and J. Braunstein, *Chem. Commun.*, 148 (1966).
74. L. E. Topol, R. A. Osteryoung, and J. H. Christie, *J. Phys. Chem.* **70**, 2857 (1966).
75. P. G. Zambonin, *J. Electroanal. Chem.* **24**, 365 (1970).
76. T. R. Griffiths and P. J. Potts, *Inorg. Chem.* **14**, 1039 (1975).
77. H. S. Hull and A. G. Turnbull, *J. Phys. Chem.* **74**, 1783 (1970).
78. T. Kozlowski and R. F. Bartholomew, *J. Electrochem. Soc.* **114**, 937 (1967).
79. H. E. Bartlett and K. E. Johnson, *J. Electrochem. Soc.* **114**, 64 (1967).
80. A. J. B. Cutler, C.E.R.L. private communication (1978).
81. I. Vogel, *A Textbook of Practical Organic Chemistry*, Longmans, Green, London (1954).
82. F. Daniels, J. W. Williams, P. Bender, R. A. Alberty, C. D. Cornwell, and J. E. Harriman, *Experimental Physical Chemistry*, 7th ed. McGraw-Hill, New York (1970).
83. G. Landresse, *Anal. Chim. Acta* **56**, 29 (1971).
84. R. C. Howie and D. W. Macmillan *Inorg. Nucl. Chem. Lett.* **6**, 399 (1970).
85. P. D. Power, Ph.D. thesis, University of Southampton (1966).
86. T. K. Mukherjee, DIC thesis, Imperial College, London University (1976).
87. J. Phillips and R. A. Osteryoung, *J. Electrochem. Soc.* **124**, 1465 (1977).
88. S. Pizzini, R. Morlotti, and E. Romer, *J. Electrochem. Soc.* **113**, 1305 (1966).
89. G. L. Holleck and J. Giner, *J. Electrochem. Soc.* **119**, 1161 (1972).
90. G. F. Warren, Ph.D. thesis, London University (1976).
91. J. W. Mellor, *Supplement to Comprehensive Treatise on Inorganic and Theoretical Chemistry*, Vol. II, Supp. II, Halsted Press, New York (1961), p. 258.
92. S. H. White, unpublished work (1975).
93. W. G. Pfann, *Zone Melting*, Wiley, New York (1958).
94. K. E. Johnson and K. F. Denning, *High Temp Sci.* **3**, 283 (1971).
95. O. Kubaschewski, E. Le Evans, C. B. Alcock, *Metallurgical Thermochemistry*, Pergamon, New York (1967).
96. W. J. Hamer, M. S. Malmberg, and B. Rubin *J. Electrochem. Soc.* **103**, 8 (1956).
97. R. Combes, J. Vedel, and B. Tremillon, *C.R. Acad. Sci. (Paris)* **273**, 1740, (1967).
98. R. Combes, J. Vedel and B. Tremillon *C.R. Acad. Sci. (Paris)*, **275**, 199 (1972).
99. R. Combes, J. Vedel, and B. Tremillon, *Electrochim. Acta* **20**, 191 (1975).
100. D. L. Manning, *J. Electroanal. Chem.* **6**, 227 (1963).

101. P. E. Field and J. H. Shaffer, *J. Phys. Chem.* **71**, 3220 (1967).
102. B. F. Hitch and C. F. Baes, Jr., *Inorg. Chem.* **8**, 201 (1969).
103. H. G. Johansen, Å. Sterten, and J. Thonstad, Paper 260, Extended Abstracts 27th Meeting ISE, Zurich (1976).
104. H. A. Laitinen and J. A. Plambeck, *J. Am. Chem. Soc.* **87**, 1202 (1965).
105. J. A. Plambeck, *J. Chem. Eng. Data* **12**, 77 (1967).
106. D. L. Maricle and D. N. Hume, *J. Electrochem. Soc.* **107**, 354 (1960).
107. J. D. Van Norman and R. J. Tivers, *Molten Salts, Characteristics and Analysis*, Ed. G. Mamantov, Marcel Dekker, New York (1969), p. 509.
108. A. Yazawa and M. Kameda, *Can. Met. Q.* **6**, 263 (1967).
109. S. H. White and R. Rudkin, unpublished work (1978).
110. H. Kojima, S. G. Whiteway, and C. R. Mason, *Can. J. Chem.* **46**, 2968 (1968).
111. P. Bowles, *Advances in Extractive Metallurgy*, I.M.M., London (1968), p. 600.
112. J. Fischer, *J. Am. Chem. Soc.* **79**, 6363 (1957).
113. R. Spencer, Ph.D. Thesis, London University (1970).
114. G. Mamantov, in *Molten Salts*, Ed. G. Mamantov, Marcel Dekker, New York (1969), p. 529.
115. J. P. Wiaux, These de Doctorat, Universite Catholique de Louvain, Belgium (1975).
116. R. Coombes, J. Vedel, and B. Tremillon, *Anal. Lett.* **3**, 523 (1970).
117. G. Delarue, *J. Electroanal. Chem.* **1**, 13, 285 (1959).
118. J. Braunstein and J. D. Brill, "Silver Complexes," *J. Phys. Chem.* **70**, 1261 (1966).
119. G. A. Mazzocchin, G. G. Bombi, and G. A. Sacchetto, *J. Electroanal. Chem.* **21**, 345 (1959).
120. R. Combes, M. N. Levelut, and B. Tremillon, *Electrochim Acta.* **23**, 1291 (1978); see *J. Electroanal. Chem.* **91**, 125 (1978).
121. S. Vire, D. Inman, and S. H. White, unpublished work (1977).
122. S. H. White and D. R. Morris, *Physical Chemistry of Process Metallurgy, Richardson Conference*, Eds. J. H. E. Jeffes and R. Tait, I.M.M., London, (1974), p. 195.
123. C. Aksaranan, V. Dosai, D. R. Morris, and S. H. White, *Can. J. Chem.* **49**, 2014 (1971).
124. D. K. Roe, Ph.D. thesis, University of Illinois (1959); University Microfilms Inc., 594560, Ann Arbor, Michigan.
125. W. J. Walsh and H. Shimotake, *International Power Sources Symposium*, Ed. 1976, D. H. Collins Academic, New York (1977), p. 725.
126. G. L. Holleck and J. Giner, *J. Electrochem. Soc.* **119**, 1161 (1972).
127. G. J. Hills, D. Inman, and L. Young, VIII Reunion CITCE, Madrid (1956).
128. G. Landress and G. Duyakaerts, *Anal. Chim. Acta* **19**, 101 (1972).
129. G. F. Warren, S. H. White, and D. Inman, *Proceedings of the International Symposium on Molten Salts*, Electrochem. Soc., Ed. J. Braustein *et al.*, Electrochemical Society, New York (1976), p. 218.
130. R. J. Armstrong, M. Phil., City University, London (1968).
131. W. J. Hamer, M. S. Malmberg, and B. Rubin, *J. Electrochem. Soc.* **112**, 750 (1965).
132. H. E. Bartlett and K. E. Johnson, *Can. J. Chem.* **44**, 2119 (1966).
133. N. P. Yao, L. A. Hereday, and R. C. Saunders, *J. Electrochem. Soc.* **118**, 1039 (1971).
134. S. D. James, *J. Electrochem. Soc.* **122**, 921 (1975).
135. J. Thonstad, *Electrochim. Acta* **15**, 1569 (1970).
136. J. De Lepinay and M. J. Barbier, *J. Electroanal. Chem.* **45**, 419 (1973); **47**, 453 (1973).
137. H. A. Laitinen and C. H. Liu, *J. Am. Chem. Soc.* **80**, 1015 (1958).
138. Y. Kanzaki and M. Takahashi, *J. Electroanal. Chem.* **90**, 305, 313 (1978).
139. A. De Haan and H. Vanderpoorten, *Electrochim. Acta.* **19**, 519 (1974).
140. D. W. Townsend, *Proceedings of the International Symposium on Molten Salts*, Ed. J. Braunstein *et al.*, Electrochemical Society, New York (1976), p. 388.
141. M. S. Foster, S. E. Wood, and C. E. Crouthamel, *Inorg. Chem.* **3**, 1428 (1964).

142. M. F. Lantratov and B. I. Skirstymonskaya, *Russ. J. Phys. Chem. (Engl.)* **36**, 1323 (1962).
143. J. Thonstad, F. Nordmo, and J. K. Rodseth, *Electrochem. Acta.* **19**, 761 (1974).
144. L. Meites and S. A. Moros, *Anal. Chem.* **31**, 23 (1959).
145. G. Boisde, G. Chauvin, and H. Coriou, *Electrochim. Acta* **11**, 375 (1966).
146. D. G. Lovering, R. M. Oblath, and A. K. Turner, *J. Chem. Soc. Chem. Commum* 673 (1976).
147. S. H. White, unpublished work (1975).
148. J. K. Brimacombe, Ph.D. thesis, London University (1970).
149. S. M. Selis, *J. Phys. Chem.* **72**, 1442 (1968).
150. M. D. Ingram, B. Baron, and G. J. Janz, *Electrochim Acta* **11**, 1629 (1966).
151. A. J. B. Cutler and C. J. Grant, Eds. Z. A. Foroulis and W. W. Smeltzer, Electrochemical Society, New York (1975), p. 591.
152. S. H. White, unpublished work (1977).
153. R. J. Heus, T. Tidwell, and J. J. Egan, *Molten Salts, Characterization and Analysis*. Ed. G. Mamantov, Mancel Dekker, New York (1969), p. 499.
154. R. Huq, D. Inman, and S. H. White, unpublished work (1976).
155. H. C. Gaur and W. K. Behl, *Electrochim. Acta* **8**, 107 (1963).
156. G. Rubel and M. Gross, *Corrosion Sci.* **15**, 261 (1975).
157. A. Withagen-Declercq, Thèse de Doctorat, Université Catholique de Louvain (1975).
158. A. G. Graves, unpublished work (1969).
159. *Encyclopaedia of the Electrochemistry of the Elements* (Editor A. J. Bard), Vol. X: Fused salt system, James A. Plambeck, Marcel Dekker, New York (1976), pp. xxi and 440.
160. W. R. Grimes, D. R. Cuneo, F. F. Blankenship, G. W. Keilholtz, H. F. Poppendiek, and M. T. Robinson, *Fluid Fuel Reactors*, Ed. J. A. Lane *et al.*, Addison Wesley, Reading, Massachusetts (1958), p. 584.
161. M. Broc, G. Chauvin, and H. Coriou, *Molten Salts, Electrolysis for Metal Production*, IMM, London (1977), p. 69.
162. S. Senderoff, G. W. Mellors, and W. J. Reinhart, *J. Electrochem. Soc.* **112**, 840 (1965).
163. F. R. Clayton, G. Mamantov, and D. L. Manning, *J. Electrochem. Soc.* **120**, 1193 (1973).
164. D. Bratland, *Trans. Metal. Soc. AIME*, Proceedings, paper 76-11 (1976).
165. P. G. Zambonin, *J. Electroanal. Chem.* **32**, App. 1 (1971).
166. F. Desimoni, F. Paniccia, and P. G. Zambonin, *J. Electroanal. Chem.* **38**, 373 (1972).
167. J. Schlegel and C. Pitak, *J. Inorg. Nucl. Chem.* **32**, 2088 (1970).
168. I.I. Vol'nov, *Peroxides, Superoxides and Ozanides of Alkali and Alkaline Earth Metals*, Transl. Z. E. Woronocow, Plenum, New York (1966).
169. *Handbook of Chemistry and Physics*, Chemical Rubber Company, Cleveland, Ohio (1971).
170. R. Spencer, D. Inman, and S. H. White, unpublished work (1969).
171. W. O'Dean and R. A. Osteryoung, *Anal. Chem.* **43**, 1879 (1971).
172. M. M. Wong and F. P. Haver, *Molten Salt Electrolysis in Metal Production*, IMM, London (1977), p. 21.
173. J. R. Selman, D. K. Denuccio, C. J. Sy, and S. K. Stennenberg, *J. Electrochem. Soc.* **124**, 1160 (1977).
174. N. A. Krasilnikova, M. V. Smirnov, and I. H. Ozeryanaya, *Tr. Inst. Elektrokhim. Akad. Nauk SSSR Uralsk. Fil.* **14**, 3 (1970).
175. V. N. Devyatkin and E. A. Ukshe, *Zh. Prikl. Khim.* **38**, 1612 (1965); *J. Appl. Chem. USSR* **38**, 1574 (1965).
176. N. Q. Minh and B. J. Welch, *J. Electroanal. Chem.* **92**, 179 (1978).
177. V. N. Devyatkin and E. A. Ukshe, *Russ. Met.*, No. 3, 42 (1966).
178. K. H. Stern and J. A. Stiff, *J. Electrochem. Soc.* **111**, 893 (1964).
179. C. S. Tedmon and W. C. Hagel, *J. Electrochem. Soc.* **115**, 151 (1968).
180. S. Senderoff, G. W. Mellors, and R. I. Bretz, *J. Electrochem. Soc.* **108**, 93 (1961).

181. J. O. M. Bockris, E. H. Crook, H. Bloom, and N. E. Richards, *Proc. R. Soc. London* **255A**, 558 (1960).
182. A. De Guibert and V. Plichon, *J. Electroanal. Chem.* **90**, 399 (1978).
183. H. C. Gaur and W. K. Behl, *Electrochem. Acta.* **8**, 107 (1963); H. C. Gaur and H. L. Jindal, *Electrochim. Acta.* **13**, 835 (1968).
184. U. Cohen and R. A. Huggins, *J. Electrochem. Soc.* **123**, 381 (1976).
185. S. Pizzini and A. Magistris, *Electrochim Acta* **9**, 1189 (1964).
186. J. Devynck, B. Tremillon, M. Sloim, and H. Menard, *J. Electroanal. Chem.* **78**, 355 (1977).
187. C. R. Boston, in *Advances in Molten Salt Chemistry*, Vol. 1, Eds. J. Braunstein, G. Mamantov, and G. P. Smith, Plenum, New York (1973), p. 129.
188. T. L. Markin, in *Electromotive Force Measurements in High Temperature systems*, Ed. C. B. Alcock, IMM, London (1968), p. 91.
189. H. Monk and D. J. Fray, *Trans. I.M.M.* **83**, C118 (1974); **82**, C161, C240 (1973).
190. S. N. Flengas and T. R. Ingraham, *Can. J. Chem.* **36**, 1662 (1958).
191. H. A. Øye and D. M. Gruen, *Inorg. Chem.* **4**, 1173 (1965).
192. K. W. Fung and G. Mamantov, *J. Electroanal. Chem.* **35**, 27 (1972).
193. L. G. Boxall, H. L. Jones, and R. A. Osteryoung, *J. Electrochem Soc.* **121**, 212 (1974).
194. C. R. Boston, *J. Chem. and Eng. Data* **11**, 262 (1966).
195. B. Tremillon, A. Bermond, and R. Molina, *J. Electroanal. Chem.* **74**, 53 (1976).
196. J. H. R. Clarke and C. Solomons, *J. Chem. Phys.* **47**, 1823 (1967).
197. E. W. Dewing, *J. Electrochem. Soc.* **123**, 1289 (1976).
198. J. P. Saget, P. Homsi, V. Plichon, and J. Badoz-Lambling, *Electrochim. Acta* **20**, 819 (1975).
199. J. P. Saget, V. Plichon, and J. Badoz-Lambling, *Electrochim. Acta* **20**, 825 (1975).
200. C. E. Thalmeyer, S. Bruckenstein, D. M. Gruen, *J. Inorg. Nucl. Chem.* **26**, 347 (1964).
201. M. Fiorani, G. G. Bombi, and G. A. Mazzocchin, *J. Electroanal. Chem.* **13**, 167 (1967).
202. B. J. Brough, D. A. Kerridge, and M. Mosley, *J. Chem. Soc. A*, 1556, (1966).
203. U. Anders and J. A. Plambeck, *J. Electrochem. Soc.* **115**, 598 (1968).
204. K. E. Johnson and H. A. Laitinen, *J. Electrochem. Soc.* **110**, 314 (1963).
205. P. K. Lorenz and G. J. Janz, *Electrochim. Acta* **15**, 1025 (1970).
206. K. S. De Haas and K. F. Fouche, *Inorg. Chim. Acta* **26**, 213 (1978).
207. J. G. V. Lessing, K. F. Fouche, and T. T. Retief, *Electrochim. Acta* **22**, 391 (1977).
208. J. Braunstein, *Inorg. Chim. Acta Rev.* **2**, 19 (1968).
209. D. L. Manning and G. Mamantov, *J. Electrochem. Soc.* **124**, 480 (1977).
210. J. Braunstein, "Statistical thermodynamics of molten salts and concentrated aqueous electrolytes," in *Ionic Interactions*, Ed. S. Petrucci, Academic, New York (1971).
211. D. Inman, D. G. Lovering, and R. Narayan, *Trans. Faraday Soc.* **63**, 3017 (1967); **64**, 2476, 2487 (1968).

Hydrogen in Ionic Liquids: A Review

Pier Giorgio Zambonin, Elio Desimoni, Francesco Palmisano, and Luigia Sabbatini

1. Introduction

The evolution of systems for the production of energy envisages hydrogen as one of the most promising new fuels. Thus, a "hydrogen economy"[1-5] appears to be one of the best alternatives for the current "fossil fuel economy." A growing number of investigations are being devoted to developing ways and means of exploiting this energy source; specialist reviews and texts consider these advances.[4-11]

The use of hydrogen in fuel cells appears, in theory, quite promising, especially when molten salt electrolytes are employed. The properties of such solvents and their influence on chemical and electrochemical processes bring about unusual reaction paths and produce enhanced reaction rates. In spite of these factors, the behavior of hydrogen in molten salts has never been the subject of review. The currently available literature on the chemistry and electrochemistry of hydrogen in various melts is, therefore, collected within this chapter.

The results of recent systematic investigations in molten nitrates receive particular attention. Work carried out in the authors' laboratory, combining several techniques, represents a unique attempt to study the various aspects of this topic in a quantitative fashion, including solubility, physicochemical interactions of hydrogen and solvent, and chemical and electrochemical mechanisms. These considerations can usefully be related to previous, extensive investigations of oxygen in the same media,[12-29] since oxygen is the ideal coreactant for hydrogen in a fuel cell. Previously

Pier Giorgio Zambonin, Elio Desimoni, Francesco Palmisano, and Luigia Sabbatini • Istituto di Chimica Analitica, Università degli Studi, Via Amendola 173, Bari, Italy.

published data are summarized in figures and tables, whilst an appendix describes preliminary attempts to characterize the upper layers of electrode surfaces which have been in contact with molten nitrates, using X-ray photoelectron spectroscopy (XPS). Examples indicate how XPS can lead to a better understanding of electrode processes.

2. Acetates

Hydrogen systems in molten acetates have been studied by Marassi *et al.*[30-32] Preliminary experiments,[30] performed at 523 K in the (Na, K) acetate eutectic melt by oscillating platinum microelectrode voltammetry, permitted the identification of waves due to the reduction of water and acetate ions, according to the following equations:

$$2H_2O + 2e = H_2 + 2OH^- \tag{1}$$

$$2CH_3COOH + 2e = H_2 + 2CH_3COO^- \tag{2}$$

In both cases, composite waves were obtained when (hydrogen + hydroxide + water) or (hydrogen + acetic acid) were added to the melt. The results were confirmed by the chronopotentiometric studies at platinum electrodes.

In the course of a subsequent investigation[31] performed in the same melt in the temperature range 523–563 K, reaction (1) was found to be a reversible, diffusion-controlled process at bright platinum electrodes, in spite of the reaction of hydroxide ions with the solvent. The Henry's law constant, the heat of solution and the standard entropy for the dissolution of water, and the water diffusion coefficient were calculated.

In a further study[32] the electrochemical behavior of hydrogen was investigated in detail in the temperature range 523–573 K. Cyclic voltammetric work performed at bright platinum electrodes in hydrogen-saturated melts established the reversibility of the electro-oxidation reaction:

$$H_2 + 2CH_3COO^- = 2CH_3COOH + 2e \tag{3}$$

as well as the inverse of reaction (1). Nonreproducible results were obtained when using vitreous carbon, gold, and tungsten electrodes, probably because of overvoltage effects. The Henry's law constant, the heat of solution, and the standard entropy were obtained from hydrogen solubility data (see Table 1). No hydrogen-consuming reaction was observed. The diffusion coefficient for hydrogen was estimated to be 5×10^{-4} cm^2 sec^{-1} at 523 K in the given solvent (see Table 2).

TABLE 1

Solvent	Species	T, K	Solubility, mol cm^{-3} atm^{-1}	ΔH, kJ mol^{-1}	$\Delta S°$, J K^{-1} mol^{-1}	Ref.
(Na, K)CH$_3$CO$_2$ (46.3–53.7) mol %	H$_2$	530	3.29×10^{-7}	8.2	-19.8 at 523 K	32
		548	3.43×10^{-7}			
		566	3.74×10^{-7}			
		573	3.76×10^{-7}			
(Li, K, Na)$_2$CO$_3$ eutectic	H$_2$	873	1.8×10^{-6}			59
(Li, K)Cl eutectic	HCl	763	$(1.20 \pm 0.14)10^{-6}$	14.65		79
		842	$(1.44 \pm 0.10)10^{-6}$			
		948	$(1.74 \pm 0.11)10^{-6}$			
		677	1.04×10^{-6}			76
		735	1.27×10^{-6}			
		793	1.56×10^{-6}			
(Na, K)Cl (50–50) mol %	HCl	1023	2.1×10^{-6}	9.80 ± 0.17		81
NaCl	HCl	1203	1.0×10^{-6}	-35.16 ± 0.50		81
KCl	HCl	1173	2.4×10^{-6}	-13.10 ± 0.33		
RbCl	HCl	1103	4.2×10^{-6}	5.80 ± 0.34		
CsCl	HCl	1020	9.8×10^{-6}	17.79 ± 0.28		
LiF–BeF$_2$ (64–36) mol %	H$_2$	873	$(4.34 \pm 0.20)10^{-8}$			95
Li$_2$BeF$_4$	H$_2$	773	$(1.78 \pm 0.13)10^{-8}$			94
		873	$(4.42 \pm 0.12)10^{-8}$			
		973	$(4.84 \pm 0.46)10^{-8}$			
	D$_2$	773	$(2.22 \pm 0.13)10^{-8}$			
		873	$(2.82 \pm 0.22)10^{-8}$			
		973	$(5.33 \pm 0.55)10^{-8}$			
	H$_2$ + D$_2$[a]	773	$(2.02 \pm 0.24)10^{-8}$	29.39 ± 7.53 at 1000 K	-14.65 ± 8.79	
		873	$(4.17 \pm 0.29)10^{-8}$			
		973	$(5.08 \pm 0.56)10^{-8}$			
(Na, K)OH (51–49) mol %	H$_2$	500	2.24×10^{-8}			112
(Na, K)NO$_3$ equimolar mixture	H$_2$	508	1.28×10^{-7}	14	-22 at 533 K	122
		533	1.50×10^{-7}			
		573	1.88×10^{-7}			
		603	2.22×10^{-7}			

[a] H$_2$ and D$_2$ results considered identical.[94]

TABLE 2

Solvent	Species	T, K	$D \times 10^4$ cm^2 sec^{-1}	Technique	Ref.
(Na, K)CH$_3$CO$_2$ (46.3–53.7) mol %	H$_2$	523	5.0	Potential step voltammetry	32
(Li, Na)$_2$CO$_3$ eutectic	H$_2$	783	4.0	Chronopotentiometry and linear sweep voltammetry	60
		823	4.4		
		873	5.3		
		923	6.2		
		973	6.7		
(Li, K)Cl eutectic	H$^+$	763	2.4	Chronopotentiometry	79
		843	2.3		
		723	2.1	Linear sweep voltammetry	78
		677	2.0	Linear sweep voltammetry	76
		735	2.4		
		793	2.0		
		677	1.8	Chronopotentiometry	
		735	2.7		
		793	2.1		
(Na, K)NO$_3$ equimolar mixture	H$_2$	513	> 0.78	Rotating disk electrode	146

3. Bisulfates

Most information in the literature on hydrogen in molten bisulfates concerns kinetic studies of hydrogen evolution and dissolution during melt electrolysis.

3.1. Hydrogen Electrodes in Bisulfate Melts

A recent paper by Tremillon et al.[33] describes a glass electrode sensitive to hydrogen ions which can be used in bisulfate melts. pH variations in molten KHSO$_4$ at 493 K have been measured and titration curves for sulfate ion–sulfuric acid are reported. The value of the autoprotolysis product for pure molten KHSO$_4$ has been evaluated.

A reversible hydrogen electrode[34] was characterized by studying the following reversible galvanic cell (without liquid junction):

$$\text{platinized Pt}_{(s)}, \text{H}_{2(p=1)}, \text{KHSO}_{4(l)}, \text{O}_{2(g)}, \text{SO}_{3(g)}, \text{Pt}_{(s)}$$

which can be formed by interrupting the electrolysis of a pure bisulfate melt performed under appropriate experimental conditions. The following half-cell reactions occur:

anodic: $$H_2 = 2H^+ + 2e \qquad (4)$$

cathodic: $$H_2SO_4 + SO_3 + \tfrac{1}{2}O_2 + 2e = 2HSO_4^- \qquad (5)$$

or $$SO_3 + \tfrac{1}{2}O_2 + 2e = SO_4^{2-} \qquad (6)$$

The authors discuss the possibility of using this half-cell as a reference electrode.

3.2. Cathodic Hydrogen Evolution

Tajima *et al.*[35] first observed that, during the electrolysis of molten potassium bisulfate, hydrogen and oxygen are formed on the electrodes in the ratio 2 : 1. The overall reaction occurring in the electrolysis cell at low overvoltage is

$$2KHSO_4 = K_2SO_4 + SO_3 + \tfrac{1}{2}O_2 + H_2 \qquad (7)$$

The cathodic hydrogen evolution occurring during electrolysis of a bisulfate melt is considered in terms of the following partial reactions:

$$M + H^+ + e = M\text{---}H \qquad (8)$$

$$M\text{---}H + M\text{---}H = H_2 + 2M \qquad (9)$$

$$M\text{---}H + H^+ + e = H_2 + M \qquad (10)$$

in which step (8) may be followed by step (9) or, alternatively, (10). Kinetic investigations of hydrogen evolution performed by Shams El Din[36] and analyzed using Tafel plots and open circuit overpotential decay measurements suggested that, on bright platinum, step (9) was rate determining. On the contrary, Arvia and co-workers deduced that on both bright platinum and graphite electrodes[37] the slow step was the initial discharge (8), while the rate-determining step is the atom recombination reaction (9) on platinum black.[34] Subsequently, Gilroy[38] stated that on bright platinum the rate-determining step was the atom–atom recombination at low overpotentials and the ion–atom electrochemical desorption [reaction (10)] at high overpotentials. The two regions were separated by a transition region in which hysteresis was observed. These findings were subsequently confirmed (mainly by potentiondynamic measurements) by Arvia *et al.*,[39-41] who additionally ascribed the changeover in mechanism to the formation, at the reaction interface, of sulfide species arising from the reduction of melt components. If the latter surmise is true, then hydrogen evolution would be appreciably hindered and this may explain the observed change in the Tafel slope from $RT/2F$ to $2RT/F$. Thus the initial ion discharge reaction would have appeared as the rate-determining step.

Further electrochemical studies have been performed on both dense[42, 43] and porous[44] graphite electrodes. The experimental cathodic Tafel slopes and the kinetic parameters deduced from cathodic overvoltage decay suggest that either reaction (9) or (10) can be rate determining in an activated adsorption process but they do not permit an unambiguous definition of the reaction mechanism.

The cathodic evolution of hydrogen on gold[36, 45] resembles that in aqueous acid solutions at ordinary temperatures and consequently it has been discussed in terms of a mechanism involving fast hydrogen ion discharge followed by an adatoms recombination reaction as rate-determining step.

Hydrogen evolution has also been studied on palladium.[46] In this case, the rate-determining step in the intermediate overvoltage region of the steady-state current–voltage curves is adatom recombination, while at higher overvoltages the ion–adatom reaction becomes the slow step.

3.3. Anodic Hydrogen Dissolution

The kinetics of hydrogen electro-oxidation in molten bisulfates was studied by conventional steady-state and transient techniques on platinum,[39, 41] graphite,[42–44] gold,[45] and iron.[47]

The results obtained on platinum electrodes at 458 K were interpreted[41] on the basis of a mechanism involving two different sites (α and β) on the electrode surface, i.e.,

$$(H_2)_b = (H_2)_e \tag{11}$$

$$(H_2)_e + xPt_\alpha + yPt_\beta = x(H)Pt_\alpha + y(H)Pt_\beta \tag{12}$$

$$(H)Pt_\alpha = (H^+)_e + Pt_\alpha + e \tag{13}$$

$$(H)Pt_\beta = (H^+)_e + Pt_\beta + e \tag{14}$$

$$(H^+)_e = (H^+)_b \tag{15}$$

$$\overline{\text{overall} \qquad (H_2)_e = 2(H^+)_e + 2e} \tag{16}$$

where b = bulk, e = electrode, $0 \leq x \leq 2$, $0 \leq y \leq 2$, and $(H)Pt_\alpha$ and $(H)Pt_\beta$ are hydrogen adatoms weakly and strongly bonded to the electrode surface. The kinetic parameters for some of the reported reactions were evaluated from anodic current–potential profiles obtained under appropriate potentiodynamic conditions.

The results obtained on dense[42, 43] and porous[44] graphite electrodes were interpreted on the basis of different reaction mechanisms. The most probable scheme involves a competition between the ionization of hydrogen atoms adsorbed on the electrode surface and the oxidation of the graphite itself. Independent support for the latter hypothesis is obtained from the mechanism deduced[48] for the anodic oxidation of graphite in the same melt.

Only preliminary, inconclusive results have been given for hydrogen dissociation on gold[45] and iron.[47]

4. Carbonates

Most information concerning hydrogen behavior in molten carbonates has been gained from high-temperature, H_2–O_2 fuel cell applications (see for example References 10, 11, 49–52).

Interesting information about the chemistry of hydrogen in molten carbonates was indirectly obtained from electrolysis studies of the melt in the presence of moisture and/or hydroxide ions.

Lorenz and Janz[53] report the results of massive electrolysis performed in the $(Li, Na, K)_2CO_3$ eutectic melt in the temperature range 833–1153 K, using a compartmented cell combined with an on-stream gas-chromatographic apparatus.

The authors confirmed that the cathodic discharge of pure carbonates occurs (as previously[54–56] postulated and thermodynamically predicted) according to the following processes:

$$CO_3^{2-} + 2e = CO + 2O^{2-} \qquad (17)$$

$$2CO_2 + 2e = CO + CO_3^{2-} \qquad (18)$$

$$CO_3^{2-} + 4e = C + 3O^{2-} \qquad (19)$$

The carbon dioxide involved in reaction (18) originates from melt dissociation:

$$CO_3^{2-} = CO_2 + O^{2-} \qquad (20)$$

or from additional equilibrium between melt anions and water impurities

$$CO_3^{2-} + H_2O = CO_2 + 2OH^- \qquad (21)$$

Gas-chromatographic analysis of carbon monoxide in the effluent gas stream indicated the presence of a very small contribution from reaction (18). Carbon monoxide, mainly produced by reaction (17), led to hydrogen evolution when water and/or hydroxide impurities were present in the melt according to the reactions

$$CO + 2OH^- = CO_3^{2-} + H_2 \qquad (22)$$

$$CO + H_2O = CO_2 + H_2 \qquad (23)$$

The effects of hydrogen-consuming reactions such as

$$CO + H_2 = C + H_2O \tag{24}$$

$$CO + 3H_2 = CH_4 + H_2O \tag{25}$$

are apparently negligible under these experimental conditions.

Arkhipov et al.[57] studied the electroactivity of hydrogen by voltamme-tric techniques, using hollow palladium tube electrodes (through which hydrogen can diffuse into the electrolyte) under the following experimental conditions: temperatures between 773 and 1073 K, hydrogen pressure be-tween 0.1 and 2.5 atmospheres, and electrode wall thickness of 20–500 μm. The diffusion of hydrogen through the palladium walls was found to be the slow step.

Bawa and Truitt[58] studied the electro-oxidation of hydrogen on wire screen nickel electrodes in the $(Li, Na)_2CO_3$ electrolyte at 933 K. They suggested that hydrogen atoms formed on the electrode surface

$$H_2 = 2H_{(ads)} \tag{26}$$

can diffuse to the reaction sites and react according to the process

$$2H_{(ads)} + CO_3^{2-} = H_2O + CO_2 + 2e \tag{27}$$

Volgin et al.[59, 60] studied the electro-oxidation of hydrogen on smooth platinum electrodes in various carbonate eutectics in the temperature range 783–973 K. Hydrogen diffusion coefficients obtained by chronopoten-tiometric and linear potential sweep methods are given at various tempera-tures (see Table 2). The solubility of molecular hydrogen in the $(Li, K, Na)_2CO_3$ eutectic at 873 K was measured[59] (see Table 1) and the validity of Henry's law verified.

Broers et al.[61] found that the polarization of hydrogen (Au) anodes in molten carbonates results primarily from the accumulation of CO_2 and H_2O which are the products of the following electrode reaction:

$$H_2 + CO_3^{2-} = H_2O + CO_2 + 2e \tag{28}$$

The authors assumed that the reaction was close to equilibrium at all current densities, contrary to the subsequent findings of Busson et al.,[62] which found discrepancies between the calculated thermodynamic poten-tials and the experimental open circuit potentials (o.c.p.). The reasons for these conflicting results were identified by Vogel and Iacovangelo,[63] who recently performed o.c.p. measurements in $(Li, K)_2CO_3$ at 923 K using gold electrodes for the electrooxidation of hydrogen and other fuel gases. Under suitable experimental conditions they found that the o.c.p. data usually agreed with the corresponding thermodynamically calculated values. Exceptions were those experiments in which equilibrium could only be attained close to the complete oxidation of excess methane in the feed gas.

5. Chlorides

The first work performed on hydrogen electrodes in chloride melt was reported in a brief review[64] by Delimarskii and Markov; it concerned some investigations in fused carnallite, aluminum chloride, and bromide, utilizing the system $H_3O^+/H_2(g)$. The work has mainly historical value and deals with the possibility of obtaining hydrogen electrode potential values useful for relating the electrochemical series in molten salts to that in aqueous solution. The experiments were performed in strongly hygroscopic salts which may retain their water when molten.

More recently, Boxall and Johnson[65] constructed an emf series suitable for relating electrode potentials in molten alkali metal salts to the hydrogen electrode at 25°C. In the same paper the potential of the $H^+(HCl)/H_2$ electrode in the (Li, K)Cl eutectic was derived versus the Cl_2/Cl^- reference half-cell. Standard potentials at different temperatures were also reported.

Takehara and Yoshizawa[66] carried out a basic study on H_2-Cl_2 fuel cells in the (Li, K)Cl eutectic. They compared the results which can be obtained at low temperature (by using hydrochloric acid as an electrolyte) with those in molten chlorides. Mechanisms for the high-temperature-type cells were formulated and discussed.

Barde *et al.*[67-69] studied hydrogen electro-oxidation on palladium diffusion electrodes at 773 K in pure chloride melts and in several molten mixtures of chlorides, hydroxides, carbonates, and oxides over a large temperature range. The authors[68] found that the rate of the electrochemical process is controlled by the diffusion of hydrogen through the palladium walls. Furthermore, they studied the effects of the electrode wall thickness and of the potassium hydroxide concentration on hydrogen current–potential profiles. According to the authors high current densities, ranging between 0.5 and 1 A/cm^2 can be obtained[69] under appropriate experimental conditions.

Polart *et al.*[70-73] compared the behavior of hydrogen electrodes at platinum, platinum–rhodium, platinum–iridium, and palladium in the (Li, K)Cl eutectic melt. For all the systems studied the potential at which the electrode reaction occurs is controlled by the diffusion of hydrogen through the metal to the reaction zone. Because of its high permeability to hydrogen, palladium offers the most efficient conversion.[70] Hydrogen electro-oxidation was found[71] to be influenced by the presence of oxyanions (i.e., O^{2-}, OH^-, CO_3^{2-}). The authors obtained current values of 800 mA cm^{-2} at 873 K under polarization of 200 mV on dipping, nonporous palladium sheets in (Li, K)Cl eutectic containing 0.5 mol/l CaO.[73]

An $H^+(HCl)/H_2$ hydrogen electrode suitable for use in anhydrous (Li, K)Cl melt was studied by Laitinen and Plambeck[74]; a purified mixture of H_2 + HCl was bubbled over platinized platinum foils dipped into the melt. The hydrogen system proved potentiometrically reversible at 723 K under totally anhydrous conditions. Previously, Littlewood and Argent[75] could not obtain equilibrium values for the same system in the (Na, K)Cl equimolar melt at 973 K, probably because of the comtamination by oxide species.

In the paper by Laitinen and Plambeck[74] some thermodynamic data were reported. For the half-cell reaction

$$HCl + e = \tfrac{1}{2}H_2 + Cl^- \tag{29}$$

the authors obtained a standard potential value $E° = -0.6935 \pm 0.0047$ V (vs. a Pt/Pt^{2+} 1.0 M reference) while for the standard e.m.f. of the overall reaction

$$\tfrac{1}{2}H_2 + \tfrac{1}{2}Cl_2 = HCl \tag{30}$$

they calculated $\Delta E^0 = 1.0161 \pm 0.0050$ V. The relevant free energy, enthalpy, and entropy values are in agreement with NBS and JANAF data.

Recently the system $H^+(HCl)/H_2$ was studied by Minh and Welch[76, 77] using chronopotentiometry and linear sweep voltammetry on both platinum and glassy carbon electrodes. According to the authors the dissolved HCl is completely dissociated and cathodic evolution of hydrogen proceeds via a mechanism consisting of a reversible, one-electron transfer step followed by a slow atom–atom recombination reaction[77]:

$$H^+ + e = H \tag{31}$$

$$H + H = H_2 \tag{32}$$

Three types of hydrogen electrodes, i.e., $H^+(HCl)/H_2$; $H_2O/H_2, OH^-$; and $OH^-/H_2, O^{2-}$ were studied by Kanzaki and Takahashi[78] by voltammetric techniques on platinum electrodes in the (Li, K)Cl eutectic melt at 723 K. They found that the three systems are voltammetrically reversible, and in all cases they formulated mechanisms characterized by one-electron transfer reactions:

$$H^+(HCl)/H_2: \qquad 2(H^+ + e = H) \tag{31}$$

$$2H = H_2 \tag{32}$$

$$\overline{}$$

$$2H^+ + 2e = H_2 \tag{33}$$

$$H_2O/H_2, OH^-: \qquad 2(H_2O + e = H + OH^-) \tag{34}$$

$$2H = H_2 \tag{32}$$

$$2H_2O + 2e = H_2 + 2OH^- \qquad (35)$$

$$OH^-/H_2, O^{2-}: \qquad 2(OH^- + e = H + O^{2-}) \qquad (36)$$

$$2H = H_2 \qquad (32)$$

$$\overline{2OH^- + 2e = H_2 + 2O^{2-}} \qquad (37)$$

The standard potentials for the three systems, versus a Ag/AgCl $(0.1 \text{ mol kg}^{-1})$ reference half-cell, are reported as well as the diffusion coefficient for $H^+(HCl)$; see Table 2. The dissociation constants for water and hydroxide under the specified experimental conditions were calculated to be

$$|H^+||OH^-|/|H_2O| = 10^{-6.08} \qquad (38)$$

$$|H^+||O^{2-}|/|OH^-| = 10^{-10.90} \qquad (39)$$

The solubility of $H^+(HCl)$ was determined by Minh and Welch[76] and by Van Norman and Tivers[79] in the (Li, K)Cl eutectic melt and by Ukshe and Devyatkin[80, 81] in single chlorides and in various (Na, K)Cl mixtures. The experimental method used in all cases was the gas-stripping technique. A guide to the solubilities obtained is given in Table 1.

The diffusion coefficient for $H^+(HCl)$ was calculated from chrono-potentiometric and linear sweep voltammetric data by Minh and Welch[76] and by Van Norman and Tivers[79] on platinum and glassy carbon, respectively. Their experimental values are in good agreement and are reported in Table 2.

Some examples of hydrogenation reactions effected by molecular hydrogen or metallic hydrides in various molten chlorides are reported in a review by Sundermeyer.[82]

6. Chloroaluminates

The behavior of hydrogen in molten sodium tetrachloraluminate has been studied by Tremillon *et al.*[83, 84] In this medium acids and bases are conveniently defined as acceptors or donors of Cl^-, respectively, and then the melt may be defined as acidic or basic when $AlCl_3$ or NaCl, respectively, are present in excess with respect to the pure equimolar mixture. The species H^+ is a strong acid which can exist only as HCl, irrespective of the acidity of the melt, because of the reaction

$$2AlCl_4^- = Al_2Cl_7^- + HCl \qquad (40)$$

for which the equilibrium lies completely to the right.

The following hydrogen systems were studied[83] at a platinum rotating microelectrode in molten $NaAlCl_4$, at a temperature of 448 K under both basic and acidic conditions:

$$\text{(basic)} \qquad H_2 + 2Cl^- = 2HCl + 2e \qquad (41)$$

$$\text{(acidic)} \qquad H_2 + 4AlCl_4^- = 2HCl + 2Al_2Cl_7^- + 2e \qquad (42)$$

On the evidence of experimental current–potential profiles the authors concluded that in both cases these hydrogen systems are voltammetrically reversible.

Cyclic and potential-step voltammetric experiments[84] performed in the same melt at 483 K showed that two cathodic waves could be obtained upon bubbling partially wet hydrochloric acid through the melt. The height of the waves could be related to the HCl and H_2O partial pressures, respectively, thus indicating that the waves were attributable to the electroreduction of dissolved HCl [see reaction (41)] and water, respectively. In this last case the electrode reaction suggested was

$$H_2O + 2e = H_2 + O^{2-} \qquad (43)$$

where the species O^{2-} certainly participates in further equilibria.

The standard potentials assigned to the systems studied are $E^\circ_{(41)} = 1.19$ V and $E^\circ_{(43)} = 0.80$ V vs. a reference electrode consisting of an aluminium wire dipped in NaCl saturated melt connected to the solution via a porous Pyrex membrane.

On the basis of experimental results the equilibrium constant for the reaction

$$AlCl_4^- + H_2O(g) = 2HCl(g) + AlOCl_2^- \qquad (44)$$

was calculated to be

$$K = \frac{[AlOCl_2^-] \cdot p_{HCl}^2}{p_{H_2O}} = 10^{-8.1} \text{ atm mol kg}^{-1} \qquad (45)$$

This value suggests the possibility of eliminating oxide species (i.e., $AlOCl_2^-$ ions) from the melt by sparging with dry HCl.

Finally, the authors verified[83] the possibility of producing molecular hydrogen directly by adding metallic aluminum to an acidic or basic tetrachloroaluminate solvent, according to the following reactions:

$$2Al + 6H^+ + 8Cl^- = 2AlCl_4^- + 3H_2 \qquad (46)$$

and

$$2Al + 6H^+ + 14AlCl_4^- = 8Al_2Cl_7^- + 3H_2 \qquad (47)$$

7. Fluorides

Hydrogen evolution from molten fluorides was studied by Pizzini *et al.*[85-87] under various experimental conditions. Overvoltage measurements at bright platinum electrodes have been undertaken in order to study[85] the reduction of hydrogen at 523 K, in molten potassium difluorides containing different water concentrations. It was found that hydrogen evolution occurs according to the following reactions:

$$(\alpha) \qquad HF + e = \tfrac{1}{2}H_2 + F^- \qquad (48)$$

$$(\beta) \qquad H_3O^+ + e = \tfrac{1}{2}H_2 + H_2O \qquad (49)$$

The limiting current observed for reaction (48) proved independent of the water content, whereas the limiting current for reaction (49) was proportional to the water concentration. In the current density region in which the transition from reaction α to β occurs, the electrode can be treated as a "bifunctional electrode"[88, 89] for which the coupling reaction in solution is

$$H_2O + HF = H_3O^+ + F^- \qquad (50)$$

A subsequent kinetic investigation of reactions (48) and (49) has been reported by Pizzini and Magistris[86] using galvanostatic overvoltage measurements. These authors found that the diffusion current for reaction (48) was limited by the slow dissociation of the solvent anions

$$HF_2^- = HF + F^- \qquad (51)$$

In turn, reaction (51) can influence the rate of the electrode process (49) because of the concomitant equilibrium (50). The overvoltage observed in the current density range where reaction β occurs was explained in terms of a Volmer–Tafel-type mechanism in which the slow charge-transfer steps were

$$H_3O^+ + e = \tfrac{1}{2}H_2 + H_2O_{ads} \qquad (52)$$

$$H_3O^+ + e = H_{ads} + H_2O_{ads} \qquad (53)$$

and the slow chemical step was the desorption of water from the electrode surface:

$$H_2O_{ads} = H_2O_{soln} \qquad (54)$$

Hydrogen (and oxygen) evolution reactions at bright platinum electrodes at 873 K were studied by Pizzini and Morlotti[87] in the (Li, K, Na)F eutectic mixture containing several oxide species (O^{2-}, OH^-, H_2O). The results accounted for the high value of the solubility of water and hydrofluoric acid[90] in molten alkali fluorides. Hydrogen evolution apparently occurs according to the same mechanistic pathways as in the difluoride melts.

Baes *et al.*[91–93] studied the HF/H_2, F^- system in the Li_2BeF_4 melt over the temperature range 773–1173 K. The system proved to be reversible on palladium tube[91] and platinum gauze[92] electrodes and it could, therefore, be employed as a "reference" in fluoride-containing melts.

The solubility of hydrogen and deuterium in molten Li_2BeF_4 has been measured by Malinauskas and Richardson[94] using a two-chamber gas-stripping apparatus.[95] These workers found that Henry's law was verified within the temperature range 773–1073 K and that the "combined solubility" (expressed as K_c = gas concentration in the dissolved state/gas phase concentration) of hydrogen and deuterium can be represented by

$$\log(10^3 K_c) = \log T - (1535/T) - 0.7684 \tag{55}$$

See Table 1 for complete solubility data.

8. Hydroxides

Literature information on hydrogen in molten hydroxides appears rather scant if one considers the potentialities that such media can have, at least, in theory, in the field of fuel cells (cf. the wide applications of concentrated hydroxide solutions, at relatively high temperature and pressure, as solvent for efficient "alkaline fuel cells"[5, 8, 10, 96–103]).

The general chemical properties and the acid–base characteristics of these melts have been described[104–106] by Goret and Tremillon; some important aspects are considered in Chapter 15 by Combes. The dissociation equilibrium of fused hydroxides can be represented by

$$2OH^- = H_2O + O^{2-} \tag{56}$$

for which the ionic product has been estimated to be

$$K_i = [H_2O][O^{2-}] = 10^{-11.6} \tag{57}$$

for the (Na, K)OH eutectic at a temperature of 500 K.

The Bronsted acid–base concept has been retained in molten hydroxides, so that a substance acting as proton donor (with respect to OH^-) and forming H_2O according to

$$HA + OH^- = H_2O + A^- \tag{58}$$

is defined as an "acid," while a proton-acceptor species

$$B + OH^- = BH^+ + O^{2-} \tag{59}$$

is defined as a "base."

Analogous to the pH (pH_3O^+) scale employed in aqueous solutions, a pH_2O scale has been formulated for molten hydroxides, so that the

(Na, K)OH eutectic melt at 500 K is "acidic" or "basic" when the pH_2O is, respectively, lower or higher than 5.8.

In the course of a voltammetric investigation at a platinum rotating wire electrode in the (Na, K)OH eutectic melt at 493 K, Goret[104, 105] identified a number of redox reactions involving hydrogen which were related to the acidity and the composition of the melt. In acidic melts, water can be reduced according to the electrode process

$$2H_2O + 2e = H_2 + 2OH^- \tag{60}$$

which is voltammetrically reversible.

Similar results have been obtained by Kern *et al.*[107, 108] on palladium tube electrodes in the same solvent. Under these experimental conditions, the limiting current was controlled by the diffusion of hydrogen through the metal walls.

In his thesis,[105] Goret reports a standard potential value of 0.78 ± 0.05 V (vs. a "standard sodium" electrode) for the reduction of water, deduced from the $E_{1/2}$ value of process (60). Subsequently, Tremillon[109] determined a value of 0.82 V, vs. the same reference, on the basis of an E vs. pH_2O diagram. For the same electrode process, Zecchin *et al.*,[110] on the evidence of chronopotentiometric data, found a value of -0.185 ± 0.02 V vs. a Cu(I)/Cu reference.[111] After conversion to the "sodium" reference scale, the value reported by Zecchin *et al.* was practically coincident[110] with the value given by Goret.

No hydrogen evolution was observed[105] due to solvent anion reduction arising from

$$2OH^- + 2e = H_2 + 2O^{2-} \tag{61}$$

because solvent cations are more easily reduced than OH^- itself. When metallic sodium is added to acidic melts, water is consumed[104] according to the process

$$2H_2O + 2Na = H_2 + 2Na^+ + 2OH^- \tag{62}$$

and the hydrogen produced can react with the excess metallic sodium to form sodium hydride. NaH may then undergo an oxidation process (following the oxidation of the residual sodium) according to the reaction

$$2NaH = 2Na^+ + H_2 + 2e \tag{63}$$

Reaction (63) was also observed when metallic sodium was added to a hydrogen-containing anhydrous melt. Additionally, cathodic evolution of hydrogen was obtained by electroreduction of ammonia dissolved in the melt according to

$$NH_3 + e = \tfrac{1}{2}H_2 + NH_2^- \tag{64}$$

The solubility of hydrogen in acidic hydroxide melts, at 500 K, was measured (see Table 1) by Eluard[112] using the coulometric method developed[113] by Vogel and Smith. His experimental value is in good agreement with the one determined[114] in aqueous KOH at 473 K.

9. Other Melts

Few papers have been published on hydrogen systems in fused organic salts. The potentiometric behavior of the half-cell

$$(Pt)H_{2(1 \text{ atm})}/(CH_3)_2NH_2^+Cl^- \tag{65}$$

in molten dimethylammonium chloride and other aliphatic ammonium chlorides has been studied by Kisza[115] at 444 K. The electrode proved potentiometrically reversible in the solvents tested so that its use as a standard reference electrode for organic melts seems feasible. The solubility product for HCl in fused dimethylammonium chloride was measured.[116]

Kisza et al.[117, 118] carried out a potentiostatic study of hydrogen evolution in alkylammonium chlorides and showed that at bright platinum electrodes the process is not strictly diffusion controlled. The relevant electrode mechanism was found to be dependent upon the applied potential and on the working temperature, the authors examined[119] how the previously obtained electrochemical results for hydrogen evolution could fit Volmer-, Heyrovsky-, and Tafel-type mechanistic models.

A platinum rotating disk electrode voltammetric investigation was performed by Vedel and Tremillon[120] in ethylpyridinium bromide to obtain the dissociation constants of some acid–base couples. The authors reported cathodic, anodic, and mixed wave profiles for the process

$$2HBr + 2e = H_2 + 2Br^- \tag{66}$$

obtained under a stream of dry HBr, H_2, and HBr + H_2 mixtures, respectively. Under their experimental conditions, the HBr/H_2, Br$^-$ system was found to be voltammetrically reversible.

10. Nitrates

Molten nitrates, widely employed[121] for a wide range of studies, have been used for investigations of hydrogen behavior only within the last few years. However, this recent work represents a systematic study covering various aspects of the problem and will now be considered in depth. All available data, relevant to the (Na, K)NO$_3$ equimolar mixture, are summarized in the present section together with some preliminary information on work in progress.

10.1. Manometric Measurements

A highly sensitive pressure gauge described in Figure 1 was used to obtain hydrogen solubility[122] and kinetic[123] data. It is derived from an apparatus previously[124-128] employed to measure the solubility of several other gases in various molten systems. The solvent (*a*) was contained in a Teflon or platinum beaker (*b*) maintained in a thermostated cell connected to a calibration flask (*c*) and having a precision micrometer (*d*) or barometric (*e*) manometer to detect low or high pressure variations. Suitable connections permitted gas influx and vacuum application to the entire apparatus. The outer part of the vacuum line was water-jacketed and thermostated at room temperature. The cold-finger (*f*) was used for kinetic studies[123] only and disconnected in the course of solubility experiments.

A typical procedure for a hydrogen solubility experiments consisted of completely deaerating the apparatus, introducing the gas, and then rapidly reading the initial pressure P_i. During vigorous magnetic stirring, the pres-

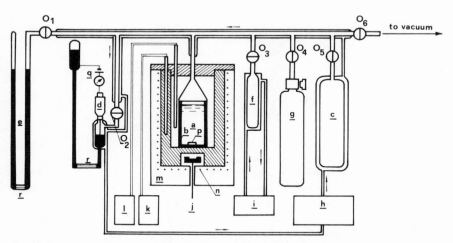

FIGURE 1. Scheme of experimental setup for solubility and kinetic manometric measurements. *a*, melt; *b*, beaker (Teflon or platinum); *c*, water-jacketed calibration flask; *d*, micrometer manometer; *e*, barometric manometer; *f*, cold water reservoir (for kinetic studies only); *g*, gas tank; *h*, *i*, thermostats; *j*, magnetic stirrer; *k*, thermoregulator; *l*, resistance thermometer; *m*, aluminium block thermostat; *n*, heating tapes; o_1–o_6, vacuum stopcocks; *p*, Teflon-coated magnetic bar; *q*, current detector; *r*, capillaries. See References 124, 125, and 129.

sure drop was recorded as a function of time. From the pressure variation, ΔP, the moles of dissolved gas, ΔN, could be calculated using the relationship

$$\Delta N = A(T)\Delta P \tag{67}$$

where $A(T)$ is a temperature-dependent calibration constant. Details of the calibration procedure are given in References 125 and 129.

The Henry's law coefficient for hydrogen dissolution at the working temperature T can be calculated from the expression

$$K_H = A(T)\Delta P/VP_f \tag{68}$$

where V is the known volume of the melt and P_f the equilibrium pressure. For the $(Na, K)NO_3$ equimolar melt in the temperature range 500–600 K, the solubility of hydrogen could be expressed[122] by the relationship

$$\log K_H/\text{mol cm}^{-3} \text{ bar}^{-1} = -5.39 - 768/T \tag{69}$$

The heat and the standard entropy of dissolution were obtained through the expressions

$$\Delta H = -R\frac{d \ln K_H}{d(1/T)} \tag{70}$$

$$\Delta S^0 = \Delta H/T + R \ln(C_s/C_g) \tag{71}$$

where C_s and C_g are the concentrations of hydrogen in solution and in the gas phase, respectively (see Table 1 for the relevant thermodynamic parameters).

The theoretical solubility constant, K_T, can be calculated on the basis of simple theory,[130, 131] which is expressed in the relationship

$$-4\pi Nr^2\gamma = RT(\ln K_T + \ln RT) \tag{72}$$

where r is the molecular radius, γ is the surface tension, and K_T is the solubility constant expressed in $\text{mol cm}^{-3} \text{ atm}^{-1}$. In the absence of appreciable solute–solvent interactions, solubilities can generally be predicted within 100% error by substituting r in equation (72) with the crystallographic radius of the gas molecules (see Reference 126 and papers quoted therein). For hydrogen, the crystallographic radius is not available and by introducing in equation (72) the radius obtained from gas-viscosity data[132] a surprising discrepancy of a factor of 3 between the theoretical and experimental values was evident.[122] Compare curves K_T and K_H in Figure 2.

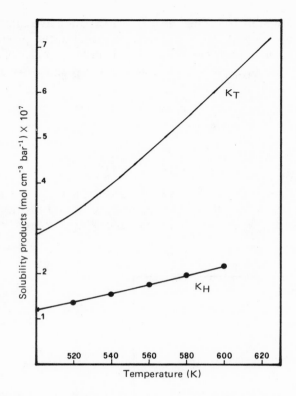

FIGURE 2. Solubility products (mol cm^{-3} atm^{-1}) vs. temperature (K) for the dissolution of hydrogen in the (Na, K)NO$_3$ equimolar melt. K_H, curve obtained from experimental values; K_T, curve calculated from equation (72). See References 122, 129, and 131.

10.2. Kinetics

In the course of solubility measurements, the presence of a slow chemical reaction between hydrogen and melt anions became apparent and was attributed to the overall process

$$H_{2(g)} + NO_3^- = NO_2^- + H_2O_{(g)} \tag{73}$$

A kinetic study[123] of reaction (73) was undertaken using the same experimental setup, applying the appropriate calibration data used for the solubility experiments. In order to maintain a constant water partial pressure over the melt, a certain amount of water was introduced into the coldfinger (f) in Figure 1, via a suitable capillary passing through the thermostating mantle. Hydrogen was then introduced into the deaerated apparatus at the desired pressure and the kinetics of the hydrogen consumption followed by recording the pressure variation as a function of time (see for example Figure 3A). Analysis of the total nitrite produced in the course of the reaction was carried out on quenched samples of the melt.

Treatment of the kinetic data according to a second-order process (73) was effected using the relationship

$$\log P/P^\circ = -Bt \tag{74}$$

where

$$B = \frac{k\,|\,NO_3^-\,|\,K_H}{2.3[K_H + A(T)/W]} \tag{75}$$

k is the rate constant for the direct process (73), K_H is the Henry's law constant for the hydrogen dissolution process,[122] $A(T)$ is the calibration factor and W the weight of the solvent, and P° and P are the initial and

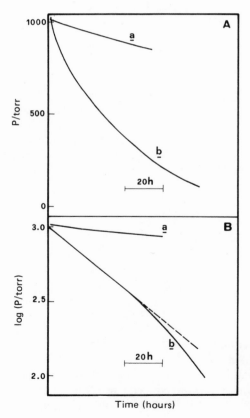

FIGURE 3. Examples of hydrogen pressure vs. time curves (A) and relevant semilogarithmic plots (B) at different temperatures; curves a, $T = 653$ K; curves b, $T = 686$ K. See Reference 123 and unpublished work.

instantaneous pressures, respectively. The rate constants derived from the slope of the log P vs. time plots (see Figure 3B) can be expressed by the general relationship

$$\log k = \log K° - \frac{E_a}{2.3RT} \tag{76}$$

where

$$E_a = 38 \pm 1 \text{ kcal mol}^{-1} \tag{77}$$

$$K° = 5.10^8 \text{ mol}^{-1} \text{ kg sec}^{-1} \tag{78}$$

As is apparent from Figure 3B, after a certain reaction time the rate of hydrogen consumption appears higher than expected on the basis of a simple second-order process. These findings have been explained[123, 129] on the basis of a hydrogen-consuming, autocatalytic mechanism paralleling the direct reaction (73)

$$NO_2^- + H_2 = X^- + H_2O \tag{79}$$

$$X^- + NO_3^- = 2NO_2^- \tag{80}$$

$$H_2 + NO_3^- = H_2O + NO_2^- \tag{73}$$

where X^- is an unstable intermediate and reaction (80) summarizes all the fast steps following the rate-determining (79). The occurrence of this kind of mechanism is confirmed by graphical analysis,[123] such as shown in Figure 4, in which the characteristic sigmoidal curve of an autocatalytic process is evident. The rate constant k_a for reaction (79) can be obtained from the relationship

$$\frac{\Delta|NO_2^-|}{K_H P \Delta t} = k|NO_3^-| + k_a|NO_2^-| \tag{81}$$

where the symbols have the previously specified meanings.

The relevant activation energy and pre-exponential factor were calculated[123] to be

$$E_a = 46 \pm 3 \text{ kcal mol}^{-1} \tag{82}$$

$$k_a = 1.10^{13} \text{ mol}^{-1} \text{ kg sec}^{-1} \tag{83}$$

The results obtained in these kinetic studies seem to indicate that, at moderate temperatures and in the absence of large nitrite concentrations, the effect of reaction (73) can be neglected when working under a hydrogen atmosphere. On the contrary, chemical interaction of hydrogen with nitrate melts becomes quite pronounced[133] in the presence of noble metals.

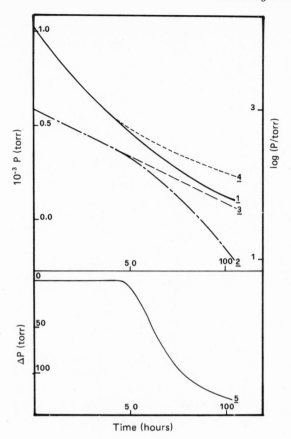

FIGURE 4. Graphical procedure: curve *1*, experimental *P* vs. time curve; curve *2*, logarithmic plot relevant to curve *1*; curve *3*, extrapolation from the straight-line portion of curve *2*; curve *4*, antilogarithm of curve *3* which represents the theoretical contribution of the "direct mechanism"; curve *5*, difference between curves *1* and *4*. See Reference 123.

10.3. Electrochemical Investigations

Potentiometric and voltammetric investigations have been conducted using an experimental setup such as the one described in Figure 5. Purification of the melt (mainly from water and nitrite) was accomplished by purging the solvent with nitric acid[134] and dry nitrogen[19] (but see Chapter 12 by White). When necessary, constant water partial pressures were maintained in the cell by bubbling the purging gases (H_2, N_2, CO_2, or their mixtures) through thermostated water saturators. The actual water content could be directly controlled *in situ* by recording the relevant voltammetric[124, 135–137] reduction wave (but see Reference 172). Precise hydroxide or carbonate additions were introduced into the melt as small drops[19, 134] of solid solutions of the appropriate material dissolved in the solvent melt. The use of low hydroxide and carbonate concentrations required considerable purification[138] of the gas stream from carbon dioxide and other acidic impurities. A knowledge[122, 124, 127] of the relevant

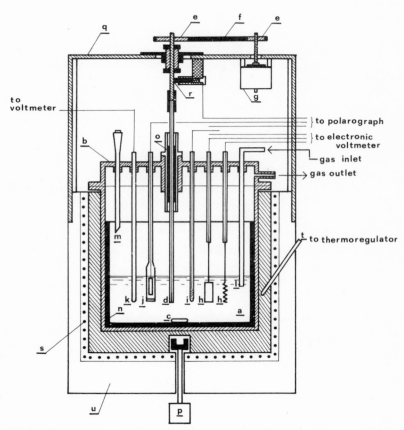

FIGURE 5. Experimental setup for electrochemical investigations. *a*, solvent; *b*, Pyrex cell; *c*, Teflon or platinum-coated magnetic bar; *d*, rotating disk electrode; *e*, cog-wheels; *f*, driving belt; *g*, synchronous motor; *h*, gold, platinum, or nickel electrodes for potentiometric experiments; *i*, reference electrode; *j*, Teflon fritted counterelectrode; *k*, thermocouple; *l*, gas bubbler; *m*, inlet for solid samples; *n*, Teflon or platinum beaker; *o*, Teflon support for rotating disk electrode (with Viton o-rings for gas-tightness); *p*, magnetic stirrer; *q*, r.d.e. assembly; *r*, graphite–brass electric contact; *s*, heating tapes; *t*, thermocouple; *u*, aluminium block thermostat. See References 19 and 129.

Henry's law coefficient permitted the estimation of the actual concentration of water, hydrogen, and carbon dioxide on the basis of their partial pressure in the cell.

Rotating disk electrodes were made by sealing platinum or gold wires into soft-glass tubes. The rotation speed was precisely controlled through a drive train of pulleys and cogs operated by a synchronous motor. Polarographic experiments were performed by substituting the rotating disk electrode with a glass capillary of suitable length. In this case a cold trap (dry ice + acetone) was connected to the cell to avoid the escape of mercury vapor.

The electrodes employed for potentiometric experiments were spirals or foils of gold, platinum, or nickel polished using standard[129, 138] procedures. All the electrochemical results are referred to the Ag^+/Ag (0.07 mol kg^{-1}) reference half-cell.

10.3.1. Potentiometry

A potentiometric investigation[138] of the $(Pt, Au)H_2O/H_2, OH^-$ system was carried out in the $(Na, K)NO_3$ equimolar melt at 503 K. The potential variations of the indicator electrodes obtained upon changing the hydrogen, water, and hydroxide concentrations indicated a marked potentiometric irreversibility of the overall electrode process

$$H_2 + 2OH^- = 2H_2O + 2e \qquad (84)$$

In fact, for both the $(Pt, Au)H_2O/H_2, OH^-$ systems studied, the relationship

$$E_{Au, Pt} = K + \frac{RT}{F} \ln \frac{[H_2O]}{[H_2][OH^-]} \qquad (85)$$

was obtained; this is quite different from the form of the Nernst equation that can be written on the basis of reaction (84), i.e.,

$$E = E^0 + \frac{RT}{F} \ln \frac{[H_2O]^2}{[H_2][OH^-]^2} \qquad (86)$$

The irreproducibility observed in the value of K in equation (85) may be attributable to the influence on the electrode potential of the activities of surface species, e.g., noble metal oxides produced[139] via reactions such as

$$Pt + NO_3^- = PtO + NO_2^- \qquad (87)$$

$$PtO + NO_3^- = PtO_2 + NO_2^- \qquad (88)$$

$$PtO_2 + Pt = 2PtO \qquad (89)$$

which are consistent with the strong oxidizing power of nitrate melts.[140–143] Similar behavior has been observed previously under comparable experimental conditions during potentiometric studies[21, 26, 29] of oxygen electrodes. The experimental findings are consistent with the following mechanistic model:

$$S + H_2 + OH^- \xrightarrow{\text{fast}} SH + H_2O + e \qquad (90)$$

$$\underline{SH + OH^- \xrightarrow{\text{slow}} S + H_2O + e} \qquad (91)$$

$$H_2 + 2OH^- = 2H_2O + 2e \qquad (84)$$

where S represents a (catalytic) species (Pt, PtO, etc.) present at the electrode surface. Nonoxidized platinum or gold atoms probably coexist with the corresponding oxides on the electrode surface; thus the activity of these species, and then K in equation (85) (which contains these activities), becomes dependent on the degree of oxidation, which is related to the metal–melt contact time. The standard potential for this "hydrogen electrode" was calculated[138] to be $E^\circ_{H_2O/H_2, OH^-} = -1.96$ V at 503 K.

A further investigation[144] was performed on the (Ni)H_2O/H_2, OH^- system at 503 K in the same (Na, K)NO_3 solvent. While the potential dependence of the hydroxide concentration was reproducible, although unexpected (slope of E vs. log[OH^-] equal to $2.3RT/2F$), no well-defined dependence of the water and hydrogen concentrations was observed. At present, the only explanation of these puzzling findings seems to involve nonreproducible variations of the activities of surface oxides due to poorly defined phenomena (chemical reactions, adsorption effects, etc.) accompanying changes in the hydrogen, hydroxide, or water concentrations. More information can probably be obtained by combining electrochemical and spectroscopic surface techniques.

In general, the electrode mechanisms proposed to explain potentiometric results, for both oxygen[21, 26, 29] and hydrogen[138, 144] electrodes, involve solid state species formed on the electrode surface arising from the oxidizing power of nitrate solvents. A further investigation was initiated[145] using X-ray photoelectron spectroscopy in order to determine the actual state of the electrode surfaces. Although work along these lines has only just begun, preliminary results seem quite promising. Some of these, which indicate the potentialities of the ESCA technique for characterizing surfaces when chemically modified by contact with molten nitrates, are described in the Appendix.

10.3.2. Voltammetry

Parallel to the potentiometric work, voltammetric studies were initiated on different "hydrogen systems" in the (Na, K)NO_3 equimolar melt in order to collect information about various aspects of hydrogen electroactivity.

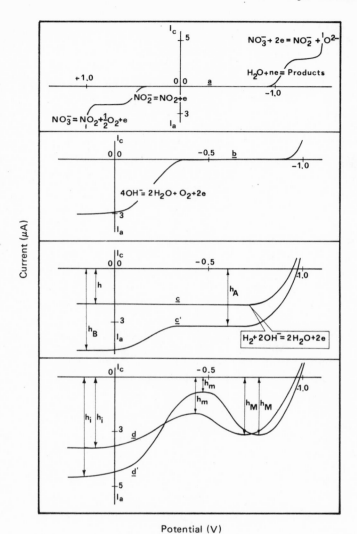

Potential (V)

FIGURE 6. Voltammetry of the H_2O/H_2, OH^- system at a platinum rotating disk electrode at 513 K. Curve a, voltammogram recorded in a pure melt under wet nitrogen or hydrogen atmosphere; curve b, hydroxide wave for a melt maintained under nitrogen atmosphere and containing $[OH^-] = 9.3 \times 10^{-4}$ mol kg^{-1}; curve c, current–potential profile obtained in a melt saturated with slightly wet hydrogen and containing $[OH^-] = 6.2 \times 10^{-4}$ mol kg^{-1}; curve c', theoretical curve for hydroxide concentrations higher than expressed by equation (94) (e.g., $[OH^-]_{c'} = 1.4 \times 10^{-3}$ mol kg^{-1}); curves d and d', current–potential profiles recorded in a melt saturated with slightly wet hydrogen and containing $[OH^-] = 1.21 \times 10^{-3}$ mol kg^{-1} and $[OH^-] = 1.68 \times 10^{-3}$ mol kg^{-1}, respectively. See References 20, 146, 149, 150, and unpublished work.

A first paper[146] describes Pt r.d.e. voltammetric studies on the H_2O/H_2, OH^- system. Some interesting current–potential profiles, recorded under the same experimental conditions as in the above paper, are reported in Figure 6. Curve *a* represents the background obtained in a melt flushed with pure nitrogen and not containing proton-acceptor species. Only the small waves due to the reduction of water[124, 135–137] and to the oxidation of nitrite[147–150] were recorded within the electroactivity range of the solvent, which is limited by the process of electroreduction[140, 151] and electro-oxidation[152, 153] of the solvent anions. The same result was obtained when, *ceteris paribus*, hydrogen was bubbled through the melt. This indicates that the direct electro-oxidation of hydrogen according to the reaction

$$H_2 = 2H^+ + 2e \tag{92}$$

does not occur. Curve *b* shows the effect of purging the melt with pure nitrogen in the presence of hydroxide ions, which are electro-oxidized according to the overall[20] reaction

$$2OH^- = H_2O + \tfrac{1}{2}O_2 + 2e \tag{93}$$

On flushing hydrogen through a melt containing hydroxide, current–potential profiles *c*, *d*, and *d'*, in the same figure, were obtained. All these voltammograms were recorded at constant hydrogen concentration (melt maintained under the pressure of 1 atm of hydrogen) whilst varying the concentration of hydroxide. The hydroxide wave is absent in curve *c*, which was obtained in the presence of an excess of hydrogen at the electrode. The diffusion of hydroxide controls the relevant limiting current, attributed[146] to the process

$$H_2 + 2OH^- = 2H_2O + 2e \tag{84}$$

When hydroxide concentration is larger than the value defined by the relationship

$$[OH^-] = 2\left[\frac{D_{H_2}}{D_{OH^-}}\right]^{2/3}[H_2] \tag{94}$$

which can be evaluated by applying the Levich[154, 155] * equation

$$I_D = 0.62nFAD^{2/3}v^{-1/6}\omega^{1/2}C \tag{95}$$

then a hydroxide wave is recorded. The relation (94) determines the conditions under which anodic currents such as *h* in Figure 6 are controlled by the diffusion of both hydrogen and hydroxide.

* For a review see Reference 156.

When hydroxide was in excess at the electrode *experimental* profiles such as *d*, characterized by a current minimum, were recorded instead of *theoretical* curves such as *c'* (in which h_A and h_B would be controlled by the diffusion of hydrogen and hydroxide, respectively).

These experimental results were rationalized[146] by the following mechanism. When hydrogen is in excess over hydroxide at the electrode, it may be consumed via a mechanism such as

$$H_2 = 2H^+ + 2e \qquad (92)$$

$$2H^+ + 2OH^- = 2H_2O \qquad (96)$$

overall $\quad H_2 + 2OH^- = 2H_2O + 2e \qquad (84)$

The H^+ ions formed in the step (92) might react with the OH^- ions present at the electrode; the constant current *h* would then be controlled by the diffusion of hydroxide. For excess hydroxide concentrations, the hydroxide ions which do not react according to the first mechanism could be electro-oxidized according to a mechanism such as

$$2(OH^- = OH^\bullet + e) \qquad (97)$$

$$2OH^\bullet = H_2O + \tfrac{1}{2}O_2 \qquad (98)$$

overall $\quad 2OH^- = H_2O + \tfrac{1}{2}O_2 + 2e \qquad (93)$

The hydroxyl radicals produced in this way on the electrode surface would be available to catalyze the *chemical* consumption of hydrogen, in the diffusion layer, according to a third mechanism

$$OH^\bullet + H_2 = H_2O + H^\bullet \qquad (99)$$

$$H^\bullet + NO_3^- = OH^\bullet + NO_2^- \qquad (100)$$

overall $\quad H_2 + NO_3^- = H_2O + NO_2^- \qquad (73)$

The chemical consumption of hydrogen could then be responsible for the observed current minimum, since the actual hydrogen concentration in the diffusion layer would be smaller than in the bulk of the melt as soon as the production of $OH^\bullet$ radicals begins. An alternative explanation of these findings, involving impurity nitrite ions, is suggested in Chapter 11 by Lovering and Oblath.

The diffusion coefficient of hydrogen estimated[146] from the present results is $D_{H_2} \geqslant 7.8.10^{-5}$ cm^2 sec^{-1} at 513 K (see Table 2).

Further experiments performed by using gold[157] in place of platinum demonstrated that the observed phenomena were independent of the electrode material. Typical curves obtained using the two electrodes are compared in Figure 7. The voltammograms obtained on the gold r.d.e.

(curves *a* and *b*) are qualitatively very similar to those on the platinum r.d.e. (curves *a'* and *b'*). The shift on the potential scale of the minimum and the broadening of the maximum (compare curves *b* and *b'*) might be explained on the basis of different discharge overvoltages of hydroxide and water on platinum and gold. This is particularly apparent in curves *c* and *c'*, which show the discharge waves of water and hydroxide on gold and platinum, respectively.

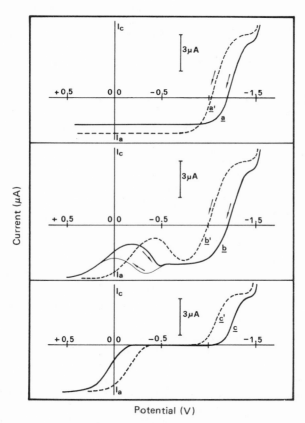

FIGURE 7. Voltammetry of the H_2O/H_2, OH^- system at gold (curves *a*, *b*, and *c*) and platinum (curves *a'*, *b'*, and *c'*) rotating disk electrodes at 513 K. Curves *a* and *a'*, voltammograms recorded in a melt saturated with slightly wet hydrogen and containing $|OH^-| = 7 \times 10^{-4}$ mol kg^{-1} and $[OH^-] = 9.3 \times 10^{-4}$ mol kg^{-1}, respectively; curves *b* and *b'*, voltammograms recorded in a melt saturated with slightly wet hydrogen and containing $[OH^-] = 1.3 \times 10^{-3}$ mol kg^{-1} and $[OH^-] = 1.35 \times 10^{-3}$ mol kg^{-1}; curves *c* and *c'*, background waves recorded in a melt maintained under wet nitrogen atmosphere and containing $[OH^-] = 1.16 \times 10^{-3}$ mol kg^{-1}. See References 20, 124, 157, and unpublished work.

No wave was obtained[157] under any conditions on the DME, probably because of the high discharge overvoltage of molecular hydrogen on mercury.

The most recent study[134] of the voltammetric behavior of hydrogen was undertaken by substituting the proton acceptor hydroxide ions with carbonate ions. Curve *a* in Figure 8 is the curve obtained at a platinum r.d.e. in a melt flushed with pure nitrogen and containing carbonate ions which can be electro-oxidized according to the overall[22] reaction

$$CO_3^{2-} = CO_2 + \tfrac{1}{2}O_2 + 2e \tag{101}$$

On changing nitrogen for hydrogen in the carbonate-containing melt, curves *b* and *c* were obtained according to the overall reaction

$$H_2 + CO_3^{2-} = H_2O + CO_2 + 2e \tag{102}$$

Curves *b* and *c* relate to an excess on the electrode of hydrogen and carbonate respectively, i.e., to carbonate concentrations lower and, respectively, higher than the value given by the relationship

$$[CO_3^{2-}] = \left[\frac{D_{H_2}}{D_{CO_3^{2-}}} \right]^{2/3} \cdot [H_2] \tag{103}$$

The limiting current *h* is controlled by the diffusion of carbonate, while *h'* is controlled by the diffusion of hydrogen.

A qualitative difference between the voltammetric behavior of hydrogen/hydroxide and hydrogen/carbonate systems is that in this last case no minimum of the anodic current is observed. Perhaps no catalyzed hydrogen-consuming reaction occurs in the proximity of the electrode. The limiting current value *h'* can be used for the analytical detection of hydrogen in molten nitrates.

On flushing wet carbon dioxide through the melt, cathodic waves such as *d* in Figure 8 were recorded and relate to the electrode process

$$H_2O + CO_2 + 2e = H_2 + CO_3^{2-} \tag{104}$$

Curves such as *d'* can be obtained on flushing *dry* carbon dioxide through the melt, the relevant electroreduction[22] process being

$$CO_2 + NO_3^- + 2e = CO_3^{2-} + NO_2^- \tag{105}$$

Both currents (h_2 and h_3) can be used for the analytical determination of carbon dioxide in wet and dry melts, respectively. Further, qualitative, support for reactions (102) and (104) has been obtained[134] from sweep voltammetry experiments. The profile shown as curve *e* in the figure was obtained upon flushing a mixture of hydrogen, water, and carbon dioxide through a carbonate-containing melt. The results indicate reversible voltam-

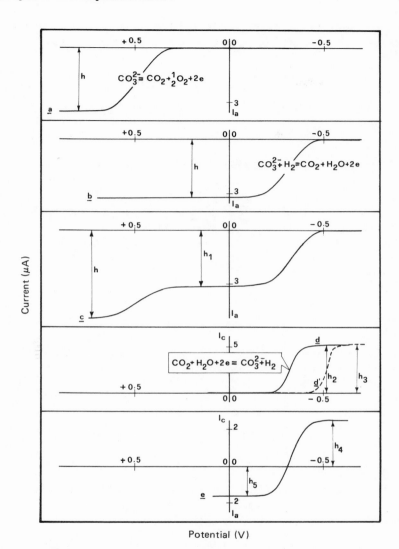

Potential (V)

FIGURE 8. Voltammetric behavior of the CO_2, H_2O/CO_3^{2-}, H_2 system at a platinum rotating disk electrode at 513 K. Curve *a*, carbonate wave obtained in a melt maintained under flux of nitrogen and containing $[CO_3^{2-}] = 5.2 \times 10^{-4}$ mol kg^{-1}; curves *b* and *c*, current–potential profiles obtained on flushing slightly wet hydrogen through a melt containing $[CO_3^{2-}] = 4.7 \times 10^{-3}$ mol kg^{-1} and $[CO_3^{2-}] = 7.3 \times 10^{-3}$ mol kg^{-1}, respectively; curves *d* and *d'*, carbon dioxide reduction wave obtained in a melt maintained under an atmosphere of wet ($p_{H_2O} = 20$ torr) or dry CO_2; curve *e*: composite wave for process (102) recorded in a melt containing $[CO_3^{2-}] = 2.7 \times 10^{-4}$ mol kg^{-1} and maintained under flux of a mixture of CO_2, H_2, and H_2O ($p_{CO_2} = p_{H_2} = 370$ torr, $p_{H_2O} = 20$ torr). See References 22, 134, and unpublished work.

metric behavior in each case. In the presence of an excess of water and carbonate h_4 and h_5 are, respectively, proportional to the concentrations of carbon dioxide and hydrogen.

The results described show a number of interesting aspects concerning hydrogen electroactivity. Further work is in progress in this area using various proton acceptor species and electrode materials.

ACKNOWLEDGMENTS

This work was carried out with the financial assistance of Consiglio Nazionale delle Ricerche (C.N.R., Rome). Permission to reproduce data in the figures is gratefully acknowledged from Analytical Chemistry, Journal of Physical Chemistry, Journal of the Chemical Society: Faraday Trans. I, Journal of Electroanalytical Chemistry, Annali di Chimica and La Chimica e l'Industria. Copyright by the American Chemical Society, The Chemical Society, Elsevier Scientific Publishing Company and Societa Chimica Italiana.

Appendix

Most metals can undergo spontaneous or potential-assisted corrosion reactions when placed in contact with ionic solvents.* A knowledge of the metal oxidation states, as well as of the passivating layer composition, is of paramount importance for elucidating electrochemical mechanisms involving surface species acting as catalysts or influencing reaction products.

X-ray photoelectron spectroscopy (XPSA or ESCA) is a spectroscopic technique capable of analyzing electrode surfaces within the top 30–40 Å (see for example Reference 159).

ESCA studies are currently in progress in this laboratory and are being employed to monitor alterations of catalytically and electrochemically significant metal–oxygen surfaces resulting from the exposure of noble[145, 160] and nonnoble[161, 162] metals to molten nitrates.

Examples of photoelectronic spectra of the platinum $4f$ levels recorded on chemically oxidized samples maintained in contact with a $(Na, K)NO_3$ equimolar melt for various lengths of time at two different temperatures are reported, in chronological sequence, in Figures 9 and 10. The presence of metallic platinum together with its $+2$ and $+4$ oxidation states can be inferred from the analysis of these spectra.

Apparently, the extent of oxidation is larger the longer the metal-melt contact time and the higher the temperature. Known values[163, 164] of bind-

---

* For a review see Reference 158.

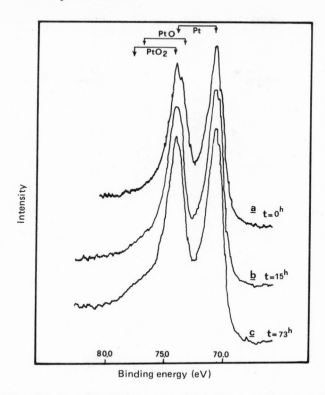

FIGURE 9. Examples of X-ray photoelectron spectra of Pt $4f_{5/2,7/2}$ levels recorded on platinum samples held at 523 K in contact with a (Na, K)NO$_3$ melt for the following times (in hours): (a) $t = 0$; (b) $t = 15$; (c) $t = 73$. The binding energies of Pt $4f$ level for Pt, PtO, PtO$_2$ species are quoted. See References 145, 160, 163, and unpublished work.

ing energies for Pt, PtO, and PtO$_2$ can be used to deconvolute the spectra and to determine the contribution of the various platinum species. Results obtained in a recent study[160] show that by plotting the sum of the peak area percentages of the two oxides as a function of time, kinetic curves (following a logarithimic law) are obtained, similar to those reported[165, 166] for the oxidation of metals by molecular oxygen in the gas phase. From the analysis of these experimental findings, in neither case does it seem probable that diffusion of molecular oxygen across the metal or the first oxide layer is responsible for the oxidation of the inner atomic layers. According to our present understanding, a mechanism involving electron tunnelling through an initial very thin oxide film (formed by direct chemical interaction between nitrate ions and the metal surface) followed by diffusion of oxide ions through the solid phase, seems to be most appro-

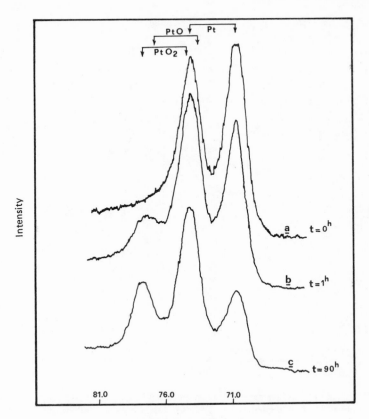

FIGURE 10. Examples of X-ray photoelectron spectra of Pt $4f_{5/2, 7/2}$ levels recorded on platinum samples held at 670 K in contact with a (Na, K)NO$_3$ melt for the following times (in hours): (a) $t = 0$; (b) $t = 1$; (c) $t = 90$. The binding energies of Pt $4f$ level for Pt, PtO, and PtO$_2$ species are quoted. See References 145, 160, 163, and unpublished work.

priate; however, because of the complexity of the system, a sound model cannot yet be fully formulated.

Information obtained by means of XPS analysis concerning the state of the platinum electrode surface provides an important contribution towards a meaningful interpretation of a series of potentiometric results obtained for hydrogen[138] and oxygen[21, 26] electrodes in molten nitrates. As mentioned above, electrode potential variations with time observed in the course of the potentiometric work appeared dependent mainly on the temperature and on the initial state of the electrode surface. Only after a certain "stabilization time" did the potential reach constancy over significant periods of time. On the basis of the ESCA evidence this behavior has been rationalized in terms of the actual composition (and, thus, the "activity") of the surface compounds being a function of the metal-melt time of contact (see Figures 9 and 10). In the case of the potentiometrically irreversible systems[21, 26, 138] involving surface species in the potential-determining step, the potential itself becomes a function of time. The "equilibrium activities" of these compounds (attained when the electrode reached constant potential) are *de facto* included in the pseudoconstant K term of Nernst relationships such as (85). Hence, small variations sometimes observed for this K term can be explained on the basis of slightly different final "surface states" related to the history, including mechanical pretreatment, of the electrodes.

ESCA studies[161, 162] have also been performed using nickel plates as electrodes for hydrogen and oxygen systems in molten nitrates. Figure 11 shows XPS spectra pertaining to Ni $2p$ and O $1s$ levels recorded for samples exposed to the $(Na, K)NO_3$ equimolar mixture at 520 K for different periods of time. As expected, the microcorrosion phenomena are faster and more pronounced for nickel than for platinum. After chemical oxidation carried out for 30 min, the oxide film thicknesses approach the ESCA escape depth and then completely obscure the underlying metal signal (see curve b'). The spectra indicate the presence of nickel in different oxidation states, although an unequivocal interpretation is, at present, difficult because of controversy in the literature[164, 167–169] concerning the assignment of binding energies in the various nickel compounds. Probably the oxides formed are NiO and Ni_2O_3; the presence of species such as $NiOOH$ and $Ni(OH)_2$ on the surface of the electrode dipped in the melt seems unlikely because of their thermal[170] instability. The initially formed oxide might reasonably be Ni_2O_3, whilst NiO would become the predominant species after longer periods of immersion (compare, for example, the spectra recorded in the O $1s$ region). These results are consistent with those obtained from the oxidation of nickel surfaces by other methods[169, 171] (e.g., direct high-temperature reaction with molecular oxygen or bombardment by O_2^+ ions). The following equilibria could well be involved in the formation of oxidized species on nickel electrodes in contact with nitrate melts:

$$Ni + NO_3^- = NiO + NO_2^- \tag{A.1}$$

$$2NiO + NO_3^- = Ni_2O_3 + NO_2^- \tag{A.2}$$

$$Ni_2O_3 + Ni = 3NiO \tag{A.3}$$

Upon this assumption, an attempt was made to explain the potentiometric behavior of systems such as $(Ni)O_2$, H_2O/OH^-, $(Ni)H_2O/H_2$, OH^-, $(Ni)CO_2$, O_2/CO_3^{2-}, and $(Ni)H_2O$, CO_2/H_2, CO_3^{2-} in molten alkali nitrates utilizing ESCA information to assess the actual composition of the

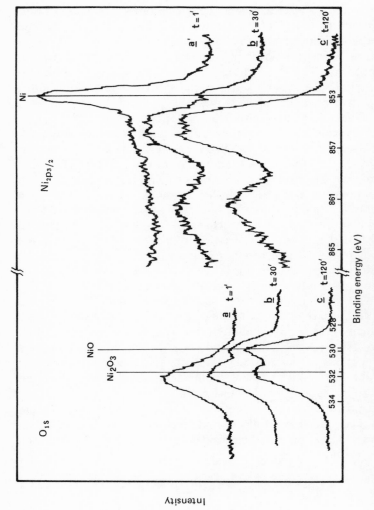

FIGURE 11. Examples of X-ray photoelectron spectra for the nickel $2p_{3/2}$ and oxygen 1s levels recorded on nickel samples held at 510 K in contact with a $(Na, K)NO_3$ melt for the following times (in minutes): (a), (a'), $t = 1$; (b), (b'), $t = 30$; (c), (c'), $t = 120$. The binding energies for metallic nickel in the $2p_{3/2}$ level and for oxides in the oxygen 1s level are quoted. See References 161, 162, 164, 169, and unpublished work.

nickel electrodes. This was successful in the case of oxygen electrodes.[161, 162] So far inconclusive results have been obtained for hydrogen systems, mainly because of the difficulty of obtaining reproducible experimental data.

References

1. D. P. Gregory, *Sci. Am.* **228**, 13 (1973).
2. D. P. Gregory, D. Y. C. Ng, and G. M. Long, in *Electrochemistry of Cleaner Environments*, Ed. J. O'M. Bockris, Plenum, New York (1972), p. 226.
3. C. Marchetti, *Euro. Spectra* **10**, 117 (1971).
4. D. P. Gregory, in *Modern Aspects of Electrochemistry*, Vol. 10, Eds. J. O'M. Bockris and B. E. Conway, Plenum, New York (1975), p. 239.
5. D. A. Mathis, in *Hydrogen Technology for Energy*, Noyes Data Corp., Park Ridge, New Jersey (1976).
6. S. Srinivasan and R. H. Wiswall, *Proceedings of the Symposium on Energy Storage*, Eds. J. B. Berkowitz *et al.*, The Electrochemical Society, Princeton, New Jersey (1976), p. 82.
7. K. W. Kordesh, in *Modern Aspects of Electrochemistry*, Vol. 10, Eds. J. O'M. Bockris and B. E. Conway, Plenum, New York (1975), p. 377.
8. Y. Breele *et al.*, *Principles, Technologie, Applications des Piles à combustible*, Editions Technip, Paris (1972), p. 24.
9. W. Mitchell Jr., *Fuel Cells*, Academic, New York (1963).
10. K. W. Kordesh, *J. Electrochem. Soc.* **125**, 77c (1978).
11. J. M. King Jr., 10th Intersociety Energy Conversion Engineering Conference, Newark, Delaware (1975), p. 237.
12. *Fused Salts*, Ed. B. R. Sundheim, McGraw-Hill, New York (1964).
13. D. Bauer and J. P. Beck, *J. Electroanal. Chem.* **40**, 233 (1972).
14. F. R. Duke, *Advan. Chem. Ser.* **49**, 220 (1965).
15. J. Jordan, W. B. McCarthy, and P. G. Zambonin, in *Molten Salts*, Ed. G. Mamantov, Marcel Dekker, New York (1969), p. 575.
16. A. F. Alabyshev, M. F. Lantranov, and A. G. Morachevski, in *Reference Electrodes for Fused Salts*, The Sigma Press Publishers, Washington, D.C. (1965).
17. K. W. Fung and G. Mamantov, in *Advanced in Molten Salt Chemistry*, Vol. 2, Eds. J. Braunstein *et al.*, Plenum, New York (1973), p. 199.
18. P. G. Zambonin, *J. Electroanal. Chem.* **24**, App. 25 (1970).
19. P. G. Zambonin, *J. Electroanal. Chem.* **33**, 243 (1971).
20. P. G. Zambonin, *Anal. Chem.* **43**, 1571 (1971).
21. P. G. Zambonin and G. Signorile, *J. Electroanal. Chem.* **35**, 251 (1972).
22. P. G. Zambonin, *Anal. Chem.* **44**, 763 (1972).
23. P. G. Zambonin, *J. Electroanal. Chem.* **45**, 451 (1973).
24. P. G. Zambonin, *J. Phys. Chem.* **78**, 1294 (1974).
25. P. G. Zambonin, *J. Am. Chem. Soc.* **97**, 4405 (1975).
26. E. Desimoni, L. Sabbatini, and P. G. Zambonin, *J. Electroanal. Chem.* **71**, 73 (1976).
27. E. Desimoni, F. Paniccia, and P. G. Zambonin, in *Reactivity of Solids*, Eds. J. Wood *et al.*, Plenum, New York (1977), p. 137.
28. L. Sabbatini, F. Palmisano, P. G. Zambonin, and B. A. DeAngelis, *Ann. Chim.* **67**, 525 (1977).
29. B. Morelli, E. Desimoni, L. Sabbatini, and P. G. Zambonin, *J. Electroanal. Chem.* **94**, 5 (1978).

30. R. Marassi, V. Bartocci, and F. Pucciarelli, *Talanta* **19**, 203 (1973).
31. R. Marassi, V. Bartocci, F. Pucciaerelli, and P. Cescon, *J. Electroanal. Chem.* **47**, 509 (1973).
32. R. Marassi, V. Bartocci, M. Gusteri, and P. Cescon, *J. Appl. Electrochem.* **9**, 81 (1978).
33. J. P. Vilaverde, G. Picard, J. Vedel, and B. Tremillon, *J. Electroanal. Chem.* **54**, 279 (1974).
34. A. J. Arvia, A. J. Calandra, and H. A. Videla, *Electrochim. Acta* **10**, 33 (1965).
35. S. Tajima, M. Soda, T. Mori, and N. Baba, *Electrochim. Acta* **1**, 205 (1959).
36. A. M. Shams el Din, *Electrochim. Acta* **17**, 613 (1962).
37. H. A. Videla and A. J. Arvia, *Electrochim. Acta* **10**, 21 (1965).
38. D. Gilroy, *Electrochim. Acta* **17**, 1771 (1972).
39. H. Urtizberea, A. J. Calandra, and A. J. Ariva, *J. Electroanal. Chem.* **55**, 145 (1974).
40. R. O. Lezna, W. E. Triaca, and A. J. Arvia, *J. Electroanal. Chem.* **57**, 113 (1974).
41. R. O. Lezna, W. E. Triaca, and A. J. Arvia, *J. Electroanal. Chem.* **71**, 51 (1976).
42. E. J. Balskus, J. J. Podesta, and A. J. Arvia, *Electrochim. Acta* **15**, 1557 (1970).
43. E. J. Balskus, J. J. Podesta, and A. J. Arvia, *Electrochim. Acta* **16**, 1663, (1971).
44. E. J. Balskus, W. E. Triaca, and A. J. Arvia, *Electrochim. Acta* **17**, 45 (1972).
45. A. J. Arvia, F. DeVega, and H. A. Videla, *Electrochim. Acta* **13**, 581 (1968).
46. A. J. Arvia, A. J. Calandra, and H. A. Videla, *Electrochim. Acta* **14**, 25 (1969).
47. R. G. Casino, J. J. Podesta, and A. J. Arvia, *Electrochim. Acta* **16**, 121 (1971).
48. A. J. Arvia, W. E. Triaca, and H. A. Videla, *Electrochim. Acta* **15**, 9 (1970).
49. M. W. Breiter, *Electrochemical Processes in Fuel Cells*, Springer-Verlag, Berlin (1969), p. 217.
50. G. H. J. Broers and M. Shenke, in *Fuel Cells*, Vol. 2, Ed. G. J. Yong, Reinhold, New York (1963), p. 6.
51. *Hydrocarbon Fuel Cell Technology*, Ed. B. S. Baker, Academic, New York (1965), Part III.
52. B. K. Andersen, *Thermodynamic Properties of Molten Alkali Carbonates*, Fysisk Kemisk Institut, The Technical University of Denmark, Lyngby.
53. P. K. Lorenz and G. J. Janz, *Electrochim. Acta* **15**, 1025 (1970).
54. H. Bartlett and K. Johnson, *J. Electrochem. Soc.* **114**, 457 (1967).
55. M. V. Smirnov, L. A. Tsiovkina, and V. A. Oleinkova, *Electrochem. Molten Solid Electrolytes* **3**, 61 (1965).
56. M. D. Ingram and G. J. Janz, *Electrochim. Acta* **11**, 1629 (1966).
57. G. G. Arkhipov, L. P. Klevtsov, and G. K. Stepanov, *Tr. Inst. Elektrokhim. Akad. Nauk. SSSR Ural'sk Filial No.* **8**, 113 (1966).
58. M. S. Bawa and J. K. Truitt, *Extended Abstracts Vol. of the Fall Meeting of the Electrochem. Soc. Inc.*, Montreal (1968), p. 177.
59. M. A. Volgin and A. L. L'Vov, *Issled. Obl. Khim. Istochnikov Toka* **2**, 26 (1971).
60. M. A. Volgin, A. L. L'Vov, and V. A. Laskutin, *Elektrokhimia* **9**, 368 (1973).
61. G. H. J. Broers and M. Schenke, in *Hydrocarbon Fuel Cell Technology*, Ed. B. S. Baker, Academic, New York (1965).
62. N. Busson, S. Palous, J. Millet, and R. Buvet, *Electrochim. Acta* **12**, 1609 (1967).
63. W. M. Vogel and C. D. Iacovangelo, *J. Electrochem. Soc.* **9**, 1305 (1977).
64. Yu. K. Delimarskii and B. F. Markov, in *Electrochemistry of Fused Salts*, Engl. Transl., Ed. R. E. Wood, The Sigma Press, Washington, D.C. (1961), p. 162.
65. L. G. Boxall and K. E. Johnson, *Electroanal. Chem.* **30**, 25 (1971).
66. Z. Takehara and S. Yoshizawa, *Abh. Saechs. Akad. Wiss. Leipzig Math. Naturwiss. Kl.* **49**, 317 (1968).
67. R. Barde, R. Buvet, and J. Dubois, *C.R. Acad. Sci.* **254C**, 1627 (1962).
68. R. Barde and R. Buvet, *J. Chim. Phys.* **60**, 1365 (1963).
69. R. Barde, R. Buvet, and N. J. Bosson, Belg. Pat., 640, 628 (May 29, 1964); Fr. Appl. (Dec. 10, 1962) and (Nov. 14, 1963).

70. J. Polart, P. Degobert, and O. Bloch, *C.R. Acad. Sci.* **225C**, 515 (1962).
71. J. Polart and P. Degobert, *C.R. Acad. Sci.* **255C**, 2103 (1962).
72. J. Polart, *C.R. Acad. Sci.* **256C**, 2159 (1963).
73. J. Polart, P. Degobert, and O. Bloch, Fr. Pat. 1, 368, 109 (July 31, 1964), Appl. (Sep. 26, 1962).
74. H. A. Laitinen and J. A. Plambeck, *J. Am. Chem. Soc.* **87**, 1202 (1965).
75. R. Littlewood and E. J. Argent, *Electrochim. Acta* **4**, 114 (1961).
76. N. Q. Minh and B. J. Welch, *Aust. J. Chem.* **28**, 965 (1975).
77. N. Q. Minh and B. J. Welch, *Aust. J. Chem.* **28**, 2579 (1975).
78. Y. Kanzaki and M. Takahashi, *J. Electroanal. Chem.* **58**, 349 (1975).
79. J. D. Van Norman and R. J. Tivers, *J. Electrochem. Soc.* **118**, 258 (1971).
80. E. A. Ukshe and V. N. Devyatkin, *Zh. Fiz. Khim.* **39**, 3074 (1965).
81. V. N. Devyatkin and E. A. Ukshe, *Fiz. Khim. Elektrokhim. Rasplav. Solei Shlakov, Tr. Vses, Soveshch.*, 3rd (1966), p. 130.
82. W. Sundermeyer, *Angew. Chem. Internat. Edn.* **4**, 222 (1965).
83. G. Letisse and B. Tremillon, *J. Electroanal. Chem.* **17**, 387 (1968).
84. B. Tremillon, A. Bermond, and R. Molina, *J. Electroanal. Chem.* **74**, 53 (1976).
85. S. Pizzini, G. Sternheim, and G. B. Barbi, *Electrochim. Acta* **8**, 227 (1963).
86. S. Pizzini and A. Magistris, *Electrochim. Acta* **9**, 1189 (1964).
87. S. Pizzini and R. Morlotti, *Electrochim. Acta* **10**, 1033 (1965).
88. K. Nagel and F. Wendler, *Z. Elektrochem.* **60**, 1064 (1956).
89. K. Nagel and F. Wendler, *Z. Elektrochem.* **63**, 213 (1959).
90. D. G. Hill, B. Porter, and A. S. Gillespie, *J. Electrochem. Soc.* **105**, 408 (1958).
91. G. Dirian, K. A. Roemberger, and C. F. Baes Jr., USAEC report No. ORNL-3789 (1965), p. 76.
92. B. F. Hitch and C. F. Baes Jr., *Inorg. Chem.* **8**, 201 (1969).
93. C. F. Baes Jr., USAEC report No. CONF-690801 (1969), p. 617.
94. A. P. Malinauskas and D. M. Richardson, *Ind. Eng. Chem. Fundam.* **13**, 242 (1974).
95. A. P. Malinauskas, D. M. Richardson, J. E. Savolainen, and J. H. Shaffer, *Ind. Eng. Chem. Fundam.* **11**, 584 (1972).
96. F. T. Bacon, *Electrochim. Acta* **14**, 569 (1969).
97. R. M. Lawrence and W. H. Bowman, *J. Chem. Ed.* **48**, 359 (1971).
98. K. Smrcek, *Collect. Czech. Chem. Commun.* **36**, 1193 (1971).
99. K. Smrcek and J. Jandera, *Collect. Czech. Chem. Commun.* **38**, 1899 (1973).
100. H. Binder, A. Kohling, W. H. Kuhn, W. Linder, and G. Sandstede, in *From Electrocatalysis to Fuel Cells*, Ed. G. Sandstede, Battelle Seattle Research Center, University of Washington Press (1972), p. 131.
101. M. A. Gutjar, in *From Electrocatalysis to Fuel Cells*, Ed. G. Sandstede, Battelle Seattle Research Center, University of Washington Press (1972), p. 143.
102. K. V. Kordesh, in *From Electrocatalysis to Fuel Cells*, Ed. G. Sandstede, Battelle Seattle Research Center, University of Washington Press (1972), p. 157.
103. K. V. Kordesh, in *Fuel Cells*, Ed. W. Mitchell Jr., Academic, New York (1963), p. 329.
104. J. Goret, *Bull. Soc. Chim. France*, 1074 (1964).
105. J. Goret, Ph.D. thesis, Faculte des Sciences de l'Universite de Paris (1966).
106. J. Goret and B. Tremillon, *Bull. Soc. Chim. France* 67 (1966).
107. G. Kern, Ph.D. thesis, Faculte des Sciences de l'Universite de Paris (1967).
108. G. Kern, P. Degobert, and O. Bloch, *C.R. Acad. Sci. Paris* **256C**, 1500 (1963).
109. B. Tremillon, *Int. Symposium on Non-Aqueous Electrochemistry, Paris, 1970*, Butterworths, London (1971), p. 395.
110. S. Zecchin, G. Schiavon, and G. G. Bombi, *J. Electroanal. Chem.* **50**, 261 (1974).
111. G. Schiavon, S. Zecchin, and G. G. Bombi, *J. Electroanal. Chem.* **38**, 473 (1972).
112. A. Eluard, Ph.D. thesis, Faculte des Sciences de l'Universite de Paris (1970).
113. W. M. Vogel and S. W. Smith, *J. Electroanal. Chem.* **18**, 215 (1968).

114. W. M. Vogel, K. J. Routsis, V. J. Kehrer, D. A. Landsman, and J. G. Tschinkel, *J. Chem. Eng. Data* **12**, 465 (1967).
115. A. Kisza, *Z. Phys. Chem. (Leipzig)* **237**, 97 (1968).
116. A. Kisza, *Rocz. Chem.* **41**, 351 (1967).
117. A. Kisza and B. Rol, *Bull. Acad. Polon. Sci. Ser. Sci. Chim.* **22**, 801 (1974).
118. A. Kisza and U. Twardoch, *Rocz. Chem.* **44**, 1503 (1970).
119. A. Kisza and B. Rol, *Bull. Acad. Polon. Sci. Ser. Sci. Chim.* **22**, 809 (1974).
120. J. Vedel and B. Tremillon, *Bull. Soc. Chim. France* 220 (1966).
121. K. W. Fung and G. Mamantov, in *Advances in Molten Salt Chemistry*, Vol. 2, Eds. J. Braunstein, G. Mamantov, and G. P. Smith, Plenum, New York (1973), p. 230.
122. E. Desimoni, F. Paniccia, and P. G. Zambonin, *J. Chem. Soc. Faraday Trans. 1*, **69**, 2014 (1973).
123. E. Desimoni, F. Paniccia, and P. G. Zambonin, *J. Phys. Chem.* **81**, 1985 (1977).
124. P. G. Zambonin, V. L. Cardetta, and G. Signorile, *J. Electroanal. Chem.* **28**, 237 (1970).
125. E. Desimoni, F. Paniccia, and P. G. Zambonin, *J. Electroanal. Chem.* **38**, 373 (1972).
126. F. Paniccia and P. G. Zambonin, *J. Chem. Soc. Faraday Trans. 1* **68**, 2083 (1972).
127. F. Paniccia and P. G. Zambonin, *J. Chem. Soc. Faraday Trans. 1* **69**, 2019 (1973).
128. R. E. Andresen, F. Paniccia, P. G. Zambonin, and H. A. Oye, *Proceedings of the 4th Nordic High Temperature Symposium, NORTHEMPS-75*, Vol. 1 (1975), p. 127.
129. E. Desimoni, F. Paniccia, L. Sabbatini, and P. G. Zambonin, *Proceedings of the International Symposium on Molten Salts*, Eds. J. P. Pemsler *et al.*, The Electrochemical Society, Princeton, New Jersey (1976), p. 584.
130. M. Blander, W. R. Grimes, N. W. Smith, and G. M. Watson, *J. Phys. Chem.* **63**, 1164 (1959).
131. H. H. Uhlig, *J. Phys. Chem.* **41**, 1215 (1937).
132. R. A. Svehla, *Estimated Viscosities and Thermal Conductivities of Gases at High Temperatures*, Tech. report No. R-132, NASA, Lewis Research Center, Cleveland, Ohio (1962), p. 35.
133. P. G. Zambonin, unpublished work.
134. E. Desimoni, F. Palmisano, L. Sabbatini, and P. G. Zambonin, *Anal. Chem.* **50**, 1895 (1978).
135. H. S. Swofford and H. A. Laitinen, *J. Electrochem. Soc.* **110**, 814 (1963).
136. G. J. Hills and P. D. Power, *J. Polarograph. Soc.* **13**, 71 (1967).
137. M. Peleg, *J. Phys. Chem.* **71**, 4553 (1967).
138. E. Desimoni, F. Paniccia, L. Sabbatini, and P. G. Zambonin, *J. Appl. Electrochem.* **6**, 445 (1976).
139. R. L. Every and R. L. Grimsley, *J. Electroanal. Chem.* **9**, 165 (1965).
140. P. G. Zambonin, *J. Electroanal. Chem.* **24**, 365 (1970).
141. P. G. Zambonin and J. Jordan, *J. Am. Chem. Soc.* **89**, 6365 (1967).
142. P. G. Zambonin and J. Jordan, *J. Am. Chem. Soc.* **91**, 2225 (1969).
143. P. G. Zambonin and A. Cavaggioni, *J. Am. Chem. Soc.* **93**, 2854 (1971).
144. P. G. Zambonin, work in progress.
145. L. Sabbatini, P. G. Zambonin, E. Desimoni, and B. A. DeAngeles, *Chim. Ind. (Milan)* **59**, 493 (1977).
146. E. Desimoni, F. Palmisano, and P. G. Zambonin, *J. Electroanal. Chem.* **84**, 323 (1977).
147. P. G. Zambonin, *Anal. Chem.* **41**, 868 (1969).
148. H. S. Swofford and P. G. McCormick, *Anal. Chem.* **37**, 970 (1965).
149. E. Desimoni, F. Palmisano, and P. G. Zambonin, *J. Electroanal. Chem.* **84**, 315 (1977).
150. F. Palmisano, L. Sabbatini, E. Desimoni, and P. G. Zambonin, *J. Electroanal. Chem.* **89**, 311 (1978).
151. G. J. Hills and K. E. Johnson, *Proceedings of the 2nd Int. Congress on Polarography, Cambridge, 1959*, Pergamon Press, London (1961), p. 974.

152. W. E. Triaca and A. J. Arvia, *Electrochim. Acta* **10**, 409 (1965).
153. J. A. A. Ketelaar and A. Dammers-de Klerk, *Rec. Trav. Chim.* **83**, 322 (1964).
154. V. G. Levich, *Acta Physicochim. URSS* **17**, 257 (1942).
155. V. G. Levich, *Acta Physicochim. URSS* **19**, 133 (1944).
156. F. Opekar and P. Beran, *J. Electroanal. Chem.* **69**, 1 (1976).
157. E. Desimoni, B. Morelli, F. Palmisano, and P. G. Zambonin, *Ann. Chim.* **67**, 451 (1977).
158. A. J. Arvia and N. R. De Tacconi, in *Thin Solid Films* **43**, 173 (1977).
159. K. L. Cheng and J. W. Prather II, *CRC Crit. Rev. Anal. Chem.* **5**, 37 (1975).
160. L. Sabbatini, F. Palmisano, P. G. Zambonin, and B. A. De Angelis, *Ann. Chim.* **67**, 525 (1977).
161. B. Morelli, L. Sabbatini, and P. G. Zambonin, *J. Electroanal. Chem.* **96**, 7 (1979).
162. L. Sabbatini, B. Morelli, P. G. Zambonin, and B. A. De Angelis, *J. Chem. Soc. Faraday Trans. 1*, **75**, 2628 (1979).
163. K. J. Kim, N. Winograd, and R. E. Davis, *J. Am. Chem. Soc.* **93**, 6296 (1971).
164. T. Dickinson, A. F. Povey, and P. M. A. Sherwood, *J. Chem. Soc. Faraday Trans. 1* **71**, 298 (1975).
165. J. Benard, *Oxidation des Metaux*, Gauthier Villars et Cie, Paris (1962).
166. P. Kofstad, *High Temperature Oxidation of Metals*, New York (1966).
167. Schon and S. T. Lundin, *J. Electron. Spectrosc.* **1**, 105 (1972–73).
168. K. S. Kim and R. E. Davis, *J. Electron. Spectrosc.* **1**, 251 (1972–73).
169. K. S. Kim and N. Winograd, *Surf. Sci.* **43**, 625 (1974).
170. C. Duval, *Inorganic Thermogravimetric Analysis*, 2nd Ed., Elsevier, London (1963).
171. K. S. Kim, W. E. Baitinger, J. W. Amy, and N. Winograd, *J. Electron Spectrosc.* **5**, 351 (1974).
172. D. G. Lovering, R. M. Oblath, and A. K. Turner, *J. Chem. Soc. Chem. Commun.* 673 (1976).

14

Water Concentration Dependence of Transport Properties in Ionic Melts

P. Claes and J. Glibert

1. Introduction

The considerable solubility of water in ionic liquids, even up to quite high temperatures, is well established. Commercially available solid compounds generally contain at least traces of water, and totally anhydrous melts may often be obtained only after drastic treatment (see Chapter 12 by White, for example). Although the role of water in the chemistry and the electrochemistry of melts has received some attention, the effect of water on transport properties in these media has largely been ignored.

The investigation of solutions of water in ionic liquids over the concentration range extending from the dry melt to aqueous solutions can yield new information about the physicochemical properties of concentrated aqueous solutions. These remain poorly characterized even though they are the most interesting from a technological point of view.

In order to avoid too elaborate experimental arrangements due to the pressures developed at high temperatures by systems containing significant concentrations of water, the choice of systems investigated has been restricted to salts or mixtures of salts whose melting point is reasonably low. Therefore, results reported in this work mainly concern only two systems studied over the entire concentration range: the eut. ($LiNO_3-KNO_3$)–water and the eut. ($NH_4NO_3-NH_4Cl-LiNO_3$)–water mixtures. In order to obtain some information about the influence of the charge and the size of the cation, the work was extended to the $Ca(NO_3)_2 \cdot 4H_2O$–water system and to aqueous solutions of various salts over restricted concentration

P. Claes and J. Glibert • Department of Inorganic and Analytical Chemistry, Catholic University of Louvain, B-1348 Louvain-la-Neuve, Belgium.

ranges. The only hydroxide–water investigated system is the eut. (NaOH–KOH)–water mixture, with mole fractions of water ranging from 0 to 0.1.

In each case, the conductivity and the density were measured as a function of the water concentration. The variation of the viscosity with water content was also investigated for some systems.

The salts investigated in this work are alkali or alkaline earth chlorides, nitrates, or sulfates; the interactions between the ions of the melts or between ions and water molecules are restricted to electrostatic ones, the ions being considered as charged spheres to a first approximation. Hydroxides are different since water molecules become involved in the autoprotolysis of the melt. Therefore molten salts and molten hydroxides will be treated separately in this chapter.

2. Molten Salts

2.1. Molar Volumes

The densities of the eut. (66.7 mol % NH_4NO_3–7.5 mol % NH_4Cl–25.8 mol% $LiNO_3$)–water and of the $Ca(NO_3)_2 \cdot 4H_2O$–water mixtures were determined by pycnometry; a volumetric method using sealed pyrex dilato-

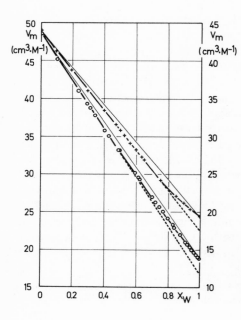

FIGURE 1. Water concentration dependence of the molar volumes of the $(NH_4NO_3–NH_4Cl–LiNO_3)$–water mixture (left side scale, $\bigcirc$) and of the $(LiNO_3–KNO_3)$–water mixture (right side scale, $+$).

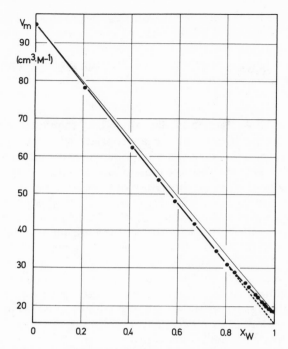

FIGURE 2. Water concentration dependence of the molar volume of the $Ca(NO_3)_2 \cdot 4H_2O - H_2O$ system.

meters was used in order to measure the density of the eut. (50 mol % $LiNO_3-KNO_3$)–water system. The molar volumes of the two eutectic mixture–water systems are plotted against the mole fraction of water (X_W) in Figure 1 while the water concentration dependence of the molar volume of the $Ca(NO_3)_2 \cdot 4H_2O$–water solution is shown in Figure 2. In the case of mixtures containing the hydrated calcium nitrate, X_W is the mole fraction of the water added to the hydrate.

In all cases, the experimental values exhibit negative departures from ideal behavior. This water contraction due to the addition of an electrolyte to the pure solvent is a well-known effect: the ion–water molecule interactions induce a breakup of the solvent structure which gives rise to a lowering of the molar volume. The same phenomena also explain the simultaneous changes of the dielectric constant and of the compressibility of water.[1]

An interesting property of these curves is the linear dependence of the molar volume of the mixture on the mole fraction of water at low water concentrations. This range extends from $X_W = 0$ to $X_W = 0.5$ for the ternary eutectic, to $X_W = 0.7$ for the binary eutectic, and to $X_W = 0.8$ for the hydrate.

The water concentration dependence of the molar volume V_m of the mixture can be described by

$$V_m = X_S V_S^0 + X_W V_W^{0'} \qquad (1)$$

where X_S is the mole fraction of the salt and V_S^0 the molar volume of the pure molten salt or hydrate at the appropriate temperature. $V_W^{0'}$ is the molar volume of water, a constant within the defined concentration range; this value, which is lower than the molar volume of pure water at the same temperature, is dependent on the nature of the electrolyte. $V_W^{0'}$ has a value of 16.9 cm^3 in the case of $(NH_4NO_3-NH_4Cl-LiNO_3)$ whereas the molar volume of pure water (V_W^0) appears to be 18.7 cm^3 at 93°C. In the case of $(LiNO_3-KNO_3)$, values of 18.3 and 19.6 cm^3 are found, respectively, for $V_W^{0'}$ and V_W^0 at 150°C, and $V_W^{0'}$ is 15.22 cm^3, V_W^0 being 18.2 cm^3 for the $Ca(NO_3)_2 \cdot 4H_2O$–water system at 50°C.

At low water concentrations, water molecules interact only with the ions of the solution. As the mole fraction of water progressively increases, the most polarizing species in the mixture initially must become hydrated by one water molecule; thereafter, solvation by more than one molecule or hydration of less polarizing ions can occur. In order to gain agreement between such a model of ion hydration and the experimental results, it must be assumed that the interaction between an ion and a water molecule modifies the volume of that molecule to an extent characteristic of the nature of the ion. The present results clearly show that the contraction of the water molecule caused by Ca^{2+} ions is more extensive than that imposed by monovalent ions.

This modified volume remains constant as long as the water concentration is low enough to preclude other interactions of the water molecules, i.e., with other ions or between water molecules attached to the same ion. At higher water concentrations, deviation from the linear relationship occurs, indicating a progressive increase in the size of the water molecule due to the advent of weaker ion–molecule interactions or of intermolecular interactions.

These results are in agreement with those reported in the literature. The additivity of the molar volumes in molten salts is well known;[2, 3] the same relationship holds good for mixtures of potassium nitrate and molten calcium nitrate tetrahydrate.[4] It may be concluded that $Ca(H_2O)_4^{2+}$ ions exhibit an ideal behavior, just as free ions do. C. A. Angell studied the conductivity of the same mixture;[5] his interpretation of the experimental results assumed that the liquid mixture consisted of NO_3^-, K^+, and $Ca(H_2O)_4^{2+}$ ions.

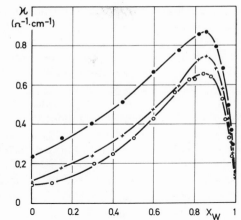

FIGURE 3. Conductivity isotherm of the eutectic mixture–water systems: $\bigcirc$, $(NH_4NO_3-NH_4Cl-LiNO_3)$–water at 93°C; $+$, $(LiNO_3-KNO_3)$–water at 150°C; $\bullet$, $(LiNO_3-KNO_3)$–water at 200°C.

2.2. Water Concentration Dependence of the Conductivity

The specific conductances of all the systems investigated were measured using capillary cells. The experimental results were extrapolated to infinite frequency.

The water concentration dependence of the conductivity of the ternary eutectic–water and of the binary eutectic–water systems is shown in Figure 3. The conductivity increases when water is added to a molten salt; a maximum value is reached at a mole fraction of water of 0.835 for the $(NH_4NO_3-NH_4Cl-LiNO_3)$–water system. A peak also occurs in the $(LiNO_3-KNO_3)$–water system at a mole fraction of water of 0.85 for all temperatures from 150 to 200°C.

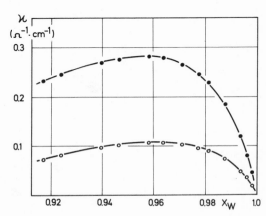

FIGURE 4. Conductivity isotherms of the $Ca(NO_3)_2$–water system; $\bigcirc$, 20°C; $\bullet$, 90°C.

As shown in Figure 4 for calcium nitrate, the conductivity maximum appears at a water mole fraction of about 0.96. Thus the mole fraction of water for the $Ca(NO_3)_2$–water system has been calculated by taking into account the four hydration water molecules. Such conductivity maxima were observed for all concentrated aqueous solutions of strong electrolytes whose solubility was high enough.[6-11] The occurrence of a maximum in the conductivity isotherm is quite normal since the conductivity is the product of the equivalent conductance and the concentration, i.e., a mobility factor multiplied by a charge concentration factor. These two factors act in the opposite sense when the concentration is varied and a maximum value of the conductivity is reached when the charge density increase is exactly counterbalanced by the corresponding decrease of the ionic mobilities.

The ratio of the number of moles of water to the number of moles of salt at the conductivity maximum (R_{max}) yields the following values for the three systems investigated: $R_{max} = 4.9$ for the $(NH_4NO_3–NH_4Cl–LiNO_3)$–water mixture, $R_{max} = 6$ for the $(LiNO_3–KNO_3)$–water system and $R_{max} = 22$ for the $Ca(NO_3)_2$–water solution. The water content at the conductivity maximum is thus strongly dependent upon the charge on the ions in the mixture which, together with the size of the charged particles, determines the strength of the interactions of the ions with water molecules. This being so, the R_{max} values may be correlated with the hydration numbers of the ions.

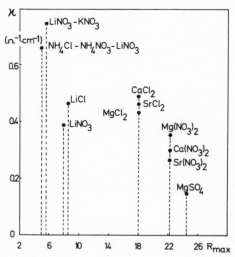

FIGURE 5. Maximum specific conductance and R_{max} values for concentrated aqueous solutions.

For a great number of salts, the conductivity maxima appear at concentrations lower than the solubility; the experimental R_{max} for alkali and alkaline earth salts are shown on the x axis in Figure 5; the ordinates give the values of the conductivity maxima.

Primary hydration numbers are defined as the mean number of water molecules bound to an ion during its translational motions; the corresponding experimental values differ according to the technique used,[12] and thus only rough estimations are available.

The number of moles of water required in order to ensure the complete hydration of one mole of a given electrolyte can be computed using the following values for the hydration numbers: K^+ and NH_4^+, 3.5; Li^+, 5.5; alkaline earth ions, 10; anions, 1. The R_{calc} values derived in this way are 6.5 for lithium salts and 12 for alkaline earth salts. Taking into account the composition of the eutectic mixtures, values of 5 and 5.3 moles of water for one mole of salt are found, respectively, for the $(NH_4NO_3-NH_4Cl-LiNO_3)$ and for the $(LiNO_3-KNO_3)$ eutectic mixtures. The R_{calc} are systematically lower than the experimental R_{max} but their order is maintained; a relationship between R_{max} and hydration numbers thus seems to be reasonable.

It is worth noting that Angell and Sare[13] report that concentrated aqueous solutions of alkali and alkaline earth chlorides separate into two phases at low temperature; the first is practically pure water, the second one is a mixture with a ratio of the number of moles of water to the number of moles of salt of 6 to 8 for the lithium chloride and of 17 to 20 for calcium chloride. Angell and Sare interpret these ratios as the total hydration numbers for the individual electrolytes. The R_{max} values reported above agree quite well with these values.

Ammonium-containing systems exhibit departures from the general behavior just described: for these solutions, R_{calc} values are slightly higher than R_{max}. For the ternary eutectic mixture, R_{calc} is 5 when R_{max} yields 4.9; Campbell and Kartzmark[6] have determined an R_{max} value of 4 in the case of ammonium nitrate, when R_{calc} yields 4.5. This difference could be explained by a lower hydration number of the ammonium ion; the value of one water molecule for one ammonium ion suggested by Peleg[14] is in agreement with this assumption.

Another interesting feature of these results is that the R_{max} values of electrolytes containing the same anion depend on the nature of the cation in the case of monovalent cations ($R_{max} = 7.9$ for $LiNO_3$ and $R_{max} = 6$ for the $LiNO_3-KNO_3$ eutectic mixture) but are the same for all the alkaline earth salts (18 for all the chlorides and 22 for all the nitrates). The reason for this may be that, owing to their high charge density, divalent cations are surrounded by a double hydration sheath. According to Berecz and Achs-Balla[15] this hypothesis is corroborated by the fact that the radii of

TABLE 1. Crystallographic Radii of Divalent
Cations Compared with the Radii of the
Hydrated Ions

Cation	Crystallographic radius (Å)	Radius of the hydrated ion (Å)
Mg^{2+}	0.65	4.28
Ca^{2+}	0.99	4.12
Sr^{2+}	1.13	4.12

the hydrated ions[16] are larger than the sum of the crystallographic radius
plus the diameter of a water molecule, 1.38 Å.[17] A comparison of these
values is shown in Table 1.

Furthermore, the radii of these hydrated ions are the same, because,
once they are surrounded by one complete hydration sheath, their charge
densities become almost equal; they then stabilize the same number of
water molecules in the second sheath.

2.3. Temperature Dependence of the Conductivity

Figure 6 shows the water concentration dependence of the activation
energy for the equivalent conductance of the $CaCl_2$–H_2O and the
$Ca(NO_3)_2$–H_2O mixtures. Figure 7 illustrates the influence of the water
concentration on the activation energy for the equivalent conductance of
the $(LiNO_3$–$KNO_3)$–H_2O solution (dotted line).

In the case of calcium nitrate and of the eutectic system, the temperature dependence of the equivalent conductance is of the Vogel–Tamman–
Fulcher (V.T.F.), type: the activation energy is thus temperature dependent.[18-21] A mean activation energy for the temperature range investigated

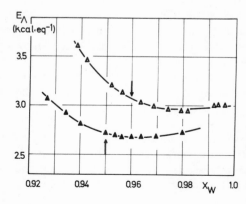

FIGURE 6. Activation energy for the
equivalent conductance of the
$Ca(NO_3)_2$–H_2O ($\triangle$) and of the
$CaCl_2$–H_2O ($\blacktriangle$) mixtures as functions of
X_W. The arrows refer to the concentrations
at the conductivity maximum.

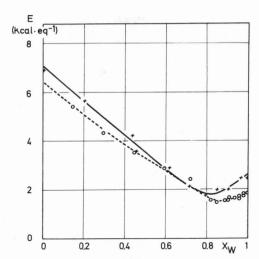

FIGURE 7. Water concentration dependence of the activation energies for equivalent conductance (– – ○ – –) and for viscosity (—— + ——) of the $(LiNO_3–KNO_3)$–water system.

was used for these two systems but the same shape is obtained for curves giving the water concentration dependence of the activation energy at a given temperature.

These diagrams suggest a change in the transport mechanism when the composition of the mixture is varied, since the activation energy decreases when water is progressively added to the molten salt. Once a given concentration is reached, further addition of water does not significantly change the activation energy any more. For the three systems investigated, the concentration at which that change occurs is very close to that at which the conductivity maximum appears.

2.4. Viscosity of Salt–Water Mixtures

As shown in Figure 8, the viscosity of the systems investigated decreases when the mole fraction of water increases.

More interesting information is gained from the water concentration dependence of the activation energy for viscous flow of the $(LiNO_3–KNO_3)$–H_2O mixture. The corresponding curve is shown in Figure 7 together with the activation energy for the equivalent conductance. Again, a marked change in behavior does occur in the vicinity of the conductivity maximum.

2.5. Conclusions

From the above results the following model can be proposed. The concentration zone in the vicinity of the conductivity maximum appears as

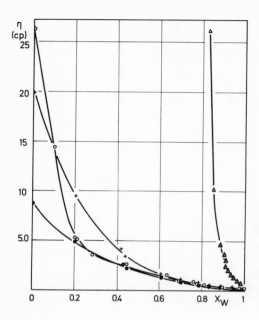

FIGURE 8. Water concentration dependence of the viscosity of salt–water system: $Ca(NO_3)_2$–H_2O at 50°C ($\triangle$); $(NH_4Cl$–NH_4NO_3–$LiNO_3)$–H_2O at 93°C ($\bigcirc$); $(LiNO_3$–$KNO_3)$–H_2O at 150°C ($+$); $(LiNO_3$–$KNO_3)$–H_2O at 200°C ($\bullet$).

a kind of "no man's land" between two regions with characteristic transport behavior. The first one extends from the pure molten salt to approximately R_{max}; these liquids are ionic liquids containing free hydrated or incompletely hydrated ions. In that region, water does not act as a dielectric medium and the electrostatic interactions are quite large. Changes in the water concentration induce modifications in the size of the moving species. The second region ranges from about R_{max} to pure water; in these mixtures, hydrated ions migrate in a dielectric medium formed by water molecules. The size of the migrating particles is practically concentration independent.

Nevertheless, the increase of conductivity and of the equivalent conductance as well as the decrease of the activation energy caused by the addition of water to the molten salt cannot be entirely explained by this size increase of the ions. As shown in Table 2, when series of electrolytes with the same anion are compared, an increase in the size of the cation lowers the specific conductivity as well as the equivalent conductance and either has no influence upon, or increases the activation energy for, conductance.[22] On the other hand, if the size of the ions was the factor governing the specific conductance of a given mixture, then the conductivities of all similar anionic salts of the alkaline earths at the conductivity maximum would be equal since the radii of their hydrated cations are the same (Table 1). Figure 5 shows that this is not true: the difference between

TABLE 2. Specific and Equivalent Conductances and Activation Energies for Equivalent Conductance of Alkali Nitrates and Chlorides

	κ $(\Omega^{-1}\,cm^{-1})$	Λ $(\Omega^{-1}\,cm^2\,eq^{-1})$	E (kcal/mol)
	$T = 640$ K		
LiNO$_3$	1.433	57.5	3.6
NaNO$_3$	1.234	56.3	3.2
KNO$_3$	0.721	39.4	3.6
RbNO$_3$	0.547	33.2	3.5
CsNO$_3$	0.447	30.1	3.7
	$T = 1080$ K		
LiCl	6.642	198	2.0
NaCl	3.596	135	3.0
KCl	2.229	110	3.4
RbCl	1.742	97	4.4
CsCl	1.588	100	5.1

the conductivities of calcium and magnesium chlorides or between magnesium and strontium nitrates at R_{max} is significant.

Conductivity increases induced by the addition of water to ionic liquids have also been observed by Campbell and co-workers.[23] A decrease in the activation energy was also evident.[24] These authors compared the effect of adding water to these systems with that of other molecular solutes such as methanol or nitrobenzene. In contrast to water additions, the two organic liquids slightly lower the specific conductance of the mixture. This effect would correspond with an increase in size of the ions of the mixture.

A more complete explanation of the influence of water on the equivalent conductance of ionic liquids requires a greater participation of water itself in current conduction by very concentrated aqueous solutions. Campbell and Williams[24] suggest the presence of ionized water molecules. Obviously, more experimental data are needed in order to support or invalidate this assumption.

3. The Molten NaOH–KOH Eutectic Mixture

The eutectic composition of this system is 51 mol % sodium hydroxide and 49 mol % potassium hydroxide.[25] The conductivity of the dry melt and of solutions of water from $X_W = 0$ to $X_W = 0.1$ was measured using a compact alumina capillary cell.[26] The specific conductances were extrapolated to infinite frequency.

3.1. Pure Eutectic Mixture

Table 3 gives a comparison of the equivalent conductance of the molten hydroxide mixture with those of fused sodium and potassium salts. These values are surprisingly comparable, though molten hydroxides differ from simple ionic liquids in many ways.

The interactions between the ions in an hydroxide melt are not strictly comparable with electrostatic interactions between ionic hard spheres: the formation of aggregates by hydrogen bonding between hydroxide ions has been deduced from IR spectroscopic investigations.[28]

In the case of molten salts, cations seem to play the major role in transport processes; in most cases, the transport number of cation is about 0.6.[29] Shvedov and Yvanov[30] measured the transport numbers of Na^+ and K^+ ions, respectively, in molten NaOH and KOH. They found $t_{Na+} = 0.10 \pm 0.03$ and $t_{K+} = 0.03 \pm 0.03$; thus current conduction is almost entirely associated with the anions.

In order to unify these results, a rather special transport mechanism has to be invoked. A proton transfer conduction mechanism similar to that proposed for water and aqueous solutions would be suitable. In molten hydroxides, proton transfer might occur through the direct involvement of water molecules and/or of oxide ions; these moieties would then take on the role of hydronium and hydroxide ions, respectively, corresponding to aqueous solutions.

The concentration of oxide ions and water molecules in the dry eutectic mixture is about 10^{-9} mol/cm^3 at 500 K.[31] This seems to be too low a concentration to explain the electrical conductance of molten hydroxides by a simple proton transfer mechanism via H_2O or O^{2-}. At 523 K, the conductivity of the neutral melt is 0.87 Ω^{-1} cm^{-1}; it rises to only 1.08 Ω^{-1} cm^{-1} for a mole fraction of water of 0.1 (i.e., for a concentration of water 10^5 higher than in the neutral melt).

TABLE 3. Comparison of the Equivalent Conductances of Molten Hydroxides and Fused Salts

T, K	Λ (Ω^{-1} cm^2 eq^{-1})			
	NaNO$_3$[a]	KNO$_3$[a]	KSCN[b]	NaOH–KOH
458	—	—	10.5	12.8
530	—	—	23.0	24.0
600	47.6	—	—	34.4
650	58.5	41.2	—	39.6

[a] Reference 22.
[b] Reference 27.

The high conductivity of the neutral melt could be explained if it is assumed that proton transfer is particularly rapid in this case; a reason for this might lie in a favorable mutual reorientation of hydroxide ions. The experimental results give no information about such an effect, but, if this interpretation is correct, the structure of the molten hydroxide should exert a large influence on the migration processes.

3.2. Acidic Melts

The influence of water on the conductivity of the NaOH–KOH mixture is shown in Figure 9.

Except at low concentrations the conductivity of the molten eutectic mixture is directly proportional to the water concentration. This effect of water is correctly described by a contribution $\mu_1 F C_{H_2O}$, where μ_1 is the conventional mobility of the proton by transfer from H_2O to OH^- ($H_2O + OH^- \rightarrow OH^- + H_2O$), F is the Faraday constant, and C_{H_2O} is the analytical concentration of water.

In order to describe the water concentration dependence at low concentration, a negative parameter must be introduced. A contribution of the form $-A(C_{H_2O})^{1/2}$, where A is a constant, is mathematically suitable; it could be the result of an influence of water molecules on the structure of the melt.

The conductivity of the acidic melt is thus described by the following equation:

$$\varkappa = K + \mu_1 F C_{H_2O} - A(C_{H_2O})^{1/2} \qquad (2)$$

K being the conductivity of the dry melt: $0.710 \ \Omega^{-1} \ cm^{-1}$ at 222°C. Introducing experimental values of the specific conductances for two different concentrations, one calculates $A = 1.025$ and $\mu_1 = 7.52.10^{-4} \ cm^2$ $sec^{-1} \ V^{-1}$ at 222°C. Equation (2) with these values of A and μ_1 exactly describes the experimental curve of Figure 9.

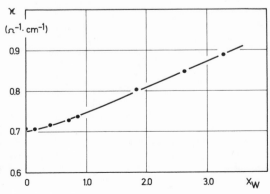

FIGURE 9. Water concentration dependence of the conductivity of the (NaOH–KOH)–water mixture at 222°C.

The mobility μ_1 is substantially larger (four times) than cation mobilities in molten salts in the same temperature range (for instance, the values of μ_{K^+} and μ_{Na^+} in $NaNO_3$–KNO_3 mixture at 275°C are,[32] respectively, $1.81.10^{-4}$ and $2.08.10^{-4}$).

The above interpretation could not, of course, be considered as a detailed description of the current conduction mechanism in molten hydroxides, but it gives an interesting starting point for the discussion of further experimental results.

References

1. J. O'M. Bockris, *Quart. Rev. Chem. Soc.* **3**, 173 (1970).
2. J. Braunstein, in *Ionic interactions*, Vol. I, Chap. IV, Ed. S. Petrucci, Academic, London (1971).
3. E. R. Van Artsdalen and I. S. Yaffe, *J. Phys. Chem.* **59**, 118 (1955).
4. J. Braunstein, L. Orr, and W. McDonald, *J. Chem. Eng. Data* **12**, 415 (1967).
5. C. A. Angell, *J. Electrochem. Soc.* **112**, 1224 (1965).
6. A. N. Campbell and E. M. Kartzmark, *Can. J. Chem.* **30**, 128 (1952).
7. A. N. Campbell, E. M. Kartzmark, M. E. Bednas, and J. T. Herron, *Can. J. Chem.* **32**, 1051 (1954).
8. A. N. Campbell, G. H. Debus, and E. M. Kartzmark, *Can. J. Chem.* **33**, 1508 (1955).
9. A. N. Campbell, J. B. Fishman, G. Rutherford, T. P. Schaefer, and L. Ross, *Can. J. Chem.* **34**, 151 (1956).
10. A. N. Campbell and W. G. Patterson, *Can. J. Chem.* **36**, 1004 (1958).
11. D. A. Lown and H. R. Thirsk, *Trans. Faraday Soc.* **67**, 132 (1971).
12. J. O'M. Bockris and A. K. N. Reddy, *Modern Electrochemistry*, Vol. I, p. 118, Rosetta Edition, Plenum, New York (1973).
13. C. A. Angell and E. J. Sare, *J. Chem. Phys.* **52**, 1058 (1970).
14. M. Peleg, *J. Phys. Chem.* **75**, 2066 (1971).
15. E. Berecz and M. Achs-Balla, *Acta Chim. Acad. Sci. Hung.* **77**, 267 (1973).
16. E. Nightingale Jr., *J. Phys. Chem.* **63**, 1381 (1959).
17. J. O'M. Bockris and A. K. N. Reddy, *J. Chem. Phys.* **52**, 1058 (1970).
18. H. Vogel, *Phys. Z.* **22**, 645 (1921).
19. G. Tamman and G. Hesse, *Z. Anorg. Allgem. Chem.* **156**, 245 (1926).
20. G. S. Fulcher, *J. Am. Ceram.* **8**, 339 (1925).
21. C. A. Angell, *J. Phys. Chem.* **68**, 218 (1964).
22. G. J. Janz, F. W. Dampier, G. R. Lakshminarayanan, P. K. Lorenz, and R. P. T. Tomkins, *Molten Salts*, Vol. I, *Electrical Conductance, Density and Viscosity Data*, National Bureau of Standards, Washington (1968).
23. A. N. Campbell, E. M. Kartzmark, and D. F. Williams, *Can. J. Chem.* **40**, 890 (1962).
24. A. N. Campbell and D. F. Williams, *Can. J. Chem.* **42**, 1984 (1964).
25. H. W. Otto and R. P. Seward, *J. Chem. Eng. Data* **9**, 507 (1964).
26. C. Dauby, J. Glibert, and P. Claes, *Electrochim. Acta* **24**, 35 (1979).
27. P. Dulieu and P. Claes, *Bull. Soc. Chim. Belges* **82**, 639 (1973).
28. J. Greenberg and L. J. Hallgreen, *J. Chem. Phys.* **35**, 180 (1961).
29. R. Oyamada, *J. Phys. Soc. Jpn.* **26**, 1068 (1969).
30. V. P. Shvedov and I. A. Yvanov, *Russ. J. Phys. Chem.* **39**, 396 (1965).
31. J. Goret and B. Tremillon, *Bull. Soc. Chim.* 1074 (1964).
32. E. P. Honig and J. A. A. Ketelaar, *Trans. Faraday Soc.* **62**, 1 (1966).

15

The Solution Chemistry of Water in Melts

Richard Combes

1. Introduction

Water is certainly the most important compound in nature. Owing to its properties and its widespread occurrences, it has much to commend it as a solvent. But, this is no longer true at high temperatures. However, molten hydrates are ionic liquids in which the water molecules are one of the components, just as in mixed solvents.

In the case of molten salts, the situation is the converse of that of diluted aqueous electrolytic solutions. Indeed, a molten salt may be considered as a solvent in which several components can be dissolved, among them, one can expect water, which is then in effect a solute.

Solute water is a very reactive compound. Its properties interfere in numerous practical situations such as the corrosive attack on materials in the presence of melts and chemical transformations in the latter, either of the solvent itself (hydrolysis) or, of the various other solutes. Thus, it appears to be of considerable interest to know more about the properties of water within the various types of molten salts and to understand its reaction processes.

These properties correspond to two main types to which the chemist has been accustomed to refer in the analysis of reactions in solution, viz., acid–base properties and redox properties. This will be the guideline in the following study for the presently important molten salts (chlorides, fluorides, hydroxides, nitrates) in Sections 3 and 4.

Richard Combes • Laboratoire d'Electrochimie Analytique et Appliquée (Associé au CNRS, LA No. 216), Ecole Nationale Supérieure de Chimie de Paris, Université Pierre et Marie Curie, 11, rue Pierre et Marie Curie, F-75231 Paris Cedex 05, France.

Naturally, the first question to be raised is that of the solubility of water vapor in media whose temperatures can reach relatively high values; this will be discussed in the next section.

In Section 5, some consequences related to the presence of water will be presented and explained through potential–acidity diagrams.

2. Solubility of Water

Like many gases, water is soluble in molten salts, which perhaps at first sight is surprising having regard to the high temperatures of these solvents. This or perhaps the technological difficulty of the corresponding measurements, may explain the relative paucity of water solubility data available in the literature.[1-4]

This solubility can be explained not only "physically" by the concept that a molten salt contains voids,[5-7] but, as described by Flengas and Block-Bolten,[3] by chemical reaction leading to the formation of one or more complex ionic species, to which more particular attention will be given in a further section of the present work.

According to the reactivity of the melt considered, experimental evidence for one or the other or both types of solubility is given. For instance, heats of mixing with Li-bearing melts have generally been found to be exothermic, because of the hydration reaction;[8-13] an excellent discussion of the corresponding results is given by Field.[4] Moreover, the presence of OH^- ions in such melts has been shown either electrochemically in chlorides,[14-17] nitrates,[12] and fluorides,[18,19] or by chemical analysis of quenched samples of LiCl–KCl.[20]

In less acidic melts, a more physical mechanism of dissolution has to be considered, for instance, recent spectroscopic studies on molten alkali chlorides demonstrate the existence of H bonds in the melts between H_2O and Cl^- ions.[21]

In fact, Burkhard and Corbett[10] had already mentioned the possibility of both kinds of dissolution when they attributed the change in the Henry's law constant, at high water vapor pressures, to hydrolysis of the melt.

Whatever the mechanism of dissolution might be, water solubilities have been determined according to two main experimental methods. In the first one, some property of the melt (gravimetric, volumetric, electrochemical) is followed while water vapor is absorbed. The second method is cryoscopy, which is based upon the decrease in the melting point of the solvent. In contradiction to other gas solubility determinations, elution or quenching techniques seem to have been scarcely used.[22,23,13] Interesting discussions on the reliability of these methods were made by Flengas and Block-Bolten[3] or by Field,[4] but it appears to us that voltammetry—

whenever possible—is the easiest and most reliable method, considering the usual high reactivity of water with melts. This point of view seems to be confirmed by the general trend which can be noticed in the more recent investigations.[12, 17–19, 24–27]

A summary of solubility data for water in fused salts is given in Table 1, where the units are those used by the various authors. Whenever possible, the Henry's law constant has been calculated in atmospheres per molality units, and the method used for the determination has been indicated.

As can be seen from this table, very few data, except those concerning molten nitrates and LiCl–KCl, on water solubility in fused salts, are available. Moreover, the water vapor pressures in equilibrium with the melt are often lacking in the other available data. Nevertheless, some remarks can be made on the following salts.

Chlorides. As a general idea, it can be said roughly that water is more soluble in carnalite than in LiCl–KCl (8 times, if we assume that atmospheric moisture was in equilibrium with the melt in the experiment quoted in Reference 30), which is 1000 times more soluble than in NaCl, allowing for the temperature difference. In fact, this difference cannot be neglected since, the higher the temperature, the less soluble is the water (a factor of 2 for 90°C in molten LiCl–KCl).

As regards LiCl–KCl melts, data given by Laitinen and co-workers[14] and those of Raynor[20] can hardly be compared because the residual moisture above the melt was not mentioned. As already pointed out by Burkhard and Corbett,[10] the solubility of water is mainly due to the presence of Li^+ ions and does not change very much with the composition, KCl functioning as a physical diluent only.

Hydroxides. The data obtained by different methods[31, 32] are very consistent and lead to a value of 0.24 mol kg^{-1} under $P_{H_2O} = 1$ atm at 753 K which is $2\frac{1}{2}$ times higher than the corresponding solubility in LiCl–KCl.

Nitrates. Most studies on water solubility have been performed on these melts and interesting comparisons can be made. There is a very good agreement between Peleg's voltammetric determinations[11] and Bertozi's gravimetric ones[8] on pure $LiNO_3$. The extrapolation of their results to the lowest available temperature in LiCl–KCl (663 K) leads to a solubility value of the same order of magnitude as in these melts (i.e., 0.3 mol kg^{-1} instead of 0.2 mol kg^{-1} for $P_{H_2O} = 1$ atm).

More care should be taken with Duke's and Doan's[9] volumetric determinations, which lead to solubility values about ten times lower for the same water pressure (1 atm) and comparable amounts of Li^+ (i.e., 9×10^{-2} mol kg^{-1} for $LiNO_3(0.467)$–$NaNO_3(0.282)$–$KNO_3(0.251)$ at 513 K instead of 0.8 mol kg^{-1} for $LiNO_3(0.25)$–$KNO_3(0.75)$ at 503 K according to Reference 8).

TABLE 1. Solubility Data of Water in Molten Salts and Experimental Technique

Melt	Composition, mol. fr.	Temperature, K	Solubility data (as given by authors)	Ref.	Henry's law const., atm mol^{-1} kg	Experimental Technique
Chlorides	NaCl	1123	1.6×10^{-7} mol cm^{-3} under $p_{H_2O} = 1$ atm	28	9.56×10^3	Equilibrium study
	NaCl(0.5)–KCl(0.5)	1013	3 μmol mol^{-1}	16		Voltammetry
	LiCl(0.5)–KCl(0.5)	663	30 μmol torr^{-1} (mol Li$^+$)$^{-1}$	10	5.14[a]	Volumetric technique
	LiCl(0.5)–KCl(0.5)	753	14.1 μmol torr^{-1} (mol Li$^+$)$^{-1}$	10	10.96[a]	Volumetric technique
	LiCl(0.53)–KCl(0.47)	753	11.8 μmol torr^{-1} (mol Li$^+$)$^{-1}$	10	12.18[a]	Volumetric technique
	LiCl(0.6)–KCl(0.4)	663	30.5 μmol torr^{-1} (mol Li$^+$)$^{-1}$	10	4.13[a]	Volumetric technique
	LiCl(0.6)–KCl(0.4)	753	11.3 μmol torr^{-1} (mol Li$^+$)$^{-1}$	10	10.70[a]	Volumetric technique
	LiCl(0.686)–KCl(0.314)	753	10.8 μmol torr^{-1} (mol Li$^+$)$^{-1}$	10	9.33[a]	Volumetric technique
	LiCl(0.59)–KCl(0.41)	723	1.9 g water in 195 g	14		Voltammetry
	LiCl(0.59)–KCl(0.41)	723	2.5×10^{-6} mol in 100 g	20		Voltammetry and chemical analysis
	MgCl$_2$(0.5)–KCl(0.5)	1008	0.03 mol %	29		Voltammetry
	MgCl$_2$(0.5)–KCl(0.5)	853–873	0.2 mol %	30		Gravimetry
Fluorides	LiF–BeF$_2$	973	$\leq 10^{-3}$ mol kg^{-1} atm^{-1}	88	$\geq 10^3$	Volumetric technique

Hydroxides / Nitrates		623–773	Linear relation between solubility pressure and the H_2O vapor		$10^{(4.88-3210/T)}$	Gravimetry
Hydroxides	NaOH	623–773		31		Gravimetry
	NaOH	640	84 μmol mol^{-1} torr^{-1}	32	0.63	Volumetric technique
	NaOH	690	44 μmol mol^{-1} torr^{-1}	32	1.20	Volumetric technique
Nitrates	NH_4NO_3	442	3 mol %	33		Cryoscopy
	$LiNO_3(0.125)$–$NaNO_3(0.464)$–$KNO_3(0.411)$	513	1.02 μmol mol^{-1} torr^{-1}	9	115.6	Volumetric technique
	$LiNO_3(0.467)$–$NaNO_3(0.282)$–$KNO_3(0.251)$	513	10.0 μmol mol^{-1} torr^{-1}	9	10.7	Volumetric technique
	$LiNO_3(0.869)$–$NaNO_3(0.069)$–$KNO_3(0.062)$	513	49.7 μmol mol^{-1} torr^{-1}	9	1.9	Volumetric technique
	$CsNO_3(0.567)$–$Ba(NO_3)_2(0.433)$	melting point > 573	630 μmol mol^{-1}	34		Cryoscopy
	$CsNO_3$	690	920 μmol mol^{-1}	34	$16 < p_{H_2O} < 21$ torr	Cryoscopy
	$LiNO_3$	527	1300 μmol mol^{-1}	34		Cryoscopy
	$NaNO_3$	580	1410 μmol mol^{-1}	34		Cryoscopy
	$NaNO_3(0.5)$–$KNO_3(0.5)$		Water solubility not affected by the presence of KCl from 0 to 5 mol %	35		
	$NaNO_3(0.5)$–$KNO_3(0.5)$	623	5×10^{-3} mol kg^{-1} for $2 < p_{H_2O} < 11$ torr	36	1.7 (mean value)	Voltammetry
	$LiNO_3$	548	161 μmol mol^{-1} torr^{-1}	11	0.56	Voltammetry
	$LiNO_3$	583	87 μmol mol^{-1} torr^{-1}	11	1.04	Voltammetry
	$LiNO_3$	610	59 μmol mol^{-1} torr^{-1}	11	1.54	Voltammetry

(continued)

TABLE 1 (continued)

Melt	Composition, mol. fr.	Temperature, K	Solubility data (as given by authors)	Ref.	Henry's law const., atm mol^{-1} kg	Experimental Technique
Nitrates (cont.)	NaNO$_3$	583	22 μmol mol^{-1} torr^{-1}	11	5.08	Voltammetry
	NaNO$_3$	606	17 μmol mol^{-1} torr^{-1}	11	6.58	Voltammetry
	KNO$_3$	610	2 μmol mol^{-1} torr^{-1}	11	66.5	Voltammetry
	LiNO$_3$	538	232 μmol mol^{-1} torr^{-1}	8	0.39	Thermobalance
	LiNO$_3$	553	165 μmol mol^{-1} torr^{-1}	8	0.55	Thermobalance
	LiNO$_3$(0.75)–KNO$_3$(0.25)	503	273 μmol mol^{-1} torr^{-1}	8	0.37	Thermobalance
		538	137 μmol mol^{-1} torr^{-1}	8	0.74	Thermobalance
	LiNO$_3$(0.25)–KNO$_3$(0.75)	503	99 μmol mol^{-1} torr^{-1}	8	1.24	Thermobalance
	LiNO$_3$(0.25)–KNO$_3$(0.75)	538	54 μmol mol^{-1} torr^{-1}	8	2.27	Thermobalance
	LiNO$_3$(0.75)–NaNO$_3$(0.25)	503	354 μmol mol^{-1} torr^{-1}	8	0.27	Thermobalance
	LiNO$_3$(0.75)–NaNO$_3$(0.25)	538	165 μmol mol^{-1} torr^{-1}	8	0.58	Thermobalance
	LiNO$_3$(0.25)–NaNO$_3$(0.75)	538	84 μmol mol^{-1} torr^{-1}	8	1.27	Thermobalance
	NaNO$_3$(0.45)–KNO$_3$(0.55)	553	50 μmol mol^{-1} torr^{-1}	13	2.47	Elution
	NaNO$_3$(0.5)–KNO$_3$(0.5)	503	8.8 $\times$ 10^{-4} mol kg^{-1} torr^{-1}	12	1.49	Voltammetry
	NaNO$_3$(0.5)–KNO$_3$(0.5)	537	5.15 $\times$ 10^{-4} mol kg^{-1} torr^{-1}	12	2.55	Voltammetry
	NaNO$_3$(0.5)–KNO$_3$(0.5)	556	3.95 mol kg^{-1} torr^{-1}	12	3.33	Voltammetry
	LiNO$_3$(0.5)–KNO$_3$(0.5)	559	2.41 $\times$ 10^{-3} mol per moles of melt	24		Voltammetry
	NaNO$_3$(0.5)–KNO$_3$(0.5)	523	0.001 < [H$_2$O] < 0.013 m 2 < p_{H_2O} < 30 torr	25	2.6	Voltammetry

^{a} The solubility follows Henry's law to p_{H_2O} = 18 torr at 753 K and to p_{H_2O} = 14 torr at 663 K.

Once more, the comparison between the various nitrates shows clearly the predominant part played by Li^+ ions.

A fair agreement can also be found between the more recent voltammetric determination made by Zambonin *et al.*[12] and Espinola and Jordan[25] on water solubility in $NaNO_3$–KNO_3 (respectively, 0.45 mol kg^{-1} atm^{-1} and 0.38 mol kg^{-1} atm^{-1} at 523 K for $P_{H_2O} = 1$ atm).

3. Acidobasicity of Water

3.1. Introduction

For a good understanding of the phenomena occurring in the study of molten salts, systematic theories on acidity, based on the laws of ionic equilibria, have been established. The first and most well known follows the Brönsted–Lowry definition to whose concept of "acceptor" and "donor" other theories will refer, considering its usefulness in low-temperature solvents. A generalization is to consider as acid–base systems the acceptor–donor couples involving other ionic particles than the proton. This is the case for the chloroacidity concept where

$$acid + Cl^- = base$$

which, coupled with the definition of the solvent system, according to Franklin,[37] enables an understanding of the chemical properties in molten chloroaluminates, for instance ($2AlCl_4^- \rightleftarrows Al_2Cl_7^- + Cl^-$, for more details, see Reference 38). This is also the case for the Lux–Flood definition[39, 40] in which the ionic particle is the oxide ion O^{2-}. Thus, every oxocomplex (i.e., compounds involving O^{2-}) are oxobases, as for instance metal oxides, hydroxides, nitrates, sulfates, carbonates, etc. Conversely, nonmetal oxides, metal cations, etc., are oxoacids. These families of compounds are among the most important ones in fused salt media as is highly emphasized in various general surveys on this field of chemistry.[41–46]

It is easy to understand that water can behave as a donor, an acceptor, or both, and the acid–base equilibria involved will be as below, according to the definition of acidity considered:

Brönsted and Lowry:

$$H_2O \rightleftarrows H^+ + OH^- \tag{1}$$

$$H_2O + H^+ \rightleftarrows H_3O^+ \tag{2}$$

Solvoacidity concept (as $AX^- \rightleftarrows A + X^-$):

$$H_2O + AX^- \rightleftarrows A + H_2OX^- \tag{3}$$

$$H_2O + 2AX^- \rightleftarrows A_2O + 2X^- + 2H^+ \tag{4}$$

Oxoacidity concept:

$$H_2O + O^{2-} \rightleftharpoons 2OH^- \tag{5}$$

$$H_2O + 2X^- \rightleftharpoons 2HX + O^{2-} \tag{6}$$

Whenever possible, a description of the study of the corresponding equilibria in the melts considered in this chapter is given below. However, it must be pointed out that very often more than one concept (i.e., oxo- and Brönsted acidities, solvo- and Brönsted acidities) could be applied to the acido-basicity of water in some salts (this is the case, for example, with molten hydroxides, where the three could be applied, or with hydrogen fluorides), but one of them has been chosen for more convenience.

3.2. Chlorides

The hydrolysis of molten or solid halides has been studied for a long time[47-61, 10, 14, 16, 20, 29, 30] and evidence was given for HCl evolution and formation for more or less soluble oxide species in the melt. As mentioned above and confirmed by the infra-red spectroscopy determinations carried out by Mignonsin and Duyckaerts on NaCl–KCl,[55] the chemical reaction involved is

$$H_2O + 2Cl^- \rightleftharpoons 2HCl + O^{2-}$$

Assuming that the activity of chloride ions remains constant and close to unity, and that solutions are diluted enough for activity coefficients to be constant, one may express the constant of the above equilibrium either by means of the partial pressures (very close to their fugacity) of the gaseous species or by their concentrations if the corresponding Henry's law relationship is known. Using thermochemical data, Edeleanu and Littlewood[62] calculated the value of this equilibrium constant, but no direct quantitative determination was possible until the problems arising in the measurement of oxide ion concentrations could be avoided.[63-65] This was mainly possible by using an oxide-ion-selective membrane electrode (for example, made of calcia-stabilized zirconia) whose behavior in various molten chlorides was tested by Combes et al.[59, 61, 66-74] In melts at temperatures lower than the one (600°C) corresponding to a sufficient conductivity of the calcia-stabilized zirconia, determinations have been performed either by means of other electrochemical methods[75] or by using various kinds of electrode.[75-79] In the latter case, special attention should be paid to the new type of membrane electrode developed by Flinn and Stern[78] in molten nitrates and recently used by Picard et al. in LiCl–KCl at 743 K.[79] According-ing to the melt, more detailed explanations and whenever possible, the constants corresponding to the two oxoacidobasic equilibria of water (K_1 and K_2)

$$2OH^- \rightleftarrows O^{2-} + H_2O, \qquad K_1 = \frac{[O^{2-}]p_{H_2O}}{[OH^-]^2}$$

$$H_2O + 2Cl^- \rightleftarrows O^{2-} + 2HCl, \qquad K_2 = \frac{[O^{2-}]p_{HCl}^2}{p_{H_2O}}$$

are given below (units for equilibrium constants are molality for nonvolatile solutes and atmospheres for gaseous species) and are also collected in Table 2.

In LiCl–KCl, both oxoacidic and oxobasic properties of water have been determined by Lysy and Combes[59] by measuring the calcia-stabilized zirconia electrode potential in melts either containing hydroxide ions and water at a given partial pressure (determination of K_1) or equilibrated with gaseous mixtures of HCl and H_2O (determination of K_2) at temperatures ranging from 915 to 1015 K. They derived the following expressions for the logarithm of the equilibrium constant:

$$\log K_1 = 6.61 - \frac{7.68 \times 10^3}{T}$$

and

$$\log K_2 = 3.54 - \frac{10.03 \times 10^3}{T}$$

which enable these results to be compared with those obtained by a different method by Kanzaki and Takahashi[75, 76] in LiCl–KCl at 723 K, and to those obtained more recently at 743 K by Picard *et al.*[79] using an yttria-stabilized zirconia tube filled with molten LiCl–KCl containing oxide and silver ions. Agreement between these sets of results is quite satisfactory as differences between the various log K_2 values are lower than 0.5 units.

In NaCl–KCl, too, both oxoacidic and oxobasic properties of water have been investigated by Combes *et al.*[57, 67, 68] using the same procedure in a 975–1090 K range, for which the following equilibrium constant values were derived:

$$\log K_1 = 5.55 - \frac{8.15 \times 10^3}{T}$$

$$\log K_2 = 41.35 - \frac{55.33 \times 10^3}{T}$$

A comparison of these results at the same temperature clearly demonstrates that water behaves as a weaker oxoacid and a stronger oxobase in LiCl–KCl than in NaCl–KCl. These differences in behavior can easily be

TABLE 2. Acid–Base Equilibria of Water and Their Corresponding Equilibrium Constant Value in Molality and Atmosphere Units

Melts	Composition	T/K	Acid–base equilibrium	$pK_2 = -\log K_2$	Ref.	Note
Chlorides	LiCl–KCl (0.45)(0.55)	1000	$2OH^- \rightleftharpoons O^{2-} + H_2O$	1.07	59	a
		1000	$H_2O + 2Cl^- \rightleftharpoons O^{2-} + 2HCl$	6.50	59	
		723	$H_2O + 2Cl^- \rightleftharpoons O^{2-} + 2HCl$	9.93	79	
		723	$H_2O + 2Cl^- \rightleftharpoons O^{2-} + 2HCl$	9.90	75	
	NaCl–KCl (0.5) (0.5)	1000	$2OH^- \rightleftharpoons O^{2-} + H_2O$	2.6	57, 67, 68	a
		1000	$H_2O + 2Cl^- \rightleftharpoons O^{2-} + 2HCl$	14.0	57, 67, 68	a
	NaCl–KCl–CaCl$_2$ (0.4) (0.4) (0.2)	1000	$H_2O + 2Cl^- \rightleftharpoons O^{2-} + 2HCl$	4.6	73	
	NaCl–KCl–CaCl$_2$ (0.35)(0.35) (0.3)	1000	$H_2O + 2Cl^- \rightleftharpoons O^{2-} + 2HCl$	4.2	73	
	NaCl–KCl–MgCl$_2$ (0.4) (0.4) (0.2)	1000	$H_2O + 2Cl^- \rightleftharpoons O^{2-} + 2HCl$	2.7	73	
	NaCl–KCl–MgCl$_2$ (0.35)(0.35) (0.3)	1000	$H_2O + 2Cl^- \rightleftharpoons O^{2-} + 2HCl$	2.3	73	
	NaCl–KCl–CeCl$_3$ (0.45)(0.45) (0.1)	1000	$H_2O + 2Cl^- \rightleftharpoons O^{2-} + 2HCl$	3.5	61	
	NaCl–KCl–CeCl$_3$ (0.4) (0.4) (0.2)	1000	$H_2O + 2Cl^- \rightleftharpoons O^{2-} + 2HCl$	2.8	61	
	NaCl–KCl–CeCl$_3$ (0.35)(0.35) (0.3)	1000	$H_2O + 2Cl^- \rightleftharpoons O^{2-} + 2HCl$	1.5	61	
Fluorides	LiF–BeF$_2$ (0.73)(0.27)	1000	$H_2O + 2F^- \rightleftharpoons O^{2-} + 2HF$	3.37	88	a
	LiF–BeF$_2$ (0.67)(0.33)	1000	$H_2O + 2F^- \rightleftharpoons O^{2-} + 2HF$	2.96	88	
	LiF–BeF$_2$ (0.6) (0.4)	1000	$H_2O + 2F^- \rightleftharpoons O^{2-} + 2HF$	2.01	88	
Hydroxides	NaOH–KOH (0.51) (0.49)	500	$2OH^- \rightleftharpoons O^{2-} + H_2O$	12.5	90 and 95–97	
	NaOH	700	$2OH^- \rightleftharpoons O^{2-} + H_2O$	7.5	98	b
	NaOH	800	$2OH^- \rightleftharpoons O^{2-} + H_2O$	6.0	98	b
	NaOH	873	$2OH^- \rightleftharpoons O^{2-} + H_2O$	5.0	98	b
	NaOH	1000	$2OH^- \rightleftharpoons O^{2-} + H_2O$	3.75	98	b
	NaOH	623	$2OH^- \rightleftharpoons O^{2-} + H_2O$	9.7	98	c
	NaOH	750	$2OH^- \rightleftharpoons O^{2-} + H_2O$	7.4	98	c
	NaOH	850	$2OH^- \rightleftharpoons O^{2-} + H_2O$	6.2	98	c
	NaOH	1000	$2OH^- \rightleftharpoons O^{2-} + H_2O$	5.0	98	c
	LiOH	750	$2OH^- \rightleftharpoons O^{2-} + H_2O$	2.5	98	c
	LiOH	850	$2OH^- \rightleftharpoons O^{2-} + H_2O$	1.8	98	c
	LiOH	1000	$2OH^- \rightleftharpoons O^{2-} + H_2O$	1.1	98	c
	KOH	750	$2OH^- \rightleftharpoons O^{2-} + H_2O$	10.3	98	c
	KOH	850	$2OH^- \rightleftharpoons O^{2-} + H_2O$	8.8	98	c
	KOH	1000	$2OH^- \rightleftharpoons O^{2-} + H_2O$	7.1	98	c
Nitrates	NaNO$_3$–KNO$_3$ (0.5) (0.5)	502	$2OH^- \rightleftharpoons O^{2-} + H_2O$	15.6	110	d

a–An equation for log K vs. T has been derived. Please see the text or the corresponding reference.
b–Combination of experimental and thermochemical data.
c–Values obtained through thermochemical data only.
d–Value obtained thanks to Zambonin and co-workers' determinations.[105, 106, 109]

explained by the stronger acidity of Li^+ compared to Na^+ due to their ionic radii, as already emphasized by the solubility data.

In $AlCl_3$–$NaCl$ mixtures, things are slightly more complicated as the oxobasic properties of water depend on the chloroacidity of the melt, as demonstrated by Tremillon and co-workers.[77] These authors showed that a nickel electrode is an indicator of oxide ion concentration in molten chloroaluminates and that O^{2-} ions were solvated in the forms $AlOCl_2^-$ and $AlOCl$ in chlorobasic and chloroacidic melts, respectively. By coupling the experimental results, obtained either with this Ni/NiO electrode or by sweep voltammetry on HCl, and thermochemical data, the following expression relating the equilibrium constant K_2 (units of molality and atmospheres) to pCl can be given in chloroaluminate melts at 483 K:

$$\log K_2 = -8.1 + \log(1 + 10^{pCl^- - 4.35})$$

These results show that traces of water may once more play a decisive role in these melts, which should be carefully treated before use.

The oxobasic properties of water have also been studied by Combes *et al.*[61] in melts containing NaCl–KCl and various proportions of $CeCl_3$ (5–30 mol %) at 1000 K. The corresponding values of $\log K_2$ are collected in Table 2 and clearly demonstrate the oxoacidic influence of cerium (III) ions.

In these latter melts, no oxoacidic properties of water are evidenced, because of the high acidity of one of the cations (Al^{3+}, Ce^{3+}).

In conclusion, a large amount of data on oxobasic properties of water in various molten chlorides is available (Table 2). Nevertheless, the pK_2 values are not a sufficient basis for comparison between various chlorides, which requires the absolute activity difference relating a given concentration of oxide ions in differently composed melts. As shown by Tremillon *et al.*,[80, 61, 73] such correlation can be made using the "oxoacidity function Ω" defined by means of the water oxobasic equilibrium, since the activities of two compounds can be expressed by their partial pressures and that of chloride ions can be assumed to be constant and close to unity. The oxoacidity scales, obtained by defining Ω in such a way that it is identical to pO^{2-} in NaCl–KCl at 1000 K chosen as solvent of reference [i.e., $\Omega = 14 + \log(p_{HCl}^2/p_{H_2O})$], are represented in Figure 1.

In this figure, the various $\log K_2$ values are obtained by subtracting 14 from the Ω value corresponding to the origin of each scale.

3.3. Fluorides

In contradiction to chlorides, quantitative determinations on the effect of water are lacking in molten fluorides. This can be ascribed to the difficulties of measurement in these melts and the lack of an accurate

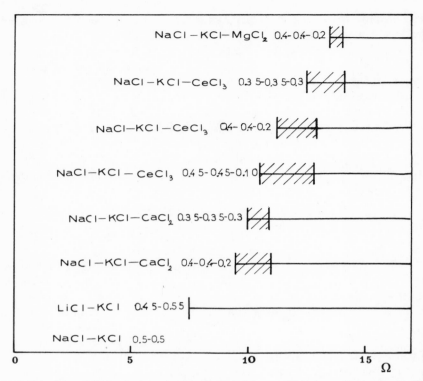

FIGURE 1. Oxoacidity scales in various molten chlorides at 1000 K (according to Combes[73]). The corresponding pK_2 values are obtained by subtracting to 14 the Ω value of the scale origin.

device for determining the oxide ion concentration. Nevertheless, some data can be found on water in KHF_2, FLINAK (LiF–NaF–KF mixture), and $LiF–BeF_2$.

Potassium hydrogendifluoride can be melted[81] and is completely dissociated into K^+ and HF_2^-. Devynck et al.[82] have shown by electrochemical measurements that HF_2^- ions are slightly dissociated at 523 K according to

$$HF_2^- \rightleftarrows HF + F^-, \qquad K_D = [HF][F^-] = 10^{-2.05} \text{ mol}^2 \text{ kg}^{-2}$$

and that the most basic* medium was obtained when the melt was saturated with KF (0.84 mol kg^{-1}).[83] Pizzini and co-workers[84, 85] studied by

_____

* In terms of the solvoacidity concept, which is the most useful in this case.

a galvanostatic method the influence of water on this melt at 523 K. Considering the acidobasic equilibrium involving water

$$H_2O + HF_2^- \rightleftarrows H_3O^+ + 2F^-$$

they explained the two reduction waves observed in the presence of water by the reactions

$$HF + e^- = \tfrac{1}{2}H_2 + F^-$$

$$H_3O^+ + e^- = \tfrac{1}{2}H_2 + H_2O$$

The latter reaction would assume that water behaves as a weak base for it is reduced at a higher potential than the solvent (HF_2^-). This hypothesis is in complete disagreement with the results of Devynck et al.[82] When adding water to a pure KHF_2 melt, the shift of the hydrogen electrode potential was the same as the one obtained for KF additions, i.e., water is a strong base in this solvent. These latter results seem the more convincing as the autodissociation constant inferred from them is close to the one calculated from thermochemical data.

Most of information on water in molten FLINAK is due to Pizzini and co-workers[86, 18] and Mamantov.[19] According to these authors, H_2O is an oxobase when involved in the equilibria

$$H_2O + F^- \rightleftarrows OH^- + HF$$

$$OH^- + F^- \rightleftarrows O^{2-} + HF$$

It must be pointed out that Mamantov et al.[87] obtained an explosive reaction by introducing hydroxide ions into molten FLINAK.

Anyway, the above-mentioned reactions have been studied thoroughly in $LiF-BeF_2$ by Mathews and Baes,[88] who found the mass balance between influent and effluent $H_2O + HF$ gaseous mixtures in a BeO-saturated melt, to be consistent with a proportionality between $[Be^{2+}]$ and P_{HF}^2/P_{H_2O} (see Table 2).

No mention of the oxobasic properties of water appears to have been made for fluoride melts.

3.4. Hydroxides

Molten alkali hydroxides are completely dissociated into Na^+, K^+, or Li^+ and OH^-. As these latter ions are submitted to a partial dissociation into H_2O and O^{2-},[39, 40, 89, 90] water will thus play a fundamental role in these media as was early emphasized by Agar and Bowden[89] and then by Lux et al.[91-94] Moreover, this role will be directly related to the oxoacidobasic properties of the melts themselves, as the autodissociation equilibrium

$$2OH^- \rightleftharpoons H_2O + O^{2-}$$

is nothing but the reverse reaction involving the oxoacid properties of water [reaction (5)]. The chemical and electrochemical properties of molten NaOH–KOH have been studied thoroughly by Tremillon et al.[95-97] By considering the dissociation constant value they determined experimentally at 500 K, one may ascribe to reaction (5) an equilibrium constant value equal to $10^{-12.5}$ in the molality scale.

Thanks to the work of Tremillon and Doisneau,[98] a correlation can be found between this value and the corresponding ones at different temperatures and compositions in molten alkali hydroxides. For example, in pure NaOH at 593 K a constant close to 10^{-10} could be found.

As a conclusion to this section, it should be pointed out that water in these media may also behave as an acid according to the Brönsted theory, for H_2O may be considered as the "solvated" proton in hydroxide melts; then it plays the same role as the H_3O^+ cation in aqueous solutions. Thus, water is also the strongest acid that can exist in the melt and, by analogy to pH, the quantity $pH_2O = -\log[H_2O]$ may be introduced. Considering the equilibrium constant determined by Tremillon et al.,[95-97] one will understand that the accessible pH_2O range in molten NaOH–KOH at 500 K is about 12 units, located between the most concentrated water solutions and the most concentrated oxide solutions (corresponding to the sodium oxide solubility in this melt, i.e., 0.05 mol kg^{-1}).

3.5. Nitrates

Results on oxoacidity in molten alkali nitrates have been quite puzzling for rather a long time, the main doubt being whether or not oxide ions could exist in these melts.[99-106] Thus, conclusions concerning the oxoacidic properties of water in these melts appeared to be inconsistent.[106-109] Attempts to clarify the situation have been made in excellent survey papers by Zambonin et al.[106, 109] and by Burke and Kerridge.[111]

Decisive experimental evidence, such as a determination of oxide ion concentration in these melts, was lacking. It has been given recently by Flinn and Stern,[78] who used an yttria-stabilized zirconia oxide ion indicator electrode. The results obtained by these authors fully support those of Zambonin[105, 106, 109] and the E–pO^{2-} diagrams given by de Haan and Vander Poorten,[110] which are mainly based on Zambonin's data. Thus, it should be kept in mind that the following equilibria are intervening (with their corresponding constant values in the molality scale) in $NaNO_3$–KNO_3 at 502 K:

$$O^{2-} + NO_3^- \rightleftarrows O_2^{2-} + NO_2^-, \qquad K = 10^{-1.86}$$

$$O_2^{2-} + 2NO_3^- \rightleftarrows 2O_2^- + 2NO_2^-, \qquad K = 10^{-10.17}$$

$$H_2O + O^{2-} \rightleftarrows 2OH^-, \qquad K = 10^{15.6}$$

$$H_2O + O_2^{2-} \rightleftarrows 2OH^- + \tfrac{1}{2}O_2, \qquad K = 10^{9.30}$$

$$H_2O + 2O_2^- \rightleftarrows 2OH^- + \tfrac{3}{2}O_2, \qquad K = 10^{3.0}$$

Through these data, some explanations concerning the apparent discrepancies between the various authors can be given. Though reacting with the melts according to the above-mentioned equilibria, oxide ions will exist to some extent except if consumed by a more reactive compound such as, oxygen, CO_2, SiO_2, or H_2O. Moreover, water is such a strong oxoacid that it will react readily on peroxide or even on superoxide. This phenomenon may be related to the very low equilibrium constant value $(3.2 \times 10^{+2})$ obtained by Fredericks and Temple[107] for equilibrium (5) in $NaNO_3$–KNO_3 at 573 K, by measuring the potentials of an oxygen gas electrode against amounts of water introduced in the melt.

As a consequence of the strong oxoacidity of water in sodium-potassium nitrates, no mention of its oxobasicity can be found. On the contrary, in molten $LiNO_3$ Peleg[11] stated that some 10^{-4} mol H_2O/mol melt is retained in spite of prolonged evacuation due to the formation of oxide species with Li^+ ions.

4. Redox Properties of Water

In addition to its acid–base properties, water can be reduced or oxidized. As seen in Section 2, these properties very often enabled the detection and/or determination of the content of water in melts.

When oxide ions are stable in the considered medium, the oxidation-reduction reactions of water can be written as below:

$$H_2O + e^- = OH^- + \tfrac{1}{2}H_2 \tag{7}$$

$$OH^- + e^- = O^{2-} + \tfrac{1}{2}H_2 \tag{8}$$

$$H_2O + 2X^- = \tfrac{1}{2}O_2 + 2HX + 2e^- \tag{9}$$

where X^- is the anionic species of the melt.

To these reactions correspond equilibrium potentials which depend on the acidity of the melt via the quantities pO^{2-} or pX^-, according to the acidity system considered to be the most useful in the molten salt. For example, in the oxoacidity system, one will obtain the following equations for the equilibrium potential corresponding, respectively, to the reduction and the oxidation of water:

$$E = E_R^\circ + \frac{2.3RT}{2F}\, pO^{2-} + \frac{2.3RT}{2F} \log \frac{p_{H_2O}}{p_{H_2}}$$

$$E = E_{OX}^\circ + \frac{2.3RT}{2F} \log K_2 + \frac{2.3RT}{2F}\, pO^{2-} + \frac{2.3RT}{4F}\, P_{O_2}$$

Very often, either the water oxidation proceeds differently via the formation of peroxide or superoxide ions (respectively, oxidation states $-I$ and $-I/2$), or it is simply not possible because of the presence in the melt of species easier to oxidize than water.

4.1. Chlorides

The first evidence of the electrochemical reduction of water was given by Laitinen *et al.*[14] and Delarue[15] when performing voltammograms on incompletely dehydrated LiCl–KCl melts. Considering that water was dissolved in the form of hydroxide ions, Laitinen *et al.*[14] attributed the limit of about -1 V [with respect to the Pt(II)Pt(O) reference electrode] to reaction (8), but Delarue,[15] who could not obtain any wave corresponding to the oxidation of dissolved OH^- ions, concluded that the corresponding electrochemical system was irreversible.

Much more recently, Kanzaki and Takahashi[75] performed a voltammetric and potentiometric study of the reduction reactions involving $H(+I)$ in molten LiCl–KCl at 723 K. The electrochemical systems corresponding to reactions (7) and (8) were found to be reversible and their standard potentials, respectively, equal to -1.69 and -2.60 V (molality scale) with respect to the standard chlorine reference electrode.

A comparison between these results and those obtained, by sweep voltammetry, by Melendres, Ackerman, and Stennenberg[17] is difficult because the temperature and the reference electrodes were different. Nevertheless, these latter authors obtained two reversible reduction peaks with $E_{1/2}$ of about 1.61 and 0.47 V versus a Li–Al reference electrode at 663 K. To attempt to make any comparison, one may use the standard potential value of the Li(I) Li(O) couple given by Laitinen and Liu[112] and obtain two $E_{1/2}$ values more negative than the former ones by some 400 mV, presumably as a result of the lower activity of Li in Li–Al.

For NaCl–KCl, no study of water reduction appears to have been carried out. The results obtained by Combes *et al.*[68] on the oxoacidity of water enable the calculation of the standard potential for the reaction (7) at 1000 K in this melt: $E^\circ = -2.4$ V (vs. Cl_2/Cl^-), taking into account the potential values corresponding to the HCl/H_2 couple.[113, 114]

In chloroaluminate melts, according to Letisse and Tremillon,[115] water reacts readily with the melt to give HCl, which gives rise to a reduction wave located at $+1.05$ V [vs. the Al(III)/Al(O) reference electrode] in a NaCl-saturated melt at 448 K. These results are in good agreement with the ones obtained by Yntema and co-workers[116, 117] in the ternary eutectic at 429 K and those more recently obtained by potentiometry by Plambeck and Skala.[118]

In a study of moisture and carbon dioxide absorption in chloride electrolytes for sodium production, Grachev and Grebenik[58, 119, 120] obtained a reduction wave for water by a voltammetric technique in NaCl–CaCl$_2$ at 873 K, roughly located at -1.3 V (vs. Cl$_2$/Cl$^-$). This value, which is in agreement with the one determined by sweep voltammetry in a MgCl$_2$–KCl melt at 973 K,[121] is much higher than the one calculable in NaCl–KCl. This clearly indicates that the reduction products of water (i.e., OH$^-$, O^{2-}) are strongly complexed in these alkaline-earth melts, as could be predicted from the oxoacidic properties discussed in the previous section.

4.2. Fluorides

Water is also electroactive in these melts, but apart from the electrochemical studies of Pizzini and co-workers[18, 84–86, 122] and those of Devynck et al.[82] nothing else appears to have been done in this field. A complete discussion on the results of the former authors has been made by Mamantov,[19] to which the reader should refer for further details as only the main results will be recalled here.

In KHF$_2$ at 523 K, two reductions resulting from the presence of water may occur:

$$HF + e^- = \tfrac{1}{2}H_2 + F^- \qquad \text{at} \simeq 0 \text{ V (vs. HF/H}_2)$$

$$H_3O^+ + e^- = \tfrac{1}{2}H_2 + H_2O \qquad \text{at} \simeq 0.4 \text{ V (vs. HF/H}_2)$$

In FLINAK at 873 K, two waves were observed, the second at -0.5 V (vs. HF/H$_2$) was attributed to hydrogen discharge from water.

Nevertheless, it should be kept in mind that the former conclusions on KHF$_2$ are in severe disagreement with the results of Devynck et al.[82] which were obtained by voltammetry and potentiometry with a hydrogen electrode in the same melt. In their view, with which it seems reasonable to agree, the first reduction wave ($\simeq 0$ V) should correspond to the reduction of HF$_2^-$ and not to that of HF as stated by Pizzini et al. The observed plateau might be due to a passivation phenomenon, as already observed in basic media. The second wave then would be ascribed not to the reduction of H$_3$O$^+$ but to the bursting of the passivating layer. To strengthen their argument, Devynck et al. remarked that it is impossible to observe the reduction of a weak acid (H_3O^+) when its conjugated base (H_2O) is introduced alone in the solvent. The conclusion would be that water is not electroactive in molten potassium hydrogendifluoride and, in order to know whether it behaves the same in molten FLINAK, a study of the same kind as that described above would be needed.

4.3. Hydroxides

Water can be reduced either by the alkali metals or electrochemically on palladium electrodes.[123] A complete study of its electrochemical reduction on rotating platinum electrodes has been performed by Goret and Tremillon,[90, 95, 96] and this led these authors to conclude that the corresponding system

$$2H_2O + 2e^- = H_2 + 2OH^-$$

is reversible in NaOH–KOH at 500 K. They found a standard potential value ($[H_2O] = 1$ mol kg^{-1}, $p_{H_2} = 1$ atm) equal to 1.04 V with respect to a standard sodium reference electrode. This value explains why the reduction of hydroxide ions can never occur since, taking into account the autodissociation constant value $(10^{-12.5})$, the standard potential of the corresponding reaction

$$2OH^- + 2e^- = H_2 + O^{2-}$$

would be equal to $1.04 - 0.1 \times 12.5 = -0.21$ V, a value more negative than that of sodium ions.

Another consequence is that dissolving sodium or potassium in this melt does not produce the strongest basic medium, since the pO^{2-} value corresponding to a zero potential value for the hydrogen electrode is $pO^{2-} = 2.1$.

On the contrary, water cannot be oxidized in molten NaOH–KOH since oxide or hydroxide ions (according to the pO^{2-} value) are easier to oxidize to peroxide and superoxide species.[90, 95, 96]

Results obtained either with thermochemical data or with experimental determinations by Doisneau and Tremillon[124] on other alkali hydroxides (particularly for pure NaOH), are in good agreement with those given above. They enable us to conclude that the water reduction is limiting the electroactivity range in all acidic or slightly acidic (i.e., water-containing melts) molten hydroxides, except maybe for LiOH, which could simultaneously be reduced in H_2. In none of them can the further reduction $(OH^- \rightarrow H_2)$ be observed, since the alkali cation is reduced first.

4.4. Nitrates

The first evidence of the redox properties of water in molten nitrates was given by the so-called "water wave" in a voltammetric study by Laitinen and Swofford.[125] But very soon afterwards evidence appeared to suggest a more complicated situation; Geckle[36] demonstrated by voltammetry and coulometry that no hydrogen was evolved at the potential corresponding to the above-mentioned wave and that 0.5 mol of NO_2^- and 0.5 mol of oxide were produced per Faraday of electricity passed.

Without making any assumptions concerning the reaction products, Peleg[11] employed the wave obtained at a Pt microelectrode at -0.7 V (vs. a massive Pt pseudo-reference-electrode) for determining the water content of pure nitrates; he demonstrated a clear interdependence of wave height with water concentration. The potential ($E_{1/2} = -1.2$ V vs. Ag(I)/Ag(O)) for this "water wave" was given by Zambonin and co-workers[12, 126] from a voltammetric study with a rotating disk electrode in $NaNO_3$–KNO_3 at 500 K. They interpreted their results in terms of the following concurrent reduction mechanisms:

$$H_2O + NO_3^- + 2e^- = 2OH^- + NO_2^-$$

$$H_2O + e^- = OH^- + 2H_2$$

with respective percentages of 84% and 16%, i.e., in direct conflict with mass spectrometric evidence[36] (see Chapter 11 by Lovering and Oblath).

An attempt to clarify this situation was made by Kerridge[127] in an interesting discussion, but more experimental evidence was obviously needed. This was given by the systematic investigations performed by Zambonin and co-workers in the last three years. In spite of the evidence given by these authors[128–131] using various electrochemical methods (potentiometry, polarography, sweep voltammetry) in molten $NaNO_3$–KNO_3 at 500 K, it seems difficult to ascribe the reduction wave to

$$2H_2O + 2e^- = H_2 + 2OH^-$$

since the evolved hydrogen would react with the melt according to

$$H_2 + NO_3^- = H_2O + NO_2^-$$

According to the evidence of voltammetry and controlled potential coulometry in $NaNO_3$–KNO_3 at 523 K reported by Espinola and Jordan[25] the overall electrode reaction could be

$$NO_3^- + H_2O + 2e^- = NO_2^- + 2OH^-$$

A standard potential of $E° = -1.96$ V [vs. Ag(I)/Ag(O)] was calculated[128] for the process

$$2H_2O + 2e \rightarrow 2OH^- + H_2$$

although some inconsistencies in the calculation and interpretation seem apparent (see Chapter 11 by Lovering and Oblath).

In agreement with the observations made in molten chlorides, this potential is shifted towards less negative values in more acidic melts. For example, in $LiNO_3$–KNO_3 at 559 K, a potential of -0.8 V [vs. Ag(I)/Ag(O)] was found by Yurkinski *et al.*[24] But, in this melt too,[25] the reduction mechanism of water involves NO_2^- ions as demonstrated by Lovering and co-workers.[132] Thus, the mechanism of the "water wave" is clearly more complicated than so far envisaged.

Nevertheless, in all cases it is clear that this reduction does not correspond to that of water since hydrogen atoms remain at the same oxidation state. Some involvement of solvent reduction

$$NO_3^- + 2e^- = NO_2^- + O^{2-}$$

possibly facilitated by the presence of water[132] and as clearly demonstrated in the E–pO^{2-} diagrams given in the excellent survey paper of de Haan and Vander Poorten[110] seems evident.

On the other hand, it may be possible to oxidize hydroxide ions in these melts,[129] according to

$$4OH^- = O_2 + 2H_2O + 4e^-, \qquad E° = -0.495 \text{ V [vs. Ag(I)/Ag(O)]}$$

5. Applications

5.1. Introduction

The applications inferred from the properties of water are too numerous in molten salt chemistry to be exhaustively reviewed in this section but they can be classified into two groups: those depending on its acidobasicity and the others depending on its oxidation–reduction properties. The former group will include the various dehydrating processes of the salts as well as some reactions involved in extractive metallurgy.

The latter group can only be understood with the help of diagrams of the Pourbaix type (representing equilibrium potentials versus the quantity used for measuring the acidity), since oxidation–reduction properties depend on the acidity of the medium both for water itself (as pointed out in the previous section) and for many other elements. Such a way of understanding phenomena has been recommended and used to a large extent by many authors[15, 41–43, 45, 67, 72, 77, 80, 95–98, 110, 133–137] either with thermochemical data or experimental determinations or a combination of both as shown by Tremillon.[43, 45] Whenever necessary, a similar procedure will be used here to explain some reactions of water with other elements. The reader should refer to the present author's work for the principles involved in the representation of the diagrams. Naturally, the corrosion phenomena related to water are to be found in this group of applications.

In this section a few examples pertaining to both groups will be discussed, in order to illustrate how the presently available data enable one to understand or predict some reactions of water in melts.

5.2. Applications Related to the Water Acidobasicity

Some dehydration processes of molten chlorides can be understood thanks to the oxoacidity scales represented in Figure 1 according to the various K_2 values obtained for equilibrium (6) by Combes *et al.*[59, 61, 75] For example, the treatment of the various melts at 1000 K by gaseous mixture of 99% HCl and 1% H_2O at atmospheric pressure, will lead to residual oxide ion concentrations (in mol kg^{-1}), respectively, equal to 10^{-16} in NaCl–KCl, $10^{-8.5}$ in LiCl–KCl, $10^{-6.5}$ in NaCl–KCl–CaCl$_2$, $10^{-2.5}$ in NaCl–KCl–MgCl$_2$, $10^{-3.5}$ in NaCl–KCl–CeCl$_3$. These figures clearly demonstrate the drastic effect of water in acidic cation melts as early emphasized by Corbett *et al.*[10]

In most extractive processes involving halides the effects of water are mainly deleterious. As an example, in the industrial electrolytic preparation of cerium, it has been possible to show, by electrochemical studies of melts containing various proportions of CeCl$_3$ in NaCl–KCl, that the formation of cerium oxychloride and insoluble Ce$_2$O$_3$ could be ascribed to the presence of water and might be avoided by using dry HCl.[61, 72] In the same kind of study, Picard *et al.*[79] have given, through a diagram represented in Figure 2, the conditions in which the various aluminum species (Al^{3+}, AlO$^+$, Al$_2$O$_3\downarrow$) could be obtained in LiCl–KCl at 743 K. Other examples of the harmful effect of water can be found in the literature related to the electrowinning of elements such as aluminum (ALCOA process), magnesium, titanium, etc.[138–142] or in the interesting survey recently made by Buckle and Finbow.[143]

The part played by water in molten hydroxides is even more fundamental as this compound is the strongest oxoacid in these melts. Thus, the solubilities of metal oxides for example will depend greatly on the concentration of H_2O in the melts as demonstrated by Tremillon *et al.*,[95–98] through reactions such as

$$M^{n+} + nOH^- \rightleftarrows MO_{n/2}\downarrow + \frac{n}{2}H_2O$$

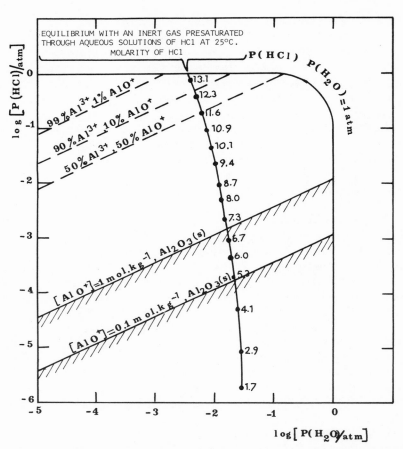

FIGURE 2. Diagram representing the stability ranges of the various aluminum species in molten LiCl–KCl at 743 K (according to Picard *et al.*[79]).

whose equilibrium constant K_A can be related to the solubility product value K_S and to the solvent autodissociation constant $K_i [K_S = K_i (K_A)^{-n/2}]$. According to these authors, the water concentration ranges corresponding to the precipitation or to the dissolution (for $[M^{n+}] = 10^{-2}$) of various metal oxides are represented in Figure 3.

In pure molten sodium hydroxide, the determinations made by Doisneau and Tremillon[144] at 623 K enabled these authors to predict that titanium could be separated from iron by varying the water content of melts in which ilmenite ($FeTiO_3$) is dissolved.

The importance of water acidobasic properties in molten nitrates has largely been discussed in the previous section in order to explain the

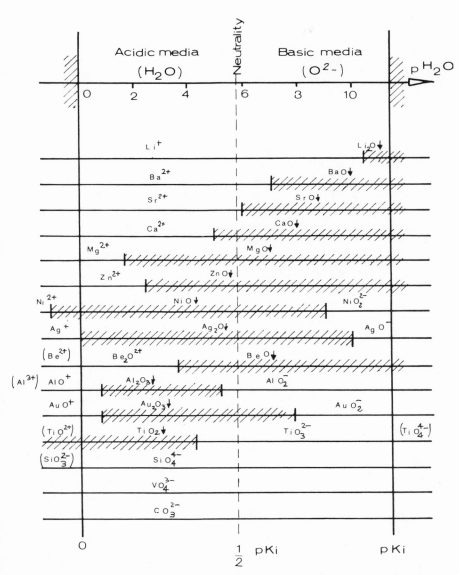

FIGURE 3. Solubility diagram of various metal oxides in molten NaOH–KOH at 500 K (according to Tremillon *et al.*[95, 96]).

numerous discrepancies in studies of these melts, and no further comment will be made now. Nevertheless, one unexpected result obtained by Flinn and Stern[78] might well result from water. When titrating superoxide ions by nickel chloride, the shift in potential of the oxide-ion indicator electrode was much larger than the one expected according to the stoechiometry of the reaction and did correspond to a consumption of oxide species greater by a factor of 2. A possible explanation is that accidental moisture would have transformed most of the superoxide in the equivalent quantity of hydroxide ions according to

$$H_2O + 2O_2^- \rightarrow 2OH^- + \tfrac{3}{2}O_2$$

then titration by $NiCl_2$ would involve these hydroxide ions $(Ni^{2+} + 2OH^- \rightarrow NiO \downarrow + H_2O)$.

The effects of water on melts are not always deleterious and, as an optimistic conclusion to this section, two examples are given. In their study on the dissociation of scheelite($CaWO_4$) in NaCl–KCl at 1000 K, Combes *et al.*[69] proposed the use of sodium hydroxide and water to dissolve this natural ore. These conditions are close to those recommended in a British Patent[145] or by Baraboshkin and Perevoskin[146] for the preparation of tungsten from its oxides in molten calcium chloride. The other example is that of the TORCO segregation process for Cu production[147, 148] in which H_2O is involved in a cycle essential to the success of the process, via the reactions (for more details, see also Reference 143):

$$2NaCl + H_2O + xSiO_2 = Na_2OSiO_{2_x} + 2HCl$$

$$Cu_2O + 2HCl = Cu_2Cl_2 + H_2O$$

$$H_2O + C = CO + H_2$$
$$Cu_2Cl_2 + H_2 = 2Cu + 2HCl$$

5.3. Applications Related to Water Oxidation–Reduction Properties

The relatively numerous data on water in molten chlorides enable the E–pO^{2-} diagrams of this compound to be plotted in terms of actual concentration scales, instead of activities determined through thermodynamical calculations, in many melts. For example, in Figure 4, such a diagram in NaCl–KCl at 1000 K is represented (in dotted lines) by considering the equilibrium constant values relating P_{HCl}, P_{H_2O}, $[OH^-]$ and $[O^{2-}]$ (P in atmospheres and concentrations in mol kg^{-1}) determined by Combes *et al.*[67, 68] and the standard potential of the $HCl(g)/H_2(g)$ couple. Its value $(-1.04\ V)$ with respect to the standard chlorine electrode (taken as zero potential, by convention) has been calculated from the standard free energy at the same temperature of the reaction between gaseous species:

$$\tfrac{1}{2}H_2(g) + \tfrac{1}{2}Cl_2(g) \rightarrow HCl(g)$$

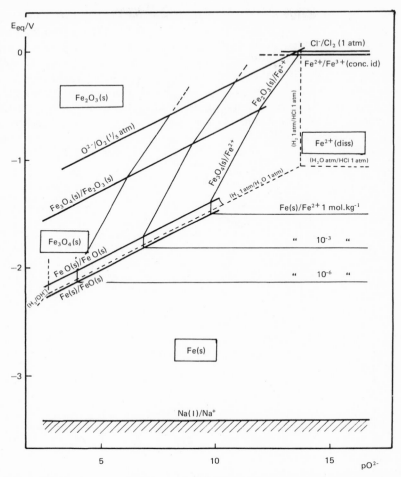

FIGURE 4. Pourbaix-type diagram for iron in molten NaCl–KCl at 1000 K (according to Combes[67]).

In order to illustrate the action of water in such a melt, the E–pO^{2-} diagram of iron has been represented in the same figure. As suggested by Trémillon[43, 45] the experimental standard potential values for Fe^{3+}/Fe^{2+} (Reference 149) and Fe^{2+}/Fe (Reference 150) have been combined with the solubility product values of FeO, Fe_3O_4, Fe_2O_3 inferred from thermochemical data and the activity coefficient of O^{2-} ions determined experimentally in this melt.[68] Comparing these two diagrams, one can see easily that iron will be oxidized to Fe^{2+} by moistened HCl. Water, under a partial pressure of 1 atm will attack iron by forming either Fe_3O_4 or FeO according to the amount of hydrogen present (i.e., the former iron oxide will be obtained for traces of hydrogen in water). For lower temperatures, the reader should refer to the work of Lewis.[151]

Such a comparison with the E–pO^{2-} diagram of zirconium, obtained by using the available experimental data,[152] leads to the prediction that water in its whole pO^{2-} stability range (4–14), would oxidize zirconium metal in ZrO_2. If the corresponding oxide layer is passivating, no further corrosion of zirconium would be observed, which is in good agreement with the conclusions of Smirnov *et al.*[153] at the end of their study on the effect of water on the corrosion of zirconium in molten NaCl–KCl.

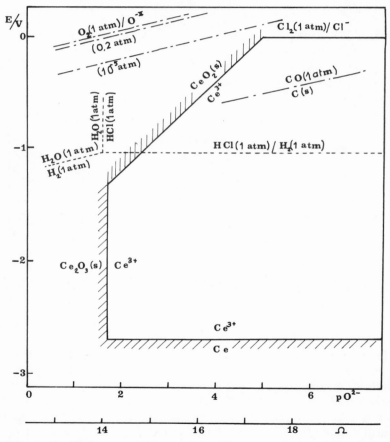

FIGURE 5. Equilibrium potentials versus pO^{2-} for various species in molten NaCl(0.35)–KCl(0.35)–CeCl$_3$(0.30) at 1000 K (according to Combes *et al.*[61]).

If water may appear as not too deleterious or even favorable in some molten chlorides[141, 153]—all those in which it is not a strong oxoacid—this is not the case in most of electrowinning processes involving these melts. In order to illustrate this fact, the E–pO^{2-} diagram of water (dotted lines) in a $CeCl_3(0.3)$–$NaCl(0.35)$–$KCl(0.35)$ melt at 1000 K has been represented in Figure 5 according to the results of Combes *et al.*[61] This diagram shows clearly the harmful effect of water above the industrial melts during the electrolytic preparation of cerium. Water appears to be able to precipitate Ce^{3+} ions into Ce_2O_3 and moreover to oxidize completely this oxide into CeO_2. It is certainly for comparable reasons that the moisture level is so drastically controlled in the new ALCOA process for the preparation of aluminum.[154]

In molten alkali hydroxides, as demonstrated by Trémillon and co-workers,[90, 95-98] the lower limit of the potential–acidity diagrams corresponds to the reduction of water except in very basic media where this limit is due to the formation of hydride.

Predictions about the effects of water on some species can be made by looking at the corresponding acidic part of the E–pO^{2-} or E–pH_2O diagrams. This has been done thoroughly by the above-mentioned authors, whose main conclusions will be recalled here. For example, copper and manganese oxides, which precipitate in neutral media, can be separated. By the progressive addition of water, only copper oxide is dissolved. Then, manganese oxide can be dissolved by the formation of MnO_4^{3-} when adding an oxidizing compound to the melt. More generally it has been shown that in concentrated water melts (acidic media), silver, gold, mercury, nickel, and platinum oxides were insoluble both in oxidizing and reducing conditions. Moreover, the use of nickel as a crucible material for hydroxide fluxes, was shown to be justified as the oxide of this element is also insoluble and adherent in oxidizing basic melts.

Some predictions have also been made for the corrosion of some metals in NaOH–KOH at 500 K through the Pourbaix-type diagrams represented in Figure 6. As can be seen, the effect of water is rather favorable. Most of the considered elements are passivated except for silver, which on the contrary is immune over a large range of acidity and potential.

A comparable study has been made by Doisneau and Trémillon[144] for iron in pure molten sodium hydroxide at 623 K. Water would have a deleterious effect on the solubilization of Fe(III) oxides, except if introduced in large quantities (very acidic media), by the formation of FeO^+. Moreover, these authors have shown that water in this melt (acidic media) would significantly corrode iron by the formation of Fe^{2+} and H_2. These latter results, obtained by electrochemical methods (potentiometry and voltammetry), are in good agreement with those of Newman *et al.*,[155] who

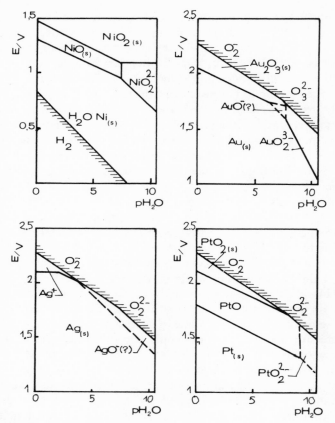

FIGURE 6. Pourbaix-type diagrams for various noble metals in molten NaOH + KOH at 500 K (according to Tremillon *et al.*[97]).

followed the hydrogen evolution on Cr–Mo steels in molten hydrated sodium hydroxide in a temperature range 1030–1270 K.

Such kinds of predictions can be made for other temperatures and for other molten hydroxides thanks to thermochemical data calculations as shown by Doisneau and Trémillon[98, 124] for LiOH, NaOH, KOH at various temperatures. These latter authors established the validity of such predictions through comparisons with some experimental determinations[156] and the main results were not found to be significantly different from those obtained in molten NaOH + KOH.

In molten alkali nitrates it has now been ascertained that the thermodynamical E–pO^{2-} diagram given by Conte and Ingram[135] was grounded on erroneous assumptions, i.e., the existence of NO_2^+ and no intermediate oxidation states between II and O for oxygen (for more detailed explanations the reader should refer to the two previous sections and to Reference 45). The first consistent E-pO^{2-} diagrams, based on the experimental determinations of Zambonin *et al.*[105, 106, 109, 128–131] in $NaNO_3$-KNO_3 at 502 K, were given by de Haan and Vander Poorten.[110] The diagram corresponding to water at a concentration of 10^{-3} mol kg^{-1} established by these authors enables us to understand the importance of this compound in molten nitrates. Extremely low oxide ion concentrations (10^{-20}–10^{-30} mol kg^{-1}) correspond to this concentration of water and thus, in carefully dried melts, it seems difficult to speak about oxide ions. Moreover, as the lower limit of the water E-pO^{2-} diagram arises from the reduction of the melt itself, water is not reducible in $NaNO_3$-KNO_3, but its presence greatly facilitates this process. On the contrary, water and hydroxide ions can be oxidized easily, and this leads to a substantial increase of pO^{-2}. This explains, for example, the purification process for molten nitrates by electrolysis recently proposed by Krongauz *et al.*[157]

Other consequences of the interactions between water and molten salts are the disruptive mechanical forces arising from the sudden vaporization of H_2O when in contact with a higher-temperature melt. These phenomena are of considerable interest in the assessment of molten salts reactors and have been studied theoretically and experimentally for the past six years.[158–165]

As a conclusion to this chapter, the parts played by water in molten salts have numerous and important consequences. In most of the industrial processes involving melts, these consequences are deleterious. For example, because of the acidobasic properties of water, sparingly soluble metal oxide species are formed and are lost for the process. The fact that water, or oxide ions generated by water, are oxidable will lead either to excessive consume of anode material or to lower electrolytical yields.

This set of effects is important both from the energetic and from the economic points of view. Therefore the increasing demand for energy-saving or raw-material-saving processes, should give a new impulse to investigation dealing with water in melts, a field which may have been underestimated for a long time.

ACKNOWLEDGMENT

The author wishes to express his deep gratitude to Professor Trémillon for his profitable suggestions and scientific support concerning this work.

References

1. M. Blander, *Molten Salts Chemistry*, Interscience, New York (1964).
2. G. Janz, *Molten Salts Handbook*, Academic, New York (1967).
3. S. Flengas and A. Block-Bolten, in *Advances in Molten Salts Chemistry*, Eds. J. Braunstein, G. Mamantov, and G. Smith, Plenum, New York (1973).
4. P. Field, in *Advances in Molten Salts Chemistry*, Eds. J. Braunstein, G. Mamantov, and G. Smith, Plenum, New York (1975).
5. H. Uhlig, *J. Phys. Chem.* **41**, 1215 (1937).
6. M. Blander, W. Grimes, N. Smith, and G. Watson, *J. Phys. Chem.* **63**, 1164 (1959).
7. H. Bloom and J. Bockris, in *Modern Aspects of Electrochemistry*, Ed. J. Bockris, Butterworths, London (1959).
8. G. Bertozzi, *Z. Naturforsch.* **A22**, 1748 (1967).
9. F. Duke and A. Doan, *Iowa State College J. Sci.* **32**, 451 (1958).
10. W. Burkhard and J. Corbett, *J. Am. Chem. Soc.* **79**, 6361 (1957).
11. M. Peleg, *J. Phys. Chem.* **71**, 4553 (1967).
12. P. Zambonin, V. Cardetta, and G. Signorile, *J. Electroanal. Chem.* **28**, 237 (1970).
13. H. Hull and A. Turnbull, *J. Phys. Chem.* **74**, 1783 (1970).
14. H. Laitinen, W. Ferguson, and R. Osteryoung, *J. Electrochem. Soc.* **104**, 516 (1957).
15. G. Delarue, *J. Electroanal. Chem.* **1**, 13 (1959).
16. D. Maricle and D. Hume, *J. Electrochem. Soc.* **107**, 354 (1960).
17. C. Melendres, J. Ackerman, and R. Steunenberg, *Proc. Int. Symp. Molten Salts* (1976), p. 575.
18. S. Pizzini, R. Morlotti, and E. Roemer, *J. Electrochem. Soc.* **113**, 1305 (1966); *Nucl. Sci. Abstr.* **20**, 43 (1966).
19. G. Mamantov, in *Molten Salts*, Marcel Dekker, New York (1969).
20. J. Raynor, *Z. Elektrochem.* **67**, 360 (1963).
21. A. Novozhilov and E. Pchelina, *Zh. Neorg. Khim.* **22** (4), 893 (1977); **22** (8), 2057 (1977).
22. F. Takanaku, S. Banya, T. Fukushima, and Y. Iguchi, *Tetsu To Hagane* **53**, 97 (1967).
23. I. Pashkeev, V. Antonenko, and V. Kozheuroy, *Fiz. Khim. Osn. Proizvod. Stali. Mater. Simp. Met. Metalloved.* 47 (1968).
24. V. Yurkinski, S. Aganesova, and A. Morachewski, *Ionnye Rasplavy* **4**, 97 (1976).
25. A. Espinola and J. Jordan, *Proc. Symp. Spectrosc. Electrochem. Charact. Solutes Nonaqueous Solvents* (1978), p. 311.
26. G. Schiavon, S. Zecchin, and G. Bombi, *J. Electroanal. Chem.* **38**, 473 (1972); **50**, 261 (1974).
27. H. Rehberg, Ger. Offen. 2, 258, 184, (1974).
28. S. Bretsznajder, *Roczniki Chemii* **10**, 729 (1930).
29. S. Rempel, *Dokl. Akad. Nauk. SSSR* **74**, 331 (1950).
30. Y. Vil'nyanskii and N. Bakina, *Zh. Priklad. Khim.* **29**, 561 (1956).
31. A. Rahmel and H. Krueger, *Z. Phys. Chem.* **55**, 25 (1967).
32. E. Hoyt, *J. Chem. Eng. Data* **12**, 461 (1967).
33. A. Keenan, *J. Phys. Chem.* **61**, 780 (1957).
34. J. Frame, E. Rhodes, and A. Ubbelhode, *Trans. Faraday Soc.* **57**, 1075 (1961).
35. H. Haug and L. Albright, *Ind. Eng. Chem. Process Res. Develop.* **4**, 241 (1965).
36. T. Geckle, U.S. AEC report No. TID 21511 (1965).
37. E. Franklin, in *The Nitrogen System of Compounds*, Reinhold, New York (1935).
38. B. Trémillon and G. Létisse, *J. Electroanal. Chem.* **17**, 371 (1968).
39. H. Lux, *Z. Electrochem.* **45**, 303 (1939); **52**, 220 (1948); **53**, 41 (1949); *Naturwissenschaften* **28**, 92 (1940).
40. H. Flood and T. Förland, *Acta Chem. Scand.* **1**, 592 (1947).

41. R. Littlewood, *J. Electrochem. Soc.* **109**, 525 (1962).
42. G. Charlot and B. Trémillon, in *Chemical Reactions in Solvents and Melts*, Gauthiers-Villars, Paris (1963), and Pergamon, New York (1969).
43. B. Trémillon, *Rev. Chim. Miner.* **3**, 767 (1966).
44. G. Mamantov, in *Molten Salts*, Marcel Dekker, New York (1969).
45. B. Trémillon, *Pure Appl. Chem.* 25 (1971).
46. W. Fung and G. Mamantov, in *Advances in Molten Salts Chemistry*, Plenum, New York (1973).
47. F. Clews and H. Thompson, *J. Chem. Soc.* 1442 (1922).
48. P. Robinson *et al.*, *J. Chem. Soc.* 836 (1926).
49. E. Berl and H. Staudinger, *Z. Angew. Chem.* **43**, 1006 (1930).
50. F. Guthrie and J. Nance, *Trans. Faraday Soc.* **27**, 228 (1931).
51. B. Neumann *et al.*, *Z. Elektrochem.* **41**, 725 (1935).
52. R. Iler and E. Tauch., *Trans. Amer. Inst. Chem. Eng.* **37**, 853 (1941).
53. E. Briner and N. Gagnaux, *Helv. Chim. Acta* **31**, 556 (1948).
54. E. Briner and P. Roth, *Helv. Chim. Acta* **31**, 1352 (1948).
55. E. Mignonsin and G. Duyckaerts, *Anal. Chim. Acta* **47**, 71 (1969).
56. N. Hanf and M. Sole, *Trans. Faraday Soc.* **66**, 3065 (1970).
57. R. Combes, J. Vedel, and B. Trémillon, *C. R. Acad. Sc. Paris* **273C**, 1740 (1971); **275C**, 199 (1972).
58. K. Grachev and V. Grebenik, *Zh. Prik. Khim.* **46**, 60 (1973).
59. R. Lysy and R. Combes, *J. Electroanal. Chem.* **83**, 287 (1977).
60. E. Savinkova, R. Lelekova, and T. Efremova, *Elektrokhim. Termodin. Svoistva Ionnykh Rasplavov* 98 (1977).
61. R. Combes, M. N. Levelut, and B. Trémillon, *Electrochim. Acta* **23**, 1291 (1978).
62. C. Edeleanu and R. Littlewood, *Electrochim. Acta* **3**, 195 (1960); *Silicates Industriels* **26**, 447 (1961).
63. H. Laitinen and B. Bhatia, *J. Electrochem. Soc.* **107**, 705 (1960).
64. R. Littlewood and B. Argent, *J. Electrochem. Soc.* **4**, 114 (1961).
65. N. Wrench and D. Inman, *J. Electroanal. Chem.* **17**, 319 (1968).
66. R. Combes, J. Vedel, and B. Trémillon, *Anal. Lett.* **3**, 523 (1970).
67. R. Combes, thesis, Paris (1973) (CNRS n°A.O. 8617).
68. R. Combes, J. Vedel, and B. Trémillon, *Electrochim. Acta* **20**, 191 (1975).
69. F. de Andrade, R. Combes, and B. Trémillon, *J. Electroanal. Chem.* **83**, 297 (1977).
70. R. Combes, F. de Andrade, and L. Carvalho, *C. R. Acad. Sci. Paris* **285C**, 137 (1977).
71. R. Combes, R. Feys, and B. Trémillon, *J. Electroanal. Chem.* **83**, 323 (1977).
72. R. Combes, M. N. Levelut, and B. Trémillon, *J. Electroanal. Chem.* **91**, 125 (1978).
73. R. Combes, to be published.
74. R. Combes, F. de Andrade, A. Barros, and H. Ferreira, *Electrochim. Acta* **25**, 371 (1980).
75. Y. Kanzaki and M. Takahashi, *J. Electroanal. Chem.* **58**, 349 (1975).
76. Y. Kanzaki and M. Takahashi, *J. Electroanal. Chem.* **58**, 339 (1975).
77. B. Trémillon, A. Bermond, and R. Molina, *J. Electroanal. Chem.* **74**, 53 (1976).
78. D. Flin and K. Stern, *J. Electroanal. Chem.* **63**, 39 (1975).
79. G. Picard, F. Séon, and B. Trémillon, to be published.
80. B. Trémillon and G. Picard, *Analyt. Chim. Acta* **82**, 273 (1976).
81. A. Opalovsky, V. Fedorov, and T. Fedotova, *J. Thermal. Anal.* **2**, 373 (1970).
82. J. Devynck, B. Trémillon, M. Sloim, and H. Ménard, *J. Electroanal. Chem.* **78**, 355 (1977).
83. G. Cady, *J. Am. Chem. Soc.* **56**, 1431 (1934).
84. S. Pizzini, G. Sternheim, and G. Barbi, *Electrochim. Acta* **8**, 227 (1963).
85. S. Pizzini and A. Magistris, *Electrochim. Acta* **9**, 1189 (1964).

86. S. Pizzini and R. Morlotti, *Electrochim. Acta* **10**, 1033 (1965).
87. F. Whiting, G. Mamantov, G. Begun, and J. Young, *Inorg. Chim. Acta* **5**, 260 (1971).
88. A. Mathews and C. Baes, *Inorg. Chem.* **7**, 373 (1968).
89. Agar and Bowden, *Proc. R. Soc. London* **169**, 206 (1939).
90. J. Goret, *Bull. Soc. Chim. Fr.* 1074 (1964).
91. H. Lux and T. Niedermaier, *Z. Anorg. Allg. Chem.* **282**, 196 (1955).
92. H. Lux and T. Niedermaier, *Z. Anorg. Allg. Chem.* **285**, 246 (1956).
93. H. Lux and T. Niedermaier, *Z. Anorg. Allg. Chem.* **298**, 285 (1959).
94. H. Lux and E. Renauer, *Z. Anorg. Allg. Chem.* **310**, 305 (1962).
95. J. Goret and B. Trémillon, *Bull. Soc. Chim. Fr.* **67**, 2872 (1966).
96. J. Goret and B. Trémillon, *Electrochim. Acta* **12**, 1065 (1967).
97. A. Eluard and B. Trémillon, *J. Electroanal. Chem.* **18**, 277 (1968); **26**, 259 (1970); **27**, 117 (1970); **30**, 323 (1971).
98. B. Trémillon and R. Doisneau, *J. Chim. Phys.* **10**, 1379 (1974).
99. F. Duke and M. Iverson, *J. Phys. Chem.* **62**, 417 (1958).
100. F. Duke and M. Iverson, *J. Am. Chem. Soc.* **80**, 5061 (1958).
101. F. Duke and M. Iverson, *Anal. Chem.* **31**, 1233 (1959).
102. R. Kust and F. Duke, *J. Am. Chem. Soc.* **85**, 3338 (1963).
103. L. Topol, R. Osteryoung, and J. Christie, *J. Phys. Chem.* **70**, 2857 (1966).
104. A. Shams El Din and A. El Hosary, *Electrochim. Acta,* **12**, 1665 (1967).
105. P. Zambonin and J. Jordan, *J. Am. Chem. Soc.* **89**, 6365 (1967); **91**, 2225 (1969).
106. J. Jordan, W. McCarty, and P. Zambonin, in *Molten Salts,* Ed. G. Mamantov, Marcel Dekker, New York (1969).
107. M. Fredericks and R. Temple, *Aust. J. Chem.* **25**, 1831 (1972).
108. M. Fredericks, R. Temple, and G. Thickett, *J. Electroanal. Chem.* **5**, 38 (1972).
109. P. Zambonin, *J. Electroanal. Chem.* **45**, 451 (1973).
110. A. de Haan and H. Vander Poorten, *Bull. Soc. Chim. Fr.* 2894 (1973).
111. J. Burke and D. Kerridge, *Electrochim. Acta* **19**, 251 (1974).
112. H. Laitinen and C. Liu, *J. Am. Chem. Soc.* **80**, 1015 (1958).
113. E. Ukshe and V. Devyatkin, *Zh. Fiz. Khim.* **39**, 3074 (1965).
114. V. Bezvoritnyi, V. Devyatkin, and A. Bezukladnikov, *Zh. Fiz. Khim.* **44**, 2105 (1970).
115. G. Letisse and B. Trémillon, *J. Electroanal. Chem.* **17**, 387 (1968).
116. W. Wade, G. Twellmeyer, and L. Yntema, *Trans. Electrochem. Soc.* **78**, 77 (1940).
117. R. Verdieck and L. Yntema, *J. Phys. Chem.* **46**, 344 (1942).
118. M. Skala, Ph.D. thesis, University of Alberta (1973).
119. V. Grebenik and K. Grachev, *Zh. Analiticheskoi Khim.* **23**, 1579 (1968).
120. K. Grachev and V. Grebenik, *Zh. Prikl. Khim.* **39**, 522 (1966).
121. A. Komura, H. Imanaga, and N. Watanabe, *Denki Kagaku* **10**, 762 (1972).
122. S. Pizzini, A. Magistris, and G. Sterheim, *Corr. Science* **4**, 345 (1964).
123. G. Kern, P. Degobert, and O. Block, *C. R. Acad. Sci. Paris* **256**, 1500 (1963).
124. R. Doisneau and B. Trémillon, *J. Chim. Phys.* **71**, 1445 (1974).
125. H. Swofford and H. Laitinen, *J. Electrochem. Soc.* **110**, 814 (1963).
126. P. Zambonin, *Anal. Chem.* **33**, 243 (1971).
127. D. Kerridge, in *Inorganic Chemistry,* Vol. 2, Butterworths, London (1972).
128. E. Desimoni, F. Paniccia, L. Sabbatini, and P. Zambonin, *J. Appl. Electrochem.* **6**, 445 (1976).
129. E. Desimoni, F. Palmisano, and P. Zambonin, *J. Electroanal. Chem.* **84**, 323 (1977).
130. E. Desimoni, F. Paniccia, and P. Zambonin, *J. Phys. Chem.* **81**, 1985 (1977).
131. E. Desimoni, B. Morelli, F. Palmisano, and P. Zambonin, *Ann. Chim.* **67**, 451 (1978).
132. D. Lovering, R. Oblath, and A. Turner, *J. Chem. Soc. Chem. Commun.* **17**, 673 (1976).
133. M. Ingram and G. Janz, *Electrochim. Acta* **10**, 783 (1965).
134. A. Rahmel, *Electrochim. Acta* **13**, 495 (1968).

135. A. Conte and M. Ingram, *Electrochim. Acta* **13**, 1551 (1968).
136. G. Bombara, G. Baudo, and A. Tamba, *Corros. Sci.* **8**, 393 (1968).
137. M. Rey, *Electrochim. Acta* **14**, 991 (1969).
138. C. Hampel in *Encyclopedia of Electrochemistry*, Reinhold, New York (1964).
139. M. Sittig, *Inorganic Chemical and Metallurgical Process Encyclopedia*, Noyes, Park Ridge, New Jersey (1968).
140. G. Milázzo, in *Electrochemistry*, Vol. 2, Dunod, Paris (1969).
141. G. Palin, in *Electrochemistry for Technologists*, Pergamon Press (1969).
142. J. Plambeck, *Fused Salt Systems* (Vol. X of the Encyclopedia of the Electrochemistry of the Elements, ed. A. Bard), Marcel Dekker, New York (1976).
143. E. Buckle and R. Finbow, *Int. Met. Rev.* **210**, 197 (1976).
144. R. Doisneau and B. Trémillon, *Bull. Soc. Chim. Fr.* **9–10**, 1419 (1976).
145. H. Slatin, British Patent, 918, 167 (1963).
146. A. Baraboshkim and V. Perevoskin, *Elektrokhimiya* **2**, 966 (1966).
147. E. Pinkney and N. Plint, *Trans. Inst. Min. Met.* **76**, c114 (1967).
148. K. Mackay and N. Gibson, *Trans. Inst. Min. Met.* **77c**, 19 (1968).
149. M. Smirrov, N. Loginov, and A. Pokrovskii, *Zh. Prikl. Khim.* **45**, 423 (1972).
150. S. Flengas and T. Ingraham, *J. Electrochem. Soc.* **106**, 714 (1959).
151. D. Lewis, Ab. Atomenergi, Stockholm, AE-424 (1971).
152. T. Sakakura, *Denki Kagaku* **36**, 306 (1968).
153. V. Volodin, I. Ozeryanaya, and M. Smirnov, in *Electrochemistry of Molten and Solid Electrolytes*, Trudy No. 8 (1966) [English translation published by Consultants Bureau, New York (1967)], p. 97.
154. ALCOA Process, U.S. Patent No. 3, 725, 222 (1973).
155. R. Newman, R. Smith, and C. Smith, *Proc. Int. Conf. Liq. Met. Technol. Energy Prod.* **2**, 841 (1977).
156. L. Antropov, D. Tkalenko, and S. Kudrya, *Ukr. Khim. Zh.* **40**, 429 (1974).
157. E. Krongauz, V. Kashcheev, and V. Busse-Machukas, *Elektrokhimiya* **8**, 1246 (1972).
158. V. Kulikov, V. Davydov, and L. Murav'ev, *Inzh- Fiz. Zh.* **23**, 435 (1972).
159. J. Hillary, P. Curry, and L. Taylor, Report U.K. Atomic Energy, TRG, 2366 (1972).
160. G. Bespalov, L. Filatova, and A. Shidlovskii, *Issled. Obl. Neorg. Tekhnol.* 31 (1972).
161. Nelson Whaton, *Tappi* **56**, 121 (1973).
162. Imai Ryukichi, *Kinzoku* **44**, 75 (1974).
163. Ogiso, Chiaki, *Yoyuen* **18**, 247 (1975).
164. J. Hightower, Oak Ridge National Laboratory report No. ORNL TM-4698 (1975).
165. R. Anderson and L. Bova, Argonne National Laboratory report No. ANL 57 (1976).

16

Oxide Species in Molten Salts

A. A. El Hosary, D. H. Kerridge,
and A. M. Shams El Din

1. Introduction

Identification of oxide species and their interactions is probably the most complex problem in molten salt chemistry at the present time. A study of the literature reveals controversy, confusion, and numerous simple discrepancies, so that rapid and conclusive answers based on a generalized theory are not to be expected. However, careful attention to the precise conditions of the experimental measurements, i.e., temperature, concentrations, the presence or absence of certain ions which did not initially feature in the equations of the reaction sequence (e.g., the nature of M^+) the container materials, etc. has shown that in the most complicated system for which data is plentiful (molten alkali metal nitrates) an attempt can now be made to construct a theory reconciling and embracing most, if not all, of the available evidence.

Such theories must make clear, in the light of the available evidence, on the one hand whether the basic species is the simple monoxide anion or whether oxidation, by melt anion or by gaseous oxygen, occurs to a significant extent. For example, in molten $NaNO_3/KNO_3$ the equilibria

$$O^{2-} + NO_3^- \rightleftharpoons NO_2^- + O_2^{2-} \tag{1}$$

$$O_2^{2-} + 2NO_3^- \rightleftharpoons 2NO_2^- + 2O_2^- \tag{2}$$

$$O_2^{2-} + O_2 \rightleftharpoons 2O_2^- \tag{3}$$

A. A. El Hosary and A. M. Shams El Din • National Research Centre, Dokki, Cairo, Egypt.
D. H. Kerridge • Chemistry Department, Southampton University, Southampton, England.

have been suggested with equilibrium constants of $K_1 \sim 3$, $K_2 \sim 6.7 \times 10^{-11}$, and $K_3 \sim 3.5 \times 10^5$ all at 229°C.[1] While on the other hand the possibility must be weighed that the oxide species may be stabilized by interaction with other solution species, for example, superoxide anions with potassium cations,[2] oxide ions with lithium,[3] silica,[4] water (to form hydroxide anions in wet melts[5]), and nitrate to form orthonitrate[6] (NO_4^{3-}) or pyronitrate[7, 8] ($N_2O_7^{4-}$), while peroxide could form peroxynitrate,[6] or peroxynitrite anions in nitrites.[9] It seems probable that similar complexities will eventually be found for other oxyanion melts, and such may now be becoming apparent in alkali metal carbonates.[10]

Despite the systemization that can thus be achieved, discrepancies still remain and much careful experimental data are still needed. However, the dangers of an oversimplistic approach may now be more widely appreciated, for example, the hope that the chemistry of any melt with the same anion can be explained on the basis of two or three simple equations, regardless of the nature of the other solvent ions, the concentrations of the solutes (whether the 10^{-5} M of the electrochemist or the 10 M of the industrial application), the occurrence of several side reactions, perhaps with secondary species not initially present, and the possibility of interaction with the container material and/or the supposedly inert atmosphere. An appreciation that such an oversimple approach will be in vain is probably a necessary prelude to the acceptance of molten salt chemistry as a discipline worthy both of academic study and of close examination for commercial possibilities.

2. Qualitative Estimation of Oxide Species in Molten Salts

Irrespective of the nature of the oxide ion species in melts, certain physical and chemical properties can be qualitatively related to their "basicity." The correlation between these properties and the basic character of the melts is mainly the result of the accumulation of data and experience in glass manufacturing. Although this information is still largely of an empirical nature, it is based on sound theoretical principles. The following are the most important of these properties.

2.1. Compound Formation and Basicity

The reaction of various metal oxides (or "oxide-ion donors") with one and the same "acid" to give compounds with definite molar ratios melting congruently (con) or incongruently (incon), can be established from a consideration of the corresponding phase diagrams. Taking SiO_2 as the common acid, the molar ratios SiO_2/metal oxide for the most acidic compound, and the nature of melting of the various glasses, are as follows[11]: ZnO, 1 (incon); Al_2O_3, 1 (incon); MgO, 1 (con); CaO, 1; SrO, 1; BaO, 1.5 (con), 2 (incon); Li_2O, 2 (incon); Na_2O, 2 (con); K_2O, 4 and Rb_2O, 4 (probable). Taking both the type and the ability of the metal oxide to bind the silica as measures of their strength, the above sequence suggests that, within the alkali metal oxides, the basicity increases with increase of the atomic weight of the metal. Alkali metal oxides are also stronger than the earth alkalies. The thermal decomposition of the alkali disilicates,[12] which decreases with the increase of the molecular weight of the disilicate, can be explained on the same basis.

Compound formation is always associated with an abrupt increase (or decrease) in the oxide-ion activity of the melt, depending on the direction whereby the compound composition is approached. Based on this fact, and by following variations in $[O^{2-}]$ using the $O_2(Pt)$ electrode, Shams El Din, Taki El Din, and El Hosary[13] found the neutralization of the acid $(NaPO_3)_3$ with the carbonates of Li, Na, K, Ca, Sr, and Ba in fused KNO_3 at 350°C to occur in steps developing at definite acid : oxide-ion donor ratios. Taking the P : O or P : e (electric charge) of the primary compound formed upon reaction of the acid with the carbonate (or vice versa) as a measure of basicity, it was concluded that this property increases as Li > Na, K and Ca > Sr > Ba.[13] This arrangement was further supported by the results of neutralization of the acids $K_2Cr_2O_7$ in KNO_3 (Reference 14) at 350°C and B_2O_3 in NaCl–KCl at 750°C.[15]

2.2. Slag Formation and Basicity

Salmang[16] found that, at 1400°C, "basic" oxides and "basic" silicates of molar ratios MO : SiO_2 larger than 1 were aggressive towards refractories. "Neutral" silicates, with molar ratios ≤ 1, attacked the refractory material to a lesser extent. Taking the attack of similar MO : SiO_2 ratios on the refractory as a measure for basicity, the oxides were arranged in their descending order of basicity as[17, 18]: CaO, PbO, BaO, FeO, SnO, Al_2O_3, Fe_2O_3, SiO_2, ZnO, Cr_2O_3. A correlation between the basicity of the oxides and the coefficients of thermal expansion and heat transfer of their silicates was also pointed out.[16]

2.3. Leachability of Glasses and Basicity

A simple silicate glass that is easily leached with water possesses, in the molten state, a high "basicity," independent of the absolute amount of the basic oxide. Potassium silicate glasses are leached to a larger extent than the corresponding sodium silicates of comparable compositions.[18] The same is also true for potassium–lead silicate and sodium–lead silicate,[19-20] when comparison is made on the basis of molar ratios. The leachability of more complex glasses has been recently described,[21-23] though of course the leachability of glasses is a measure of basicity only for those glasses whose basic components are water soluble.

2.4. Gas Uptake and Basicity

Molten glasses can absorb certain gases to extents depending on their basicity. Thus, for example, Weyl[24] found that the metasilicates of Li, Na, and K, at 1150°C, dissolved CO_2, at 750 atm pressure, to the extents of 0.2, 6.3, and 7.9%, successively. Since the uptake of CO_2 by the systems Na_2O/SiO_2 and K_2O/SiO_2 increased also with the alkali oxide content (i.e., basicity), the above results were taken to support the conclusion that within the three metasilicates the basicity increases in the direction $Li \rightarrow Na \rightarrow K$. This conclusion is partly supported and extended by the results of Niggli,[25] who studied the evolution of CO_2 in the systems $Na_2CO_3(K_2CO_3)/SiO_2$, on the one hand, and $Na_2CO_3(K_2CO_3)/TiO_2$, on the other. Working at $\sim 960°C$, the CO_2 evolution (in equivalents) was: Na_2CO_3/SiO_2, 1.35 mol; K_2CO_3/SiO_2, 0.85 mol; Na_2CO_3/TiO_2, 0.82 mol, and K_2CO_3/TiO_2, 0.64 mol. The above results show that, SiO_2 is a stronger acid than TiO_2. This last conclusion is in harmony with the fact that crystalline salts with the highest acid content possess the formulas $Na_2O \cdot 2SiO_2$ and $Na_2O \cdot 3TiO_2$, respectively.[25]

The solubility of CO_2 in binary alkali borate and alkali silicate glasses at different temperatures was similarly studied by Pearce.[26] Equilibrium was considered to be represented by

$$CO_{2(gas)} + O^{2-}_{(melt)} \rightleftharpoons CO^{2-}_{3(melt)}$$

For low concentrations, the activity of $CO^{2-}_{3(melt)}$ could be replaced by concentration, and that of CO_2 by its pressure; which was equal to unity when working at 1 atm. For the above reaction, the equilibrium constant, K, is given as

$$K = \frac{\%CO_3^{2-}}{[O^{2-}]} \quad \text{and} \quad [O^{2-}] = \frac{\%CO_3^{2-}}{K} = K'\%CO_3^{2-}$$

Since K' could not be calculated, the assumption was made that it was independent of melt composition. It could be eliminated by determining the ratios of $[O^{2-}]$ of two different melts, the $[O^{2-}]$ of pure Na_2O being taken as the standard state; and hence

$$\log[O^{2-}]_{(melt)} = \log \frac{\%CO_{3(melt)}}{\%CO^{2-}_{3(Na_2O)}}$$

On this basis, the oxygen ion activities in some binary alkali borates and silicates were estimated in the temperature range 900–1200°C.[26] The technique of measuring the quantity of CO_2 evolved upon the addition of Na_2CO_3 was applied by van Norman and Osteryoung to determine acid oxyanion compositions in nitrate and chloride melts.[27]

The solubility of H_2O vapor in silicate melts was studied by Franz and Scholze.[28] Equilibrium was considered to be represented as

$$H_2O_{(gas)} + O^{2-}_{(melt)} \rightleftharpoons 2OH^-_{(melt)}$$

The presence of OH^- in glass, either as $\geqslant$Si—OH or $\geqslant$Si—OH—O—Si$\leqslant$ groupings was proved by infrared spectroscopy.[29, 30] As expected, the solubility of water vapor in glass was found to depend upon the square root of its partial pressure in the gas phase, when enough time was allowed for the two phases to come to equilibrium. The solubility of water vapor in silicate melts of various compositions also increases with their basicity (i.e., with $[O^{2-}]$ in the melt).[28, 31, 32]

Alkali borate melts at 750 to 1050°C also dissolved measurable quantities of H_2O.[33] The solubility was similarly linearly proportional to the square root of the water vapor partial pressure. A minimum in solubility was recorded at 25 mol % K_2O, at 35–40 mol % Na_2O, and in the Li_2O/B_2O_3 system. The minima were considered to represent the "neutral point" of the corresponding glasses. H_2O solubility minima in silicate glasses were similarly described by Kurkjian and Russell.[32]

The interaction between O^{2-} in binary and ternary silicate and aluminate melts and gaseous sulfur was studied by Fincham and Richardson.[34] Depending upon whether reducing or oxidizing atmospheres prevail, the following equilibria were assumed to be established:

$$\tfrac{1}{2}S_{2(gas)} + O^{2-}_{(melt)} = \tfrac{1}{2}O_{2(gas)} + S^{2-}_{(melt)}$$

when

$$O^{2-} = \frac{S^{2-} \cdot P^{1/2}_{O_2}}{K_4 \cdot P^{1/2}_{S_2}} \tag{4}$$

or

$$\tfrac{1}{2}S_{2(gas)} + \tfrac{3}{2}O_{2(gas)} + O^{2-}_{(melt)} = SO^{2-}_{4(melt)}$$

when

$$O^{2-} = \frac{SO^{2-}_4}{K_5 \cdot P^{1/2}_{S_2} \cdot P^{3/2}_{O_2}} \tag{5}$$

Relations (4) and (5) could not, however, be used directly to calculate the O^{2-} in glasses since the equilibrium constants are composition dependent.[34, 35] Nevertheless, at a constant oxygen pressure in the gas phase, the solubility of S in glass increased with the increase in the $[O^{2-}]$ of the melt.[34, 36]

2.5. Acid–Base Indicators

A number of polyvalent metal ions, when admixed with molten glasses, impart different colors, depending upon the "basicity" of the system. This reversible color change involves also redox equilibria in such a manner that the higher valencies (of more acidic character) form in basic melts, while the lower valencies (more basic) are stable in acid melts. A detailed review of the different acid–base indicators (and of colored glasses) falls outside the scope of the present review; for more information see References 37–41. A few examples are here cited, however, to illustrate the concept. Chromic oxide, Cr_2O_3, was proposed as indicator for the basicity of glasses.[37, 38, 42, 43] Acid melts favor the existence of green Cr_2O_3, whereas basic glasses produce yellow chromate ion,

$$2Cr^{6+} + 3O^{2-} \rightleftharpoons 2Cr^{3+} + \tfrac{3}{2}O_2 \tag{6}$$

Valency changes, reaction (6), are accompanied by the evolution of O_2. The equilibrium depends, therefore, on the partial pressure of O_2 above the melt. Glasses colored by iron,[44–46] blue–green (acidic)/green–yellow (basic), are said to involve equilibria between Fe^{2+} cations and FeO_2^- anions,[47]

$$2Fe^{2+} + 3O^{2-} + \tfrac{1}{2}O_2 = 2FeO_2^- \tag{7}$$

Since the indicator changes its color with the amount of the oxide-ion donor in the glass, one would anticipate a similar response to the "strength" of the oxide-ion donor at equal concentrations. Indeed, Csaki and Dietzel[47] found the alkali metal oxides to produce the following colorations with the iron indicator:

$$K_2O + 3SiO_2 \quad \text{yellow}$$

$$Na_2O + 3SiO_2 \quad \text{yellow–green}$$

$$(\tfrac{1}{2}Na_2O + \tfrac{1}{2}Li_2O) + 3SiO_2 \quad \text{blue–green}$$

indicating an increase in basicity in the succession: $Li_2O < Na_2O < K_2O$.

Other typical color indicators are V_2O_5 (yellow–green),[42, 48] selenium (rose–brown–colorless),[49, 50] sulfur (blue–colorless),[51] MnO_2 (colorless–violet),[52] and Cu.[38, 53] Apparently, color acid–base indicators are not suitable to establish the basicity of melts which are intrinsically either oxidizing or reducing.

Color indicators have two obvious drawbacks. They change their color over a narrow range of basicity much in the same way as does an acid–base indicator in an aqueous titration. Also, their color change depends on some specific conditions, such as the O_2 partial pressure above the glass melt. The two difficulties were ingeniously surmounted through the use of *p*-block metal ions as probe ions. These are metal ions in oxidation states two units less than the number of the group in the Periodic Table to which they belong, e.g., Tl^+ (Group III), Pb^{2+} (Group IV), and Bi^{3+} (Group V). The electronic configurations of these ions are such that there is a pair of electrons in the outermost occupied (6*s*) orbital. When small concentration of the ions are dissolved in glass, the $6s \to 6p$ electronic transition gives rise to an intense ultraviolet band with sharp maximum. The frequency of the adsorption bands is considerably lowered by the increased basicity of the glass in a direct and gradual manner.[39-41] The "optical" basicity, Ω, of a certain glass, as measured by a *p*-block metal ion, M^{n+}, is defined as[31]

$$M^{n+} = \frac{v_{\text{free ion}} - v_{\text{glass}}}{v_{\text{free ion}} - v_{O^{2-}}} = \frac{\Delta v_{\text{glass}}}{\Delta v_{O^{2-}}}$$

where $v_{\text{free ion}}$ is the frequency of the unperturbed, free probe ion, $v_{O^{2-}}$ is the frequency of the same ion in an ionic oxide (e.g., CaO) and v_{glass} is the same in the glass melt. The optical basicities of glasses as determined by Tl^+, Pb^{2+}, and Bi^{3+}, as probe ions, were found to be comparable.[39, 54]

With salts of lower melting point, more conventional organic acid–base indicators can be used. For example, phenolphthalein is colorless in acidic $LiNO_3/KNO_3$ solutions at 210°C but pink (with similar absorption bands at 17,800 and 27,200 cm^{-1} to those in aqueous solutions) in basic melts (e.g., with Na_2O, Na_2O_2, or NaOH). In these conditions this indicator became unstable after 30 min, but in the less oxidizing potassium thiocyanate melt, seven organic indicators were both stable and gave color changes, on additions of acids or bases, closely similar to those found in aqueous solutions.[55]

3. Quantitative Estimation of Oxide Species in Molten Salts

3.1. Potentiometry

A turning point in the history of acidity and basicity of fused salts was achieved through Lux's modification of Bronsted's definition for acidity,[56] namely, acid + O^{2-} = base. Acid–base equilibria can thus be established through the estimation of the oxide-ion activity of the melt. Electrochemically this can be accomplished by measuring the potential of either an oxygen– or a metal–metal oxide electrode relative to a suitable reference electrode. The potential of an oxygen electrode, E, whose reaction is $O_2 + 4e = 2O^{2-}$, is given as

$$E = E° + RT/4F \ln p_{O_2}/[O^{2-}]^2 \tag{8}$$

Working at an O_2 pressure of 1 atm, the above relation reduces to

$$E = E° - RT/2F \ln [O^{2-}] \tag{9}$$

where $E°$ refers to the standard O_2/O^{2-} redox potential relative to a suitable reference.

The first trial to use the O_2 electrode for the determination of O^{2-} activities in melts dates back to Lux,[56, 57] whose cell had the form

$$O_2(Pt) \left| \begin{matrix} Na_2SO_4 \\ K_2SO_4 \end{matrix} \right\} eutectic \left| \begin{matrix} Na_2SO_4 \\ K_2SO_4 \end{matrix} \right\} eutectic \left| (Pt)O_2 \right.$$

$$\left. Na_2O \ (N) \right| \quad \left. NaPO_3 + Na_3PO_4 \right|$$

$$T = 950°C \qquad\qquad\qquad reference\ electrode$$

The cell was worked by successively adding increasing amounts of Na_2O to the left-hand side electrode. Owing to the rapid volatilization of Na_2O, and to the attack on the alumina crucible used as container, rapid changes in the emf of the cell with time were observed. For this reason, the potentials measured at increasing time intervals were extrapolated to zero times (times of addition of Na_2O). The curve relating these potentials to pO^{2-} was not linear, and at its middle a tangent was drawn whose slope amounted to $2.303RT/F$ rather than $2.303RT/2F$. Lux assumed that the potential was determined by $Na = Na^+ + e$, and that the sodium was formed by the thermal decomposition of Na_2O, $Na_2O = 2Na + Na_2O_2$. Lux's work was severely criticized on the basis of failure to account for the observed irreversible electrode reaction, low buffer capacity of the melt, and volatilization of Na_2O.[58] On the whole, the work was ascribed as being the result of "an unfortunate cell arrangement combined with not a very fortunate choice of salt system".[58]

The credit for establishing the applicability of the O_2 electrode for oxide-ion determination in melts goes to Flood, Förland, and Motzfeldt,[58] who measured the emf of concentration cells involving O_2 electrodes. Two systems were examined, namely, the $Na_2CO_3–Na_2SO_4$ and the $K_2CrO_4–K_2Cr_2O_7$ systems. The first system was found to behave almost ideally, and the measured emf values corresponded practically to those calculated from theory. The more acid system, $K_2CrO_4–K_2Cr_2O_7$, gave rise to emf values about 15% lower than expected. This was attributed to the presence of a considerable liquid junction potential and/or electronic conductivity which short-circuited the cell.[58] Concentration cells involving oxygen electrodes and $PbO \cdot SiO_2$ as solvent were examined by Didtschenko and Rochow[59] at 800, 850, and 900°C. To one electrode compartment was added increasing amounts of the oxides of the alkali metals, alkaline earth metals, and those of Tl, Pb, Cd, Zn, Bi, B, Al, Ti, Si, and Ge. The emf values were found to depend on the nature and concentration of the added oxide, as well as on the partial pressure of O_2 in the feed gas. Conclusions regarding the basicities of the various oxides were drawn. The work was, however, criticized by Förland and Tashiro.[60] The reversibility of the O_2 electrode in silicate melts was also established by Minenco, Petrov, and Ivanova,[61] and Ranford and Flengas.[62] The effects of oxide additions, O_2 gas pressure, and temperature were in accordance with that expected from the Nernst equation.

Kust and Duke attempted to establish the $E°$ value of the O electrode in fused $NaNO_3$-KNO_3 by two independent procedures.[63, 64] On the assumption that O_2 gas could be electrolytically reduced in the melt, a cathodic current of the order of a few microamperes was used to generate the assumed O^{2-} ion, and the potential of the O_2 electrode was measured relative to a $Ag/Ag(I)/glass$ membrane reference electrode. Oxide ion concentrations ranging between 10^{-7} and 10^{-5} mol/kg were used. It is most probable, however, that the O^{2-} ion resulted from the reduction of the supporting electrolyte,[65] $NO_3^- + 2e = NO_2^- + O^{2-}$ rather than from the dissolved O_2. The second procedure depended on following the potential of the O_2 electrode as a function of time after the addition of small quantities of Na_2CO_3 until steady state values were established. CO_2 resulting from $CO_3^{2-} \rightleftharpoons CO_2 + O^{2-}$ was considered to be swept away by O_2, and the steady state potentials denoted the production of Na_2O from Na_2CO_3. Within the temperature range 530–639 K, both procedures were said to give the same $E°$ value, which amounted to $E° = 0.7759 - 2.537 \times 10^{-4}T$ (K) V, relative to the $Ag/Ag(I)$ reference half cell. This value was used to evaluate the dissociation constant of the carbonate.[66] The effect of the Na^+/K^+ ratio of the melt on $E°$ was studied.[67]

The behavior of the O_2 electrode in NaOH melt was examined by Rose, Davis, and Ellingham[68] and by Afanasev and Gamazov.[69] Janz and Seagusa[70] evaluated the Pt/O_2 and Au/O_2 systems as reference electrodes in molten carbonates.

The $O_2(Pt)$ electrode was extensively used by Shams El Din and his collaborators as indicator electrode in acid–base titrations in fused salts. Following Lux's definition, the acids $K_2Cr_2O_7$,[14, 71, 72, 75, 76, 80] $NaPO_3$,[13, 73–76, 78, 80, 81] $NaAsO_3$,[78] $NaVO_3$,[74–76, 78] $Na_4P_2O_7$,[73–76, 78] $Na_4As_2O_7$,[78] $Na_4V_2O_7$,[77] NaH_2PO_4,[73, 75, 76, 78] NaH_2AsO_4,[78] Na_2HPO_4,[73, 75, 76, 78] Na_2HAsO_4,[77, 78] and the acid oxides V_2O_5,[77, 78] P_2O_5,[78] As_2O_5,[78] CrO_3,[71–79] MoO_3,[79] WO_3,[79] and B_2O_3[15] were titrated mainly in KNO_3 at 350°C. A variety of oxide-ion donors, viz., KOH,[71, 80] Na_2CO_3,[13, 14, 72, 81] $NaHCO_3$,[72] Na_2O_2,[72–80] and electrolytically generated O^{2-} (Reference 80) were employed. Sharp end points were recorded at definite, simple acid : O^{2-} ratios which enabled the characterization of the reaction products. In conformity with the results of other authors, the O_2 electrode in unbuffered KNO_3 at 350°C did not behave reversibly.[82] Although it varied its potential linearly with $\log[O^{2-}]$, it did so at a rate of $2.303RT/F$ rather than $2.303RT/2F$. This interfered with the evaluation of the $E°$ value, necessary for the calculations of acid–base equilibrium constants. On the other hand, the theoretical analysis of the titration curves revealed that the O_2 electrode behaved reversibly in well-buffered melts.[71, 72] The determination of the $E°$ value of the O_2/O^{2-} electrode in KNO_3 was made possible through the comparison of the neutralization curves of the acid $K_2Cr_2O_7$ as determined with the O_2 electrode and the Cu/CuO electrode, under the same condition[75] and the knowledge of the $E°$ value of the latter electrode.[75] In this manner, the first true acid–base equilibrium constants in fused KNO_3 were computed. These are grouped in Table 1.

The technique of acid–base potentiometric titrations in fused salts, using an $O_2(Pt)$ electrode, was similarly employed by Coumert, Porthault, and Merlin[83] to establish the acidity of some condensed phosphates in molten KNO_3, and by Rahmel[84] in alkali sulfate melts. The suitability of the $MoSi_2$ electrode to act as an indicator electrode in acid–base titrations in melts was similarly established.[85] Shams El Din and El Hosary[86] showed that a number of bimetallic electrode combinations, some of them involving gaseous O_2, were successful in determining points of equivalence in potentiometric titrations. One particular aspect of titrations involving the O_2 electrode is the differentiation of the acidity of the various cations when used with one and the same oxide-ion-donor anion. The acids $NaPO_3$,[13] $K_2Cr_2O_7$,[14] and B_2O_3 (Reference 15) were titrated with the

TABLE 1. Acid–Base Equilibrium Constants in Fused KNO_3 at 350°C

Reaction	K	Ref.
$Cr_2O_7^{2-} + O_2^{2-} = 2CrO_4^{2-} + \frac{1}{2}O_2$	1.8×10^{12}	72, 75, 76
$2PO_3^- + O_2^{2-} = P_2O_7^{4-} + \frac{1}{2}O_2$	3.1×10^{15}	73–76, 78
$P_2O_7^{4-} + O_2^{2-} = 2PO_4^{3-} + \frac{1}{2}O_2$	6.6×10^5	73–76, 78
$2H_2PO_4^- + O_2^{2-} = 2HPO_4^{2-} + \frac{1}{2}O_2 + H_2O$	3.9×10^{13}	73, 75, 76, 78
$2HPO_4^{2-} + O_2^{2-} = 2PO_4^{3-} + \frac{1}{2}O_2 + H_2O$	2.8×10^5	73, 75, 76, 78
$2HAsO_4^{2-} + O_2^{2-} = 2AsO_4^{3-} + \frac{1}{2}O_2 + H_2O$	2.6×10^2	75, 76, 78
$V_2O_7^{4-} + O_2^{2-} = 2VO_4^{3-} + \frac{1}{2}O_2$	4.5×10^6	74, 76, 78
$2AsO_3^- + O_2^{2-} = As_2O_7^{4-} + \frac{1}{2}O_2$	4.8×10^{13}	78
$As_2O_7^{4-} + O_2^{2-} = 2AsO_4^{3-} + \frac{1}{2}O_2$	4.3×10^4	78
$2H_2AsO_4^- + O_2^{2-} = 2HAsO_4^{2-} + \frac{1}{2}O_2 + H_2O$	7.4×10^{10}	78
$2CrO_3 + O_2^{2-} = Cr_2O_7^{2-} + \frac{1}{2}O_2$	1×10^{25}	79
$CrO_3 + O_2^{2-} = CrO_4^{2-} + \frac{1}{2}O_2$	4.2×10^{18}	79
$3MoO_3 + O_2^{2-} = Mo_3O_{10}^{2-} + \frac{1}{2}O_2$	1×10^{24}	79
$MoO_3 + O_2^{2-} = MoO_4^{2-} + \frac{1}{2}O_2$	4.6×10^{15}	79
$Mo_3O_{10}^{2-} + 2O_2^{2-} = 3MoO_4^{2-} + O_2$	9.9×10^{22}	79
$WO_3 + O_2^{2-} = WO_4^{2-} + \frac{1}{2}O_2$	9.5×10^{13}	79

carbonates of Li, Na, K, Ca, Sr, Ba, and Pb. Both "forward" and "backward" titration experiments were carried out, and differences in the molar ratios acid : O^{2-} at the equivalence point were correlated to the acid character of the cations. A somewhat similar technique was used by Lyalikov *et al.*[87] and Bombi and Fiorani[88] for some heavy metal cations. The basic character of different oxide-ion-donor anions was established by Shams El Din and El Hosary[80] by titrations with the two acids $NaPO_3$ and $K_2Cr_2O_7$. The various anions were found to fall into two distinct groups with widely varying basicities. The first group, the so-called "carbonate group," included carbonate, bicarbonate, oxalate, formate, and acetate. The second, more basic anions, the "oxide group" included O_2^{2-}, OH^-, and electrogenerated O^{2-}.[65, 89] This difference in the basicity of the two groups of oxide-ion donors led to the discovery of the anion pyronitrate, $N_2O_7^{4-}$.[7, 8]

A semiquantitative acid–base scale for oxyanions in molten KNO_3 was proposed by Shams El Din and Gerges.[82] This was based on the measurement of the steady state potential of the oxygen electrode in nitrate melts containing 10^{-2} M concentrations of the different oxyanion. Anions giving rise to more positive potentials than in pure KNO_3 were acids, while those giving more negative potentials were basic to the pure ground electrolyte. Assuming a reversible behavior for the oxygen electrode in these melts, the measured potentials were drawn on a straight line having a slope of $2.303RT/2F$ V. Each $2.303RT/2F$ V was reflected on the x coordinate as a unit of acidity or basicity. Pure molten KNO_3 was assigned zero acidity. Acid oxyanions fell to the left-hand side of the zero and were given a negative sign denoting decreased oxide-ion activity. Bases and basic oxyanions, on the other hand, lay to the right-hand side of the zero and were given positive signs, signifying an increase in oxide-ion activity. According to this concept the acidity (basicity) number should be related to the acid–base equilibrium constants in such a manner that the plot of the logarithm of the equilibrium constant K (given in Table 1) versus the acidity number, N, of the acid component of the reaction should give a straight line with a slope of -1. The line making the best fit to the points had a slope of -0.83.[75]

In a nonoxyanion melt (LiCl–KCl), the oxygen electrode did not respond to additions of either oxide ion or acid, although measurable inflexions were recorded at the points of equivalence.[74] The same was also confirmed by other authors.[90, 91] On the other hand, Wrench and Inman[92] reported that the O_2 electrode in the same melt responded to additions of Li_2O, but at a rate of $2.303RT/F$. This was attributed to the formation of peroxide anions.

Metal–metal oxide electrodes vary their potentials also in accordance with equation (9). Their response to $p_{O^{2-}}$, and also their E° values, depend on the ionization of the oxide: $MO_{n/2} = M^{n+} + (n/2)O^{2-}$. Förland and Tashiro[60] and Smith and Rindone[93] compared the basicities of various glasses by measuring the emf of the concentration cell of the form

$$Ag, Ag_2O \,|\, glass \,(I)\|glass \,(II)\,|\, Ag_2O, \,Ag$$

The concentration of one of the glasses was kept constant while that of the other was varied. Metal/metal electrodes were successfully employed by Hill, Porter, and Gillespie[94] and by Selis, Elliot, and McGinnis.[95] Laitinen and Bhatia[91] extensively studied the behavior of a number of metal/metal oxide electrodes in molten LiCl–KCl eutectic at 450°C. The Cu/CuO, Pt/PtO, Pd/PdO, and Bi/BiOCl were found to respond theoretically to variations in $p_{O^{2-}}$. Irreversibility was noted, however, in the behavior of Ni/NiO and Bi/Bi$_2$O$_3$ half-cells, and was attributed to the formation of a higher oxide in the first case and to a reaction with the melt in the second. Rolin and Gallay[96] used sintered oxide electrodes of Cr_2O_3 and SnO_2 for the determination of dissolved alumina, Al_2O_3, in molten cryolite. Shams El Din and Gerges[75] successfully applied the Cu/CuO electrode as indicator for O^{2-} in molten KNO_3. The electrode was found to respond reversibly to additions of O^{2-}, and when used in potentiometric titrations gave sharp inflexions at the points of equivalence.

3.2. Voltammetry

Voltammetry is a good tool for studying oxide species in fused salts. The subject has been repeatedly reviewed.[97–101] Numerous investigations have been reported on the voltammetric behavior of molten nitrates, with and without added species.[65, 102–110] Particular attention has been paid to the electroreduction of nitrate ions in alkali nitrate melts. The primary reduction products of the nitrate have been identified[65, 102] as nitrite and oxide: $NO_3^- + 2e = NO_2^- + O^{2-}$. Other processes involving the production of nitrogen and nitrogen oxides were also postulated.[102, 106] At more negative potentials, and on a Pt cathode, an unstable green solution is formed,[102, 106, 107] which is said to be a solution of the alkali metal. This is, however, not in conformity with the results of Arvia and his co-workers.[104, 105]

There is considerable controversy on the voltammetry of oxide species in fused nitrates. Swofford and McCormick,[111, 112] on the assumption that oxalate changes in molten nitrates into oxide ion, reported on a one-electron oxidation process to produce peroxide. Shams El Din and El Hosary[80, 113] proved, however, that oxalate changes into CO_3^{2-} rather than O^{2-}. Notoya and Midorikawa[114] studied the voltammetric behavior of KOH in NaNO$_3$–KNO$_3$ melt. Well-developed waves for oxygen evolution at a Pt electrode were obtained. Linearity between the limiting current and concentration of KOH was established over the concentration range 10^{-4}–10^{-2} mole fraction. Topol, Osteryoung, and Christie[109] studied the voltammetric and chronopotentiometric behavior of O^{2-}, OH^-, O_2^{2-}, and CO_3^{2-} in molten $NaNO_3$–KNO_3 at 273–350°C. Solutions of Na_2O_2 yielded upon oxidation two approximately equal waves. The oxide ion gave a single wave at $+0.3$ V relative to their reference electrode. The Na_2CO_3 and $NaOH$ solutions each produced one wave at about $+0.4$ and -0.3 V, respectively. On reversal of current after these waves, single reduction waves at -0.6 to -0.9 V were found in all cases. Francini and Martini[115] studied the reduction of gaseous O_2 in $NaNO_3$–KNO_3 melt in the absence and presence of OH^- and O^{2-} at a dropping mercury electrode. O_2 gas gave rise to a primary reduction wave, while in the presence of O^{2-} or OH^-, an oxidation wave was recorded. All waves were irreversible. OH^- and not O^{2-} was assumed to oxidize. Both species were strictly related by an equilibrium involving H_2O, $O^{2-} + H_2O \rightleftharpoons 2OH^-$, so that in the presence of water it is not feasible to distinguish between the two species. This conclusion is supported by the results given by El Hosary and Shams El Din,[110] who studied the chronopotentiometric behavior of OH^-, O_2^{2-}, and electrogenerated O^{2-} in the same melt. In a carefully dried melt both O_2^{2-} and O^{2-} ions gave ill-defined and poorly reproducible waves. However, when the melts were freely equilibrated with the atmosphere and/or the feed gas was not fully dried, the oxidation step was properly developed. No difficulty was experienced in reproducing the oxidation step when $NaOH$ was used as oxide-ion donor. The product $it^{1/2}$ of the $|O^{2-}|$ step decreased linearly with time denoting the transformation of $|O^{2-}|$ into a bound, inactive form, presumably pyronitrate, $N_2O_7^{4-}$ (References 7, 8, and 116).

Voltammetric studies carried out with a rotating Pt electrode, in Pt cells, and under carefully dried conditions revealed that the O^{2-} ion was unstable in pure $NaNO_3$–KNO_3 eutectic.[1, 117, 118] It changed through a fast reaction with the nitrate ion into peroxide, $O^{2-} + NO_3^- \rightleftharpoons NO_2^- + O_2^{2-}$, and later through a slow reaction to superoxide, $O_2^{2-} + 2NO_3^- \rightleftharpoons 2NO_2^- + 2O_2^-$. The presence of superoxide ion in the melt was confirmed by EPR signals.[119] The oxygen (Pt or Au) electrode in $NaNO_3$–KNO_3 melts is described as a superoxide electrode,[5] $O_2 + e \rightleftharpoons O_2^-$. Zambonin's results were recently confirmed by Johnson and Zacharias,[107] who showed further that in presence of Na or Li in the melt, a cathodic peak is recorded in the voltammograms. This peak was absent in KNO_3. This difference was attributed either to corresponding disparities in the solubility of Na_2O, Li_2O, and K_2O,[65, 106] or to the fast production of O_2^-.[120]

3.3. Other Techniques

Evidence on the equilibria between oxide species have been obtained by a variety of methods other than electrochemical which may be broadly divided into spectroscopy, physical measurements, and conventional chemical analysis.

Spectroscopic techniques have been limited by the extremely corrosive nature of basic melt solutions, particularly if not completely anhydrous, but a Raman line at 1107 cm^{-1} was observed when NaO_2 or KO_2 was dissolved in thoroughly dry $LiF/NaF/KF$ eutectic held in quartz at 500°C, which decreased to zero intensity within 10 min. These solutions had an electronic absorption band at 254 nm (also characteristic of O_2^-). Proof of the presence of superoxide anions in the solidified melt was obtained from the X-ray powder pattern.[121, 122] Similar esr signals were obtained when either monoxide, peroxide, or superoxide were dissolved in $NaNO_3/KNO_3$, which were identified as due to paramagnetic O_2^- ions and were qualitatively similar to those of O_2^- in Na_2O_2.[119]

Cryoscopic measurements have been reported but in most cases the experimental results do not discriminate between peroxide and its thermal decomposition product, monoxide. For example, Na_2O_2 dissolved in $LiNO_3/KNO_3$ gave a freezing point depression indicative of three additional ions per solute molecule,[123] as would Na_2O. However, this result does rule out further oxidation of peroxide to superoxide by reactions (2) or (3), since six additional ions would be present if the oxidation was complete and oxidation was by nitrate anions, or four additional ions if oxidation was by oxygen gas. Similarly cryoscopy on the postulated 1:1 oxide:nitrate compounds (orthonitrates) does not distinguish between free and bonded monoxide [e.g., Na_3NO_4 and $Na_2O \cdot NaNO_3$ would both provide four additional ions in $LiNO_3$ (Reference 6)].

The related technique of phase diagram determination has indicated no compound formation between Na_2O_2 and $NaNO_3$ but possible 1:1 compound formation between Na_2O and $NaNO_3$ or $NaNO_2$ and between K_2O and KNO_3 (Reference 6).

Direct determinations of solubility have demonstrated the individuality of oxide and peroxide ions (i.e., 0.3 M Na_2O and 0.016 M Na_2O_2 in $LiNO_3/KNO_3$ at 160°C)[124] unlike the implications of equation (1).

Thermogravimetry has shown sodium peroxide to be appreciably more stable in $NaNO_3/KNO_3$ than in $LiNO_3/KNO_3$, while the slow weight gain displayed by sodium peroxide solutions in $NaNO_2/KNO_2$ (Reference 9) provides support for the supposed catalytic role of peroxide in the aerial oxidation of molten nitrite.[120] Thermogravimetry and measurement of evolved oxygen showed the decomposition temperatures of potassium superoxide to be greatly affected by the alkali metal nitrate in which it was dissolved. As expected these increased with increasing size of cation, the temperature at which peroxide was formed being 120°C in $LiNO_3$, 190°C in $NaNO_3$, and 370°C in KNO_3. Peroxide showed some stability in concentrated solutions.[125–127]

Among the more conventional analytical methods which have been used to provide information on oxide equilibria is volumetric analysis. For example, iodometric titration of aqueous solutions of quenched LiCl/KCl containing Li_2O indicated some oxidation of oxide to peroxide at 450°C,[92] and titration of aqueous solutions gave evidence of the stability of peroxide in LiF/NaF/KF at 600°C,[128] while volumetric estimation of residual peroxide enabled the equilibrium position of reaction (1) to be determined in molten $NaNO_2/KNO_2$.[9] Permanganate and acidimetric titrations of quenched melts have been used to deduce thermodynamic data on equation (3) in alkali metal carbonates[129] and combined titrimetric and manometric methods for similar data in alkali metal nitrates.[130]

Lastly a manometric determination of the equilibrium constant of reaction (3) in $NaNO_3/KNO_3$ was of crucial importance in elucidating the apparent discrepencies between different workers (see Section IV) and in giving support to thermodynamic calculations.[131]

4. The Nature of Oxide Species in Nitrate Melts

The earlier universally held view that the basic species in nitrate melts must be monoxide[109, 132, 133] seemed completely contrary to the claim[117] that monoxide was oxidized by the melt oxyanions to peroxide, and superoxide, i.e., equations (1) and (2). The situation was further confused by conflicting reports as to the form of the Nernst equation for the oxygen electrode (both one- and two-electron processes being reported) and the position of the hydroxide versus oxide plus water equilibrium (references given in Reference 4).

Moreover, the equilibrium constant given by Zambonin and Jordan[1] for equation (1) rested not only on the value obtained by those authors from voltammetric studies on basic nitrate melts but also on the value of the equilibrium constant for the reaction

$$3O_2^{2-} \rightleftharpoons 2O^{2-} + 2O_2^- \qquad (10)$$

The latter had in fact been determined for the species dissolved not in a nitrate melt but in an alkali metal hydroxide melt[134] and had been assumed to be identical, an assumption which was soon strongly challenged.[135]

However, support for the values advanced by Zambonin and Jordan[1] has now been obtained, firstly by a similar value (6×10^{-6}) for the equilibrium constant of equation (3) obtained by independent authors using a manometric technique[131] and a value of 0.5×10^{-6} derived by Zambonin[5] from potentiometric measurements. Secondly, using the manometric value for equation (3) and a value for the decomposition of nitrate

$$NO_3^- \rightleftharpoons NO_2^- + \tfrac{1}{2}O_2 \tag{11}$$

determined by chemical analysis,[136] it was possible to calculate a value for the equilibrium constant of equation (2) quite independently of any electrochemical measurements. The good agreement obtained (6.7×10^{-11} and 11×10^{-11}) allayed some fears that either electrode kinetics or attack of the platinum container by peroxide (when a higher concentration of nitrite is present, peroxide anions in a nitrate melt can attack platinum[9]) introduced errors into the value obtained from the voltammetric measurements. Thirdly, some support for the value of the equilibrium constant of equation (1), which is of course the essential pivot of the whole discussion, was obtained both from the fact that a thermodynamic calculation of the equilibrium constant for equation (10) in the solid state reaction[4] gave a similar value to that found experimentally for the species dissolved in the hydroxide melt, and that the values for equation (11) and the calculated thermodynamic value for the solid state reaction $O^{2-} + \tfrac{1}{2}O_2 = O_2^{2-}$ gave a figure of 1.2 for the equilibrium constant of equation (1), which was perhaps fortuitously close to the experimental figure[1] of 3.

Nevertheless, despite this interconnecting but consistent set of equilibrium constants supporting equations (1) and (2), the discrepancies in the reported Nernst equations for the oxygen electrode remained. An attempt to explain these differences was made by assuming that oxide ions were stabilized by anion solvation, forming, for example, either orthonitrate (NO_4^{3-})[6, 63, 66] or pyronitrate $(N_2O_7^{4-})$,[7, 8, 80] so that, depending on the value of the equilibrium constants for such stabilization, the significant basic species in nitrate melts would be, depending upon the actual concentration of base, either superoxide (giving a one-electron slope) or the stabilized oxide ion (giving a two-electron slope). However, the concentration ranges over which such different slopes had been reported by different workers showed that this was not a sufficient explanation.[4]

A further hypothesis was required, namely, that additonal oxide stabilization was occurring owing to silicate formation, i.e.,

$$O^{2-} + xSiO_2 \rightleftharpoons (SiO_2)_x \cdot O^{2-} \tag{12}$$

in melts held in silica or glass containers (for which two-electron slopes have been reported) which did not occur in melts held in inert containers (of platinum or p.t.f.e.) for which one-electron slopes had been reported. Fortunately this hypothesis could be tested immediately, since a value for the equilibrium constant of equation (12) could be calculated from the equilibrium constants of reaction (13),

$$NO_3^- \rightleftharpoons NO_2^+ + O^{2-} \tag{13}$$

which had opportunely already been measured as 10^{-35} in a platinum vessel,[137] and could be deduced, from two acid–base equilibria, i.e.,

$$NO_3^- + Cr_2O_7^{2-} \rightleftharpoons NO_2^+ + 2CrO_4^{2-} \qquad \text{(Reference 138)}$$

and

$$Cr_2O_7^{2-} + O^{2-} \rightleftharpoons 2CrO_4^{2-} \qquad \text{(Reference 139)}$$

which had both been measured in glass vessels, as 10^{-19}, thus yielding a value of 10^{16} for equation (12). A second value for the same equilibrium constant could similarly be obtained from the formerly discrepant values reported for the equilibrium

$$H_2O + O^{2-} \rightleftharpoons 2OH^- \tag{14}$$

where 10^{18}, in units of concentration, had been found when platinum had been used[5] and 3×10^{-2} in units of water vapor pressure when glass vessels were used.[140] After allowing for the solubility of water in the nitrate melt,[141] a value of 2×10^{15} for the equilibrium constant of equation (12) was obtained. The good agreement between these two independent series of experimental values was considered as a very satisfactory test of the silicate-stabilized oxide hypothesis, as well as a resolution of the discrepancies reported for equation (14), even though it was noted that both values for equations (13) and (14) determined in platinum containers included the value measured for equation (10) in hydroxide melts.

It may be noted that using a value of 10^{15} for the equilibrium constant of equation (12) more than 95% of initial oxide will be present, *at equilibrium*, as the silicate-stabilized species, when the initial oxide concentration is 10^{-8} *m* or more. At low concentrations of initial oxide, equilibrium does seem to be reached within the timescale of many electrochemical measurements, though at higher concentrations of base this equilibrium is only attained slowly (e.g., a 0.3 *m* solution of Na_2O_2 in equimolar $NaNO_3/KNO_3$ at 290°C in glass still contained peroxide after 24 hours, though this was quickly destroyed if powdered silica was added[4]), and thus fresh molar concentration melt solutions of monoxide and peroxide have as expected an individual identity in chemical reactions.[142]. The interconversion of superoxide and oxide has been shown to be slow and the general validity of equations (1)–(3) was confirmed though a number of differences were found, according to the nature of the base originally added.[125–127]

Besides the interactions outlined above it seems that lithium cations interact strongly with monoxide, a value of $K \approx 70$ for $Li^+ + O^{2-} \rightleftharpoons LiO^-$

was obtained at 300°C in $NaNO_3/KNO_3$.[3] Conversely potassium super-oxide is destabilized in $LiNO_3$ as compared to $NaNO_3$ and KNO_3.[125-127] More recently potassium cations were shown to stabilize superoxide anions and sodium cations to stabilize peroxide anions.[2]

5. The Nature of Oxide Species in Other Melts

In many different melt systems the only available information is that certain oxides dissolve, e.g., in ammonium acetate,[143] and zinc chloride[144] melts while oxygen and other oxidizing agents convert molten cyanide to cyanate[145] and in molten thiocyanate oxide ions take part in an equilibrium[146] $O^{2-} + SCN^- \rightleftharpoons S^{2-} + OCN^-$. However, in molten lithium perchlorate there is the report of the loss of oxygen from KO_2 and $M_2O_2(M = Li, Na, or K)$ solutions.[147, 148]

Compared to the studies carried out in single or mixed nitrates, comparatively little has been done on the nature of the oxide species even in the other commonly used melts. However, the solubility and reactivity of different oxide species in various chloride melts have been recently reviewed by Kerridge.[149] The solubility of gaseous CO and CO_2 in chloride melts[150, 151] increases in the presence of oxide anions due to the formation of carbonate. Although water solubility in chloride melts is very small, traces can lead to hydrolysis, $H_2O + 2Cl^- = HCl + O^{2-}$.[152, 153] Peroxide anions are stable in LiCl–KCl eutectic at 400–500°C,[154] but not in NaCl–KCl at 700°C.[155, 156] Peroxide has been found in melts to which oxide anions were originally added, e.g., in LiCl/KCl at 450°C[92] in contact with oxygen and Stern[157] is quoted (in Reference 92) as "finding up to 1% peroxide in oxide solution in NaCl at 900°C," whereas he reported after heating impure Na_2O (containing Na_2O_2) in NaCl at 900°C for an unstated time most of the excess O_2 was evolved but $\sim 1\%$ peroxide remained in the quenched melt.[157] In addition later measurements have indicated the position of the equilibrium to be somewhat variable.[158] The equilibria in LiCl–KCl, NaCl–KCl, LiCl–NaCl–KCl at 585°C were assumed to be.[159]

$$O_2 + Cl^- \rightleftharpoons O_2^- + \tfrac{1}{2}Cl_2$$

$$\Updownarrow O_2$$

$$O_2 + 2Cl^- \rightleftharpoons O_2^{2-} + Cl_2$$

The behavior of the O^{2-}/O_2 electrode in chloride melts has been repeatedly studied.[74, 80–92, 160–162] Agreement seems to be established that the Nernstian plot E–log a_O^{2-} has a slope of $2.303RT/F$. It is not possible to differentiate chronopotentiometrically between the two anodic reactions: $O^{2-} + MO \overset{-2e}{\rightleftharpoons} MO-O$, $2O = O_2$, and $O_2^{2-} \overset{-2e}{\rightleftharpoons} O_2$.[161] The oxidation of O^{2-} ion by cyclic voltammetry[125] on Pt, Pd, and Rh occurs in one step, corresponding to $O^{2-} \rightarrow O + 2e$, $O + O \rightarrow O_2$. On a gold anode, however, a two-step process is recorded favoring the mechanism: $O^{2-} \rightarrow O^- + e$, $O^- \rightarrow O + e$, $O + O \rightarrow O_2$.

In accord with the RT/F slope, superoxide has been reported not to disproportionate in LiCl/KCl at 400–500°C,[163] but to do so in NaCl/KCl at 700°C.[156, 164] The interaction of oxide with a number of acids, $CaWO_4$,[165, 166] Ca^{2+},[167] Ce^{3+},[168] CO_2,[169] and H_2O,[170, 171] has been studied in NaCl/KCl.

In $NaAlCl_4$ the oxide ion has been found to be a strong dibase, $O^{2-} + AlCl_4^- \rightleftharpoons AlOCl_2^- + 2Cl^-$, though the oxydichloride is also a weak base in the presence of sufficiently strong acid.[172]

In molten alkali hydroxides, free O^{2-}, O_2^{2-}, O_2^- and dissolved O_2 are said to be in equilibrium.[134] O^{2-} is the strongest base and reductant. It oxidizes to O_2^{2-} and further to O_2^-:

$$2O^{2-} - 2e \rightarrow O_2^{2-}, \qquad O_2^{2-} - e \rightarrow O_2^-$$

The O_2^{2-} is stable only in basic media. It reduces to O^{2-} or oxidizes to O_2^-. The O^{2-}/O_2^{2-} system is slow on Pt, but establishes readily on a NiO electrode. The system O_2^{2-}/O_2^- is rapid and reversible. In the presence of H_2O, O_2^{2-} changes into OH^- and O_2^-:

$$3O_2^{2-} + 2H_2O \rightarrow 4OH^- + 2O_2^-$$

The O_2^- ion exists only in neutral and acid media, and is reduced to O_2^{2-} by O^{2-} ion. It is not, however, oxidized to O_2. In the absence of water, O_2^- reduces along two steps:

$$O_2^- + e \rightarrow O_2^{2-}, \qquad O_2^{2-} + 2e \rightarrow 2O^{2-}$$

In the presence of H_2O, however, OH^- ions are produced instead:

$$O_2^- + 2H_2O + 3e \rightarrow 4OH^-$$

Burrows and Hills[173] studied the behavior of the oxygen electrode in Li_2SO_4–K_2SO_4 eutectic. The O^{2-} ion concentration was varied through small additions of CaO. Both bright and platinized Pt were used. Both electrodes responded sluggishly to changes in O_2 partial pressure, but the response to O^{2-} ion concentration was immediate. The negative potential drifted, however, slowly back to more positive values owing to the attack on the silica of the container. Again the Nernstian slope of $2.303RT/F$ was obtained. Coulometric acid–base titrations in $(Li, Na, K)_2SO_4$ melts at 625°C were carried out by Rahmel.[174] The titration involved the neutralization of the base Na_2CO_3 by the acid SO_3 released at the anode according to

$$SO_4^{2-} = SO_3 + \tfrac{1}{2}O_2 + 2e$$

and was followed potentiometrically by an $O_2(Pt)$ indicator electrode. Much earlier, peroxide was shown to give equivalent solutions of monoxide and evolve oxygen in Na_2SO_4/K_2SO_4 at 950°C.[175]

The equilibria

$$H_2O + F^- \rightleftharpoons OH^- + HF \quad \text{and} \quad H_2O + 2F^- \rightleftharpoons O^{2-} + 2HF$$

in molten $LiF–BeF_2$ were studied by Mathews and Baes.[176] Raman and uv spectroscopy have shown O_2^- to be unstable in molten $LiF–NaF–KF$ (References 121 and 122) though peroxide has been reported to be stable at 600°C.[128] Cyclic voltammetric and chronopotentiometric curves of O^{2-} ion in molten $LiF–BeF_2–ZrF_4$ and $LiF–BeF_2–ThF_4$ on Au, Ir, and glassy C electrodes in the temperature range 500–710°C were recently reported.[177] Well-defined and reproducible results were obtained only at a gold electrode. The following sequence of reactions were considered to occur:

$$O^{2-} = O + 2e, \quad O + O = O_2, \quad O + O^{2-} = O_2^{2-}$$

O_2^{2-} ions are further oxidized producing a postwave increasing with Na_2O_2 additions. O_2^- ions are unstable in these melts.[177]

In fluoride melts containing aluminum species, such as cryolite or MF/M_3AlF_6, complexes containing Al_2O were shown to form at M_2O/M_3AlF_6 ratios of 0.2 and AlO complexes at higher ratios.[178]

Potentiometric studies in Na_2CO_3/K_2CO_3 melts in contact with oxygen have shown the electroactive species to be peroxide and superoxide ions produced by the chemical equilibria:[10]

$$O_2 + 2CO_3^{2-} \rightleftharpoons 2O_2^{2-} + 2CO_2 \tag{13}$$

$$3O_2 + 2CO_3^{2-} \rightleftharpoons 4O_2^- + 2CO_2 \tag{14}$$

though in Li_2CO_3 only the peroxide anion was formed because of the greater stabilization of oxide by lithium cations.[179] However, Andersen,[129] who derived equilibrium constants on the basis of chemical analysis on quenched melts, concluded that such equilibria as (13) and (14) were unlikely to be significant.

In molten $NaNO_2–KNO_2$, the equilibrium constant of the reaction $O_2^{2-} + NO_2^- \rightleftharpoons O^{2-} + NO_3^-$ amounts to 0.028 at 250°C and 0.115 at 300°C.[9] These values compare with 0.33 at 229°C given by Zambonin and Jordan,[1] but were much less than that computed by Temple[135] ($> 10^6$).

References

1. P. G. Zambonin and J. Jordan, *J. Am. Chem. Soc.* **91**, 2225 (1969).
2. K. E. Johnson, P. S. Zacharias, and J. Matthews, in *Proceedings of the International Symposium on Molten Salts, 1976*, Electrochemical Society, New York (1976), p. 603.
3. J. P. Pemberton, *Diss. Abs.* **32B**, 3241 (1972).
4. J. D. Burke and D. H. Kerridge, *Electrochim. Acta* **19**, 251 (1974).
5. P. G. Zambonin, *J. Electroanal. Chem.* **33**, 243 (1971).
6. R. Kohlmuller, *Ann. Chim.* **4**, 1183 (1959).
7. A. M. Shams El-Din and A. A. El-Hosary, *J. Inorg. Nucl. Chem.* **28**, 3043 (1966).
8. A. M. Shams El-Din and A. A. El-Hosary, *Electrochim. Acta* **12**, 1665 (1967).
9. S. S. Al Omer and D. H. Kerridge, *J. Inorg. Nucl. Chem.* **40**, 975 (1978).
10. A. J. Appleby and S. B. Nicholson, *J. Electroanal. Chem.* **83**, 309 (1977).
11. W. Stegmaier and A. Dietzel, *Glastechn. Ber.* **18**, 297 (1940).
12. G. W. Morey and N. L. Bowen, *J. Phys. Chem.* **28**, 1177 (1924).
13. A. M. Shams El-Din, H. D. Taki El-Din, and A. A. El-Hosary, *Electrochim. Acta* **13**, 407 (1968).
14. A. M. Shams El-Din and A. A. El-Hosary, *J. Electroanal. Chem.* **16**, 551 (1968).
15. A. A. El-Hosary, H. D. Taki El-Din, and A. M. Shams El-Din, *Phys. Chem. Glasses* **12**, 111 (1971).
16. H. Salmang, *Glastechn. Ber.* **7**, 278 (1929/1930).
17. H. Salmang and J. Kaltenbach, *Feuerfest.* **7**, 161 (1931).
18. C. J. Peddle, *J. Soc. Glass Technol.* **4**, 20, 46, 59 (1920).
19. C. J. Peddle, *J. Soc. Glass Technol.* **5**, 195 (1921).
20. H. Karmous, *Sprechsaal Keram.*, 59 (1926).
21. H. Ohta and Y. Suzuki, *Bull. Am. Ceram. Soc.* **57**, 602 (1978).
22. T. M. El-Shamy and A. A. Ahmed, Proceedings of the 11th International Congress on Glass, Prague, July 1977, Vol. 3, pp. 181–195.
23. T. M. El-Shamy, S. E. Morsi, H. D. Taki El-Din, and A. A. Ahmed, *J. Non-Cryst. Solids* **19**, 241 (1976).
24. W. A. Weyl, *Glastech. Ber.* **9**, 641 (1931).
25. P. Niggli, *Z. Anorg. Chem.* **98**, 214 (1914).
26. M. L. Pearce, *J. Am. Ceram. Soc.* **47**, 342 (1964); **48**, 175 (1965).
27. J. D. Van Norman and R. A. Osteryoung, *Anal. Chem.* **32**, 398 (1960).
28. V. H. Franz and H. Scholze, *Glastech. Ber.* **36**, 347 (1963).
29. H. Scholze, *Glastech. Ber.* **32**, 81, 142, 278, 314 (1959).
30. R. V. Adams and R. W. Douglas, *J. Soc. Glass Technol.* **43**, 147 (1959).
31. J. M. Uye and T. B. King, *Trans. AIME* **227**, 492 (1963).
32. C. R. Kurkjian and L. E. Russell, *J. Soc. Glass Technol.* **42**, 130 (1958).
33. H. Franz, *J. Am. Ceram. Soc.* **49**, 473 (1966).
34. C. J. B. Fincham and F. D. Richardson, *Proc R. Soc. London* **A223**, 40 (1954).
35. G. E. Toop and C. S. Samis, *Trans. AIME* **224**, 878 (1962).
36. E. J. Turkdogan and M. L. Pearce, *Trans. AIME* **227** 940 (1963).
37. W. A. Weyl and F. Thümen, *Sprechsaal Keram.* **66**, 197 (1933).
38. W. A. Weyl and E. C. Marboe, *Constitution of Glasses, A Dynamic Interpretation*, Vol. 1, Wiley and Sons, New York (1962).
39. J. A. Duffy and M. D. Ingram, *J. Chem. Phys.* **52**, 3752 (1970); *J. Am. Chem. Soc.* **93**, 6448 (1971).
40. A. Paul, *Phys. Chem. Glasses* **11**, 46 (1970); **13**, 144 (1972).
41. J. A. Duffy and M. D. Ingram, *J. Non-Cryst. Solids* **21**, 373 (1976).
42. A. Paul, *Trans. Indian Ceram. Soc.* **28**, 63 (1969).

43. J. Krogh-Moe, *Phys. Cer. Glasses* 1, 26 (1960).
44. C. Anderson-Kraft, *Glastech. Ber.* 9, 557, 597 (1931).
45. W. A. Weyl, *Coloured Glasses*, Society of Glass Technology, Sheffield, England (1951).
46. R. W. Douglas, P. Nath, and A. Paul, *Phys. Chem. Glasses* 6, 216 (1965).
47. P. Csaki and A. Dietzel, *Glastech. Ber.* 18, 35, 65 (1940).
48. W. A. Weyl, B. Badger, and A. Pincus, *J. Am. Ceram. Soc.* 22, 374 (1939).
49. W. Höfler, *Glastech. Ber.* 12, 117 (1934).
50. W. Hirsch and A. Dietzel, *Sprechsaal Keram.* 68, 243 (1935).
51. A. Dietzel, *Glastech. Ber.* 16, 292 (1938).
52. W. D. Bancroft and R. L. Nuggent, *J. Phys. Chem.* 33, 481 (1929).
53. A. A. Ahmed, T. M. El-Shamy, and G. M. Ashour, *Egypt. J. Chem.* 18, 875 (1975).
54. J. A. Duffy and M. D. Ingram, *J. Inorg. Nucl. Chem.* 36, 43 (1974).
55. B. J. Brough, D. H. Kerridge, and M. Mosley, *J. Chem. Soc. A*, 1556 (1966).
56. H. Lux, *Z. Elektrochem.* 45, 303 (1939).
57. H. Lux, *Z. Elektrochem.* 52, 220 (1948); 53, 43, 45 (1949).
58. H. Flood, T. Förland, and K. Motzfeldt, *Acta Chim. Scand.* 6, 257 (1952).
59. R. Didtschenko and E. G. Rochow, *J. Am. Chem. Soc.* 76, 3291 (1954).
60. T. Förland and M. Tashiro, *Glass Ind.* 37, 381 (1956).
61. V. I. Minenko, S. M. Petrov, and N. S. Ivanova, *Zh. Fiz. Khim.* 35, 1534, (1961).
62. R. E. Ranford and S. N. Flengas, *Can. J. Chem.* 43, 2879 (1965).
63. R. N. Kust and F. R. Duke, *J. Am. Chem. Soc.* 85, 3338 (1963).
64. R. N. Kust, *J. Phys. Chem.* 69, 3662 (1969).
65. H. S. Swofford and H. A. Laitinen, *J. Electrochem. Soc.* 110, 814 (1963).
66. R. N. Kust, *Inorg. Chem.* 3, 1035 (1964).
67. R. N. Kust, *J. Electrochem. Soc.* 116, 1137 (1969).
68. B. A. Rose, G. J. Davis, and H. J. T. Ellingham, *Farad. Soc. Discuss.* 4, 154 (1948).
69. A. S. Afanasev and V. P. Gamazov, *Russ. J. Phys. Chem.* (Eng. Ed.), 38, 1537 (1964).
70. G. J. Janz and F. Saegusa, *Electrochim. Acta* 7, 393 (1962).
71. A. M. Shams El-Din, *Electrochim. Acta* 7, 285 (1962).
72. A. M. Shams El-Din and A. A. A. Gerges, *J. Electroanal. Chem.* 4, 309 (1962).
73. A. M. Shams El-Din and A. A. A. Gerges, *Electrochim. Acta* 9, 123 (1964).
74. A. M. Shams El-Din, A. A. El-Hosary, and A. A. A. Gerges, *J. Electroanal. Chem.* 6, 131 (1963).
75. A. M. Shams El-Din and A. A. A. Gerges, *Electrochim. Acta* 9, 613 (1964).
76. A. M. Shams El-Din and A. A. A. Gerges, *J. Inorg. Nucl. Chem.* 25, 1537 (1963).
77. A. M. Shams El-Din and A. A. El-Hosary, *J. Electroanal. Chem.* 7, 464 (1964).
78. A. M. Shams El-Din, A. A. El-Hosary, and A. A. A. Gerges, *J. Electroanal. Chem.* 8, 312 (1964).
79. A. M. Shams El-Din and A. A. El-Hosary, *J. Electroanal. Chem.* 9, 349 (1965).
80. A. M. Shams El-Din and A. A. El-Hosary, *Electrochim. Acta* 13, 135 (1968).
81. A. M. Shams El-Din, H. D. Taki El-Din, and A. A. El-Hosary, *Electrochim. Acta* 13, 407 (1968).
82. A. M. Shams El-Din and A. A. A. Gerges, *Electrochemistry*, Eds. A. Friend and F. Gutmann, Pergamon, New York (1964), p. 562.
83. N. Coumert, M. Porthault, and J. C. Merlin, *Bull. Soc. Chim. Fr.*, 910 (1965).
84. A. Rahmel, *J. Electroanal. Chem.* 61, 333 (1975).
85. A. M. Shams El-Din, A. A. El-Hosary, and R. M Saleh, *Metalloberflaeche* 25, 425 (1971).
86. A. M. Shams El-Din and A. A. El-Hosary, *J Electroanal. Chem.* 8, 139 (1964).
87. Lu. S. Lyalikov, E. A. Levinson, T. I. Todorova, and V. V. Nikolaeva, *Uch. Zap. Kishinev. Gos. Univ.* 56, 91 (1960).
88. G. G. Bombi and M. Fiorani, *Talanta* 12, 1053 (1965).
89. G. J. Hills and K. E. Johnson, *J. Electrochem. Soc.* 108, 1013 (1961).

90. R. Littlewood and E. J. Argent, *Electrochim. Acta* 4, 114 (1961).
91. H. A. Laitinan and B. B. Bhatia, *J. Electrochem. Soc.* 107, 707 (1960).
92. N. S. Wrench and D. Inman, *J. Electroanal. Chem.* 17, 319 (1968).
93. G. S. Smith and G. E. Rindone, *Glass Ind.* 37, 437 (1956).
94. D. G. Hill, B. Porter, and A. S. Gillespie, *J. Electrochem. Soc.* 105, 408 (1958).
95. S. M. Selis, G. R. B. Elliot, and L. P. McGinnis, *J. Electrochem. Soc.* 106, 134 (1959).
96. M. Rolin and J. J. Gallay, *Electrochim. Acta* 7, 153 (1962).
97. Yu. K. Delimarskii and B. F. Markov, *Electrochemistry of Fused Salts*, (Engl. Trans.), Sigma, Washington, D.C., (1961), Chap. 8.
98. H. C. Gaur and R. S. Sethi, *J. Electroanal. Chem.* 7, 474 (1964).
99. H. A. Laitinen and R. A. Osteryoung, in *Fused Salts*, Ed. B. R. Sundheim, McGraw-Hill, New York (1964), Chap. 4.
100. C. H. Liu, K. E. Johnson, and H. A. Laitinen, in *Molten Salt Chemistry*, Ed. M. Blander, Wiley-Interscience, New York (1964).
101. H. A. Laitinen, *Pure Appl. Chem.* 15, 227 (1967).
102. G. J. Hills and K. E. Johnson, *Advances in Polarogrophy*, Pergamon, London (1961), p. 974.
103. H. S. Swofford and C. L. Holifield, *Anal. Chem.* 37, 1503 (1965).
104. A. J. Arvia, A. J. Calandra, and W. E. Triaca, *Electrochim. Acta.* 9, 1417 (1964).
105. A. J. Calandra and A. J. Arvia, *Electrochim. Acta* 10, 474 (1965).
106. H. E. Bartlett and K. E. Johnson, *J. Electrochem. Soc.* 114, 64 (1967).
107. K. E. Johnson and P. S. Zacharias, *J. Electrochem. Soc.* 124, 448 (1977).
108. G. J. Hills and P. D. Power, *J. Polarg. Soc.* 13, 71 (1967).
109. L. E. Topol, R. A. Osteryoung, and J. H. Christie, *J. Phys. Chem.* 70, 2857 (1966).
110. A. A. El-Hosary and A. M. Shams El-Din, *Electrochim. Acta* 16, 143 (1971).
111. H. S. Swofford and P. G. McCormick, *Anal. Chem.* 37, 970 (1965).
112. P. G. McCormick and H. S. Swofford, *Anal. Chem.* 41, 146 (1969).
113. A. A. El-Hosary and A. M. Shams El-Din, *J. Electroanal. Chem.* 30, 33 (1971).
114. T. Notoya and R. Midorikawa, *Bull. Fac. Engn. Hokkaido Univ.* 36, 868 (1968).
115. M. Francini and S. Martini, *Electrochim. Acta* 13, 851 (1968).
116. A. A. El-Hosary and A. M. Shams El-Din, *J. Electroanal. Chem.* 35, 35 (1972).
117. P. G. Zambonin and J. Jordan, *J. Am. Chem. Soc.* 89, 6365 (1967).
118. P. G. Zambonin, *J. Electroanal. Chem.* 24, 365 (1970).
119. P. G. Zambonin, *J. Phys. Chem.* 78, 1294 (1974).
120. P. G. Zambonin and A. Cavaggioni, *J. Am. Chem. Soc.* 93, 2854 (1971).
121. F. L. Whiting, G. Mamantov, G. M. Begun, and J. P. Young, *Inorg. Chim. Acta* 5, 260 (1971).
122. F. L. Whiting, G. Mamantov, and J. P. Young, *J. Inorg. Nucl. Chem.* 34, 2475 (1972).
123. P. Allamagny, *Rev. Chim. Min.* 2, 645 (1965).
124. B. J. Brough and D. H. Kerridge, *J. Chem. Eng. Data* 11, 260 (1966).
125. Ya. Ya. Sauka and V. Ya. Bruner, *Izv. Akad. Nauk. Latv. S.S.R. Ser. Khim.* 6, 615 (1966).
126. V. Ya. Bruner and Ya. Ya. Sauka, *Russ. J. Inorg. Chem.* 12, 696 (1967).
127. V. Ya. Bruner and Ya. Ya. Sauka, *Russ. J. Inorg. Chem.* 13, 1727 (1968).
128. S. Pizzini and R. Morlotti, *Electrochim. Acta* 10, 1033 (1965).
129. B. K. Andersen, *Acta Chem. Scand.* A31, 242 (1977).
130. D. R. Flinn and K. H. Stern, *J. Electroanal. Chem.* 63, 39 (1975).
131. J. M. Schlegel and D. Priore, *J. Phys. Chem.* 76, 2841 (1972).
132. F. R. Duke and M. L. Iverson, *J. Am. Chem. Soc.* 80, 5061 (1958).
133. R. N. Kust and F. R. Duke, *J. Am. Chem. Soc.* 85, 3338 (1963).
134. J. Goret and B. Tremillon, *Bull. Soc. Chim. Fr.*, 67 (1966).
135. M. Fredericks, R. B. Temple, and G. W. Thickett, *J. Electroanal. Chem.* 38, App. 5 (1972).

136. R. N. Kust and J. D. Burke, *Inorg. Nucl. Chem. Lett.* **6**, 333 (1970).
137. P. G. Zambonin, *J. Electroanal. Chem.* **32**, App. 1 (1971).
138. F. R. Duke, *J. Chem. Ed.* **39**, 57 (1962).
139. R. N. Kust, *Inorg. Chem.* **6**, 2239 (1967).
140. M. Fredericks and R. G. Temple, *Austrl. J. Chem.* **25**, 1831 (1972).
141. P. G. Zambonin, V. L. Cardetta, and G. Signorile, *J. Electroanal. Chem.* **28**, 237 (1970).
142. D. A. Habboush and D. H. Kerridge, *Thermochim. Acta* **10**, 187 (1974).
143. J. B. Reed, B. S. Hopkins, and L. F. Audrieth, *J. Am. Chem. Soc.* **57**, 1159 (1935).
144. G. F. Pinaev, V. V. Pechkovskii, and L. M. Vinagradov, *Zh. Neorg. Khim.* **16**, 827 (1971).
145. W. Sundermeyer, *Angew. Chem. Int. Ed.* **4**, 222 (1965).
146. D. H. Kerridge and M. Mosley, *J. Chem. Soc. A*, 352 (1967).
147. A. Bruners, A. Salta, and I. I. Vol'nov, *Tezisy Dokl. Vses. Sov. Khim. Neorg. Perekisnykh. Soedin*, 87 (1973).
148. D. G. Lemesheva and V. Ya. Rosolovskii, *Tezsy Dokl. Vses. Sov. Khim. Neorg. Perekisnykh. Soedin*, 88 (1973).
149. D. H. Kerridge, *The Chemistry of Nonaqueous Solvents*, Vol. VB, Ed. J. J. Lagowski, Academic, New York (1978), Chap. 5.
150. J. P. Zezyanov and V. A. Il'ichev, *Zh. Neorg. Khim* **17**, 254 (1972).
151. A. B. Bezukladnikov, V. N. Devyatkin, and O. N. Il'icheva, *Zh. Fiz. Khim.* **44**, 253 (1970).
152. E. P. Mignonsin and G. Duyckaerts, *Anal. Chim. Acta* **47**, 71 (1969).
153. R. Combes, J. Vedel, and B. Tremillon, *C.R. Acad. Sci. Ser. C* **275**, 199 (1972).
154. V. I. Shapoval and V. A. Vasilenko, *Ukr. Khim. Zh.* **40**, 868 (1974).
155. V. I. Shapoval, Y. K. Delimarski, and D. G. Tsiklauri, *Ukr. Khim. Zh.* **40**, 734 (1974).
156. Y. Kaneko and H. Kojima, *Denki Kagaku* **41**, 347 (1973).
157. K. H. Stern, *J. Phys. Chem.* **66**, 1311 (1962).
158. K. H. Stern, R. Panayappan, and D. R. Flinn, *J. Electrochem. Soc.* **124**, 641 (1977).
159. E. P. Mignonsin, L. Martinot, and G. Duychaerts, *Inorg. Nucl. Chem. Lett.* **3**, 511 (1967).
160. G. Delarue, *J. Electroanal. Chem.* **1**, 285 (1959–1960).
161. D. Inman and M. J. Weaver, *J. Electroanal. Chem.* **51**, 45 (1974).
162. Y. Kanzaki and M. Takahashi, *J. Electroanal. Chem.* **58**, 339 (1975).
163. V. I. Shapoval and V. A. Vasilenko, *Ukr. Khim. Zh.* **40**, 868 (1971).
164. V. I. Shapoval, Yu. K. Delimarskii, and O. G. Tsiklauri, *Ukr. Khim. Zh.* **40**, 736 (1974).
165. F. de Andrade, R. Combes, and B. Tremillon, *C.R. Acad. Sci. Ser. C* **280**, 945 (1975).
166. R. Combes and B. Tremillon, *J. Electroanal. Chem.* **83**, 297 (1977).
167. R. Combes, J. Vedel, and B. Tremillon, *Anal. Lett.* **3**, 523 (1970).
168. R. Combes, M.-N. Levelut, and B. Tremillon, *J. Electroanal. Chem.* **91**, 125 (1978).
169. R. Combes, R. Feys, and B. Tremillon, *J. Electroanal. Chem.* **83**, 383 (1977).
170. R. Combes, J. Vedel, and B. Tremillon, *C.R. Acad. Sci. Ser. C* **275**, 199 (1972).
171. R. Lysy and R. Combes, *J. Electroanal. Chem.* **83**, 287 (1977).
172. B. Tremillon, A. Bernard, and R. Molina, *J. Electroanal. Chem.* **74**, 53 (1976).
173. B. W. Burrows and G. J. Hills, *Electrochim. Acta* **15**, 445 (1970).
174. A. Rahmel, *Electrochim. Acta* **21**, 181 (1976).
175. H. Lux, *Naturwissenschaften* **28**, 92 (1940).
176. A. L. Mathews and C. F. Baes,Jr.,*Inorg. Chem.* **7**, 373 (1968).
177. D. L. Manning and G. Mamantov, *J. Electrochem. Soc.* **124**, 480 (1977).
178. K. Ratkje and T. Förland, *Arch. Huttn.* **22**, 159 (1977).
179. A. J. Appleby and S. Nicholson, *J. Electroanal. Chem.* **53**, 105 (1974).

17

Studies of Molten Mixtures of Halides and Chalcogenides

H. C. Brookes

1. Introduction

Sulfide ores are amongst the major nonferrous base metal minerals mined today, and have been for many decades. Many sulfides are semiconductors in the solid state and considerable electrochemical investigation of this property at room temperature has been described.[1] Despite the fact that many of the industrially important reactions of extractive metallurgy concern sulfide minerals and involve high-temperature chemistry (400–1200°C) in the molten state, few overviews are available concerning the chemistry of molten chalcogenides when mixed with other melts, as is usually the case industrially.

Since 1970 much use has been made of fused salt media as electrolytes in fuel cells and batteries and, in this context also, the investigation of the chemistry of chalcogens and their compounds in the molten state is of considerable technological importance. This chapter describes the molten salt chemistry of halide + chalcogenide mixtures with emphasis on the concentrated solutions of the chalcogen. Some indication of the sort of chalcogen species that can exist in halide–halometallate solutions would be valuable, and hence reference to dilute chalcogen solution ($< \sim 0.1$ mol %) studies will be made, since many of the productive investigations of this nature have been with dilute solutions. It is a moot point, however, as to whether results on the nature of the species present at low concentration can always be extrapolated to more concentrated mixtures.

H. C. Brookes • Department of Chemistry, University of Natal, Durban, South Africa.

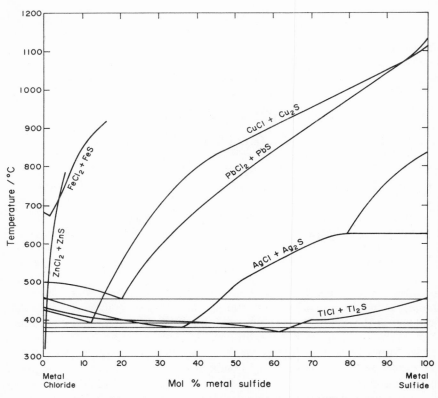

FIGURE 1. Phase diagrams for metal chloride + metal sulfide mixtures.

Investigations of one or more of the thermodynamic, structural, electrochemical, or cryoscopic properties of the following binary molten salt systems, *inter alia*, have been reported up to July 1978:

$TlCl + Tl_2S$ (References 2 and 29), $AgX + Ag_2Y$ * (References 2, 3–5, 14, 20, 21, 23–26, 42, 43, 51, and 77), $PbX_2 + PbY$ (References 6, 7, 53–58, 67, and 68), $CuX + Cu_2Y$ (References 8, 9, 31, and 69), $SnX + SnY$ (References 52, 59, and 60), $FeCl_2 + FeS$ (Reference 8), $GeS + GeI_2$ (Reference 61).

A comparison of chloride + sulfide binary phase diagrams is shown in Figure 1.

A number of reciprocal systems will be discussed when relevant to the systems mentioned above. As far as elemental chalcogens and so-called unusual oxidation states are concerned, a recent review[32] has described

* X = Cl, Br, I; Y = S, Se, Te.

work on chalcogens in the unusual positive oxidation states, only, in chloroaluminate melts. Hence, chloroaluminate solutions will only be mentioned with regard to chalcogenides (i.e., negative oxidation states) in these melts.

2. The Nature of Chalcogenide Species in the Melts

Considerable evidence is available concerning the precise nature of the chalcogenide ions in halide melts. The Te^- ion is present in LiCl + KCl (59 : 41 mol %) and $LiF + BeF_2$ (66 : 34 mol %) melts, although it is only slightly soluble in the chloride melt as Li_2Te.[33] This contrasts with the work of Gruen et al.,[27] who found Te^{2-}, whereas Bamberger et al.[34] suggested that the species present in Li_2BeF_4 is Te_3^-, both groups of workers using spectroscopic evidence around 475 nm in LiCl,CsCl, and LiCl + LiF (Reference 27), $LiF + BeF_2$ (Reference 34). Toth and Hitch's spectroscopic investigations[33] agree with electrodic work by Manning and Mamantov[32] on a tellurium electrode in fluoride melts, which also indicated the presence of Te^-. In addition, higher tellurides are strongly indicated by blue shifts on the 497-nm band found by Toth and Hitch, and the Te_2^{2-} ion has been prepared in fluoride melts.[32]

As far as sulfide species are concerned the well-known blue color of dilute sulfur/sulfide in LiCl + KCl has sparked numerous investigations leading to several reports[17, 18, 36, 37, 44] suggesting that the blue color concerns species such as S, S_2^-, S_4^{2-}, S^{2-}, S_3^-, or S_n^- with $n > 2$. It seems clear that both elemental sulfur as well as anionic species are prerequisites for the coloration.

Bernard et al.[38] reported, from a spectrophotometric and chronopotentiometric study, that S^{2-} is oxidized in three steps to S^+ in LiCl + KCl and S^{2-}, S_3^{2-}, and S_2^- or S_3^- were detected. Delarue[19] investigated redox reactions in LiCl + KCl melts and found that Tl_2S, MnS, and the alkali and alkaline earth sulfides were soluble; that CuS, Fe_2S_3, Au_2S, and HgS reacted with the cations being reduced and S^{2-} oxidized to S; and that sulfides of Zn^{2+}, Cd^{2+}, Fe^{2+}, Co^{2+}, Ni^{2+}, Cu^+, Pt^{2+}, Pd^{2+}, Sn^{2+}, Pb^{2+}, Sb^{3+}, Bi^{3+}, and Ce^{3+} were insoluble. Cleaver et al.,[39] in contrast to Delarue, found Na_2S almost insoluble in LiCl + KCl at 420°C. $Na_2S_{2.2}$ gave lime–green solutions from which sulfur slowly deposited and from cyclic voltammetric studies it was concluded that $S_2^{2-} \rightarrow S_2^- + e$ occurred at cathodic potentials on gold. Anodic sweeps produced S as product. Liu et al.[40] found electrodically that NiS was almost insoluble in LiCl + KCl ($K_{sp} = 1.6 \times 10^{-15}$ at 400°C) and Li_2S to be soluble to 0.02 mol/l at 400°C. Interaction of S with sulfide clearly greatly increases sulfide solubility,[17] which may explain Delarue's results.

Electrolysis studies of PbS in chloride melts[41] showed that the major contribution to the lowering of cathodic current efficiency for lead production was dissolution of S. When the limiting current density for sulfur evolution was reached the anode potential rapidly increased to that for evolution of S_2Cl_2 according to

$$2S + 2Cl^- \rightarrow S_2Cl_2 + 2e$$

which tends to confirm the results of Bodewig and Plambeck.[17]

In a chronopotentiometric study of up to 2 mol% of Tl_2S in NaCl + Na_2S (95 : 5 mol%) from 820 to 900°C at glassy carbon Belous *et al.*[48] found the reduction process to be diffusion controlled and that the diffusion coefficient of Tl^+ in the melt varied from 7.5×10^{-5} cm^2 sec^{-1} at 820°C to 13.0×10^{-5} cm^2 sec^{-1} at 900°C. The solubility product of Cu_2S in basic NaCl + $AlCl_3$ melts has been determined as 6.8×10^{-15} at 175°C.[28]

Electrochemical studies of the chalcogenide anions[16] in acidic sodium chloroaluminate melts at 175°C have shown the chalcogenides to be trichlorobases. The S^{2-} activity varies with melt acidity owing to the equilibrium

$$3Al_2Cl_7^- + S^{2-} \rightleftharpoons AlSCl + 5AlCl_4^-$$

which pertains at pCl > 2.8 on the molal scale. In less acidic systems, the chalcogenide ions, Y^{2-}, are dibasic yielding $AlYCl_2^-$ according to $2Al_2Cl_7^- + Y^{2-} \rightleftharpoons AlYCl_2^- + 3AlCl_4^-$. Similar results were obtained for selenides and tellurides, and the solubility of sulfides and selenides was found to be very dependent on melt acidity, as expected from the equilibria just described, sulfides being very soluble in acidic melts.[35] It is well known[45] that the sulfide ion acts as a strong base in those melts and sulfide ion activity is several orders of magnitude smaller in acid than in basic melt. Reduction of elemental sulfur yielded S^{2-} and no evidence for polysulfide ions was found,[35] in contrast with Bodewig and Plambeck's findings.[17] The relative strength of the tribases found was Te^{2-}, $Se^{2-} < S^{2-} < O^{2-}$. Similar reactions to those in the NaCl : $AlCl_3$ melts have been found from reaction of chalcogenides, such as ZnS, with $AlCl_3$ in closed tubes at 300°C to produce AlSCl,[10] AlSeCl,[11] and AlTeCl.[12] Chalcopyrite has been found, in an exploratory cyclic voltammetric study, to be soluble in NaCl + $AlCl_3$ melts and both iron and copper could be deposited from the melt.[35] Further details regarding sulfur and sulfide species when dilute in halide melts are given elsewhere in this book.

As far as information on the nature of selenium anionic species in halide melts is concerned, studies in LiCl + KCl melts showed that, as for S, Se can be oxidized to give Se_2Cl_2 and reduced to Se^{2-}.[15] Free telluride, on the other hand, was not detected in the melt since Te was oxidized to Te(II).

3. Germanium, Tin, and Lead Systems

A short résumé of some Group IV systems investigated will be given, with emphasis on those which lead to structural conclusions.

The densities, phase relations, and conductances of the $PbCl_2$ + PbS system were investigated by the Toronto school.[6, 7] The cryoscopic data, obtained from cooling curves, indicated that the pure molten PbS was dissociated. Pure $PbCl_2$ apparently contained ions other than Cl^- and Pb^{2+}. Molar volumes were almost ideal and it was concluded that $PbCl_2$ + PbS melts were structurally simpler than AgCl + AgS mixtures (Section 4). The electrical conductivities exhibited similar sorts of behavior to the analogous silver system, with conductivity changing from ionic to semiconducting as the sulfide fraction increased, and as the temperature rose, thus increasing the number and mobility of conducting electrons. The more metallic temperature dependence of molar conductivity observed for their AgCl + Ag_2S (Reference 4) system was not found for this lead system, the sulfide-rich melts exhibiting typical semiconductive behavior. This difference is associated with the more ionic nature of PbS melts in contrast to Ag_2S melts. PbS + $PbCl_2$ and PbSe + $PbCl_2$ were investigated in 1970[56] using differential thermoanalysis (dta). Compounds were also not detected in PbS + $PbCl_2$ mixtures on slow cooling but on quenching from 1000°C the compound $Pb_7S_2Cl_{10}$ was obtained. PbSe + $PbCl_2$, on the other hand, produced no compounds, being a simple eutectic with eutectic point at 478°C, 7 mol % PbSe.

In a study[67] of PbS + $PbCl_2$, $PbBr_2$, and PbI_2 binary mixtures the peritectic compounds $Pb_7S_2Br_{10}$ and $Pb_5S_2I_6$ were found, with PbS + $PbCl_2$ being a eutectic in good agreement with Bell and Flengas.[6] In all three systems a region of undetermined liquidus occurred, as was the case for TlCl + Tl_2S (Reference 2) and AgCl + Ag_2S (Reference 3). However, the compound Pb_4SCl_6 was reported,[67] confirming the suggestion of the possible compound made by Bell *et al.* earlier[6] but not the same compound described above.[56] These results on PbS mixtures were substantially confirmed by Novoselova and co-workers[68] together with details of PbSe and PbTe phase diagrams and X-ray analysis of binaries with $PbCl_2$, $PbBr_2$, and PbI_2. As regards the sulfide + halide binaries the only disagreement was for the mp of $Pb_7S_2Br_{10}$, being 394°C in the Soviet Union[68] compared to 381°C in West Germany.[67] The compounds $Pb_7Se_2Br_{10}$ (mp 383°C), and $Pb_7S_2Cl_{10}$ (d.180°C) (see above) were characterized by dta and X-ray analysis.

It was concluded[68] that PbS + PbX_2 (X = Cl, Br, I) melts were "ionic" in character, whereas the other six lead binary mixtures formed so-called "molecular" melts. This is in agreement with Bell and Flengas[6] as regards PbS being dissociated, but these workers[6] found $PbCl_2$ to be partly associated.

In another study[57] $PbS + PbI_2$ showed an incongruently melting compound $Pb_5S_2I_6$, mp 415°C. $PbS + PbBr_2$ and $PbSe + PbBr_2$ systems were again reported[58] to yield $Pb_7S_2Br_{10}$ and $Pb_7Se_2Br_{10}$, whereas $PbTe + PbBr_2$ produced no compounds at all. The $PbTe + PbI_2$ phase diagram[53] shows a simple eutectic, with eutic point 398°C, 6 mol % PbTe, and $PbTe + PbCl_2$ was also reported by the Novoselova school[54] to exhibit no compound formation. In the $PbSe + PbI_2$ section of the $Pb + Se + I$ ternary[55] a eutectic point occurred at 384°C, 14 mol % PbSe, and several heat effects due to undetermined phase transitions were noted on the cooling curves at temperatures near the liquidus.

Binary tin mixtures exhibit similar general behavior to the corresponding lead mixtures, $SnS + SnI_2$ (Reference 60) melts producing the new incongruently melting compounds Sn_3SI_4 (mp 330°C) and $Sn_7S_3I_8$ (mp 410°C). The entire range of $SnX_2 + SnY$ (X = halogen, Y = chalcogen) molten mixtures were investigated using dta by Blachnik and Kasper.[52] In $SnSe + SnI_2$ melts a new phase was also found and the incongruently melting compound Sn_3SeI_4, mp 330°C defined, whereas $SnSe + SnBr_2$ is a simple eutectic. Blachnik's dta technique avoided supercooling to a far greater extent than Morozov and Li[59] could using cooling curves, and the liquidus lines for $SnCl_2 + SnS$ are correspondingly higher by the dta technique. For $SnBr_2 + SnS$ a new phase of approximate composition $2SnS \cdot 7SnBr$ was detected by X-ray analysis. This compound melts incongruently at 250 ± 5°C. Miscibility gaps in the liquid phase were found for all the telluride mixtures, i.e., for $SnTe + SnI_2$, $SnTe + SnBr_2$, and $SnTe + SnCl_2$ as well as for $SnSe + SnCl_2$.

From the Canadian, Soviet, and West German work on these tin and lead systems it is apparent that association, and resulting immiscibility, increases from sulfide to telluride and from iodide to chloride. Compound formation would thus be expected to increase in the reverse order. The thermal stability of the compounds formed increases as

$$\Delta E_N = \{E_N(\text{halogen}) - E_N(\text{chalcogen})\}$$

decreases, which is perhaps not unexpected, E_N being electronegativity. The complexity of the situation as regards the influence of the metal concerned is illustrated in $GeS + GeI_2$ molten mixtures,[61] where the melting point of GeI_2 was determined for the first time as 459°C, with a eutectic point at 55 mol % GeS, 398°C, and *no* compounds were found. Two-liquid phase separation from 25–50 mol % GeS occurred. A conductivity study[13] of SnS and GeS showed these compounds to be semiconductors in the liquid state, like PbS, and for both SnS and GeS the molecular character was reported to increase on going from the solid to the liquid state. Further understanding of these trends requires additional viscosity, density, and conductivity as well as thermodynamic data not yet generally available.

Mozorov and Li[59] investigated reciprocal $SnCl_2$ + PbS mixtures, finding that $SnCl_2$ + PbS do not coexist in any phase field. Clearly the metathetical reaction

$$SnCl_2 + PbS \rightleftharpoons SnS + PbCl_2$$

proceeds completely to the right. A eutectic point at 440°C, 17.4 mol % SnS was found. Evidence had been presented[50] of a miscibility gap in the Ag_2S + $PbCl_2$ system where a positive $\Delta G°$ for the metathetical reaction

$$Ag_2S + PbCl_2 \rightleftharpoons 2AgCl + PbS$$

of 46 kJ/mol was found, which is apparently sufficiently positive to result in immiscibility.

Pelton and Flengas[49] determined liquidus temperatures for PbS + ACl reciprocal solutions (A = Na to Cs) over the range 0–5 mol % PbS and 96–100 mol % PbS. Monotectic temperatures of 1096°C were found for all mixtures and were independent of composition. PbS and the alkali chlorides formed true quasibinary systems, and a trend towards increasing solubility of PbS as the solvent changed from NaCl to CsCl was found. Cooling curve procedures were used for PbS-rich solutions and decantation and analysis procedures were used in alkali chloride-rich solutions. All mixtures showed very large regions of liquid–liquid immiscibility. The cryoscopic calculations suggested that alkali chlorides dissolve in PbS partly as dissociated ions and partly as associated ion pairs. The solubility of PbS in the four alkali chlorides was so small that the solutions were not expected to conduct electronically. The solubility variation of various metal sulfides in alkali halide melts clearly could be utilized in solvent extraction processes for separating sulfide ores. For example, the solubility of Ag_2S in alkali chlorides is far less than that of PbS.

4. Silver Halide + Silver Chalcogenide Systems

Blachnik and Hoppe[5] measured H^E values for liquid mixtures of Ag_2S with AgCl, AgBr, and AgI by calorimetry and concluded that these melts do not form regular solutions. The excess enthalpy data for Ag_2S + AgCl mixtures determined by Thompson and Flengas[20] disagree with that of Blachnik *et al.* The recent values appear more reasonable since the latter vary drastically over small composition ranges. The excess enthalpies found (Figure 2)[70] were much more endothermic than is generally the case in mixtures such as those of alkali halides and it appears that additional structural effects, besides those well known for simpler systems, contribute to H^E, which increased from Cl to I.

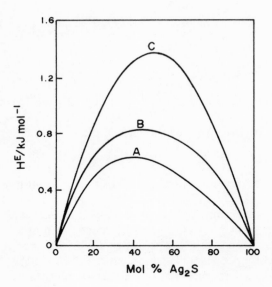

FIGURE 2. Excess enthalpies for $Ag_2S + AgX$ mixtures. A, $X \equiv Cl$; B, $X \equiv Br$; C, $X \equiv I$.[70]

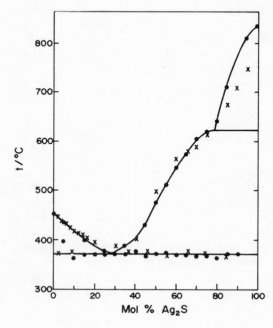

FIGURE 3. $AgCl + Ag_2S$ phase diagram. ●, References 23 and 70; ×, Reference 3.

The phase diagram for $AgCl + Ag_2S$ molten salt mixtures was determined by the Flengas group,[3] using the method of cooling curves, and by Blachnik's group[23] using dta resulting in reasonable agreement (Figure 3) on the AgCl-rich side, but higher liquidus temperatures, by up to 45°C were found in the Ag_2S-rich melts by dta techniques, probably due to well-known difficulties experienced in determining precise melting points by the cooling curve technique when the liquidus is steep. There is evidence for ionic interactions in the chloride-rich melts and for covalent association in the sulfide-rich solutions, as has been found for $TlCl + Tl_2S$ mixtures.[2] An incongruently melting compound at 50 mol % Ag_2S (presumably Ag_3SCl) suggested by Bell and Flengas[3] and subsequently not confirmed by Thompson and Flengas[42] was also not detected by Blachnik and Kahleyss,[23] who attributed the relevant heat effect to a known phase transformation of Ag_2S.[22] The $Ag_2S + AgBr$ (Reference 24) and $Ag_2S + AgI$ (Reference 25) phase diagrams show the peritectic compounds Ag_3SBr (mp 434°C) and Ag_3SI (mp > 600°C), respectively.

Densities of the $AgCl + Ag_2S$ (Figure 4) and $AgBr + Ag_2S$ melts (Reference 26) indicated ideal behavior. The molar volumes of binary molten mixtures of AgCl with Ag_2S, Ag_2Se, and Ag_2Te, together with $AgI + Ag_2Te$, $AgI + Ag_2S$, and $AgBr + Ag_2Te$ were determined dilatometrically.[26] All the systems behaved ideally except for $AgI + Ag_2Te$ and $AgI + Ag_2S$, which showed negative deviations from ideality. Bell and Flengas,[3] however, obtained molar volume isotherms for the $AgCl + Ag_2S$ system which showed (Figure 5) negative deviations from ideality above 750°C, and their temperature dependence for the density of pure fused

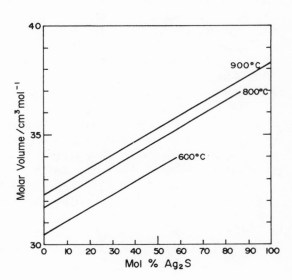

FIGURE 4. Molar volumes of $AgCl + Ag_2S$.[26, 70]

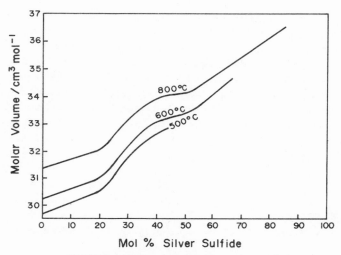

FIGURE 5. Molar volumes of AgCl + Ag$_2$S.[3]

Ag$_2$S is far greater than that reported by Blachnik *et al.* and by Glazov *et al.*,[14] who also studied this system. Negative deviations from ideality are often accepted as reflecting an increased ionic character of the melt and calculations by Blachnik[26] show that an increase in polar character for AgCl + Ag$_2$S mixtures is not unexpected. However, experimentally the greatest negative deviations from ideality for sulfide mixtures occur with AgI + Ag$_2$S molten mixtures, which tends to suggest that negative deviations are not associated solely with an increase on the electrostatic contribution to bonding interactions during mixing of these components. Denser packing and hence smaller than ideal molar volumes do not seem to depend primarily on the degree of ionicity involved. Further investigations of these apparent anomalies are necessary, since more covalency, rather than electrovalency, is expected in AgI than in AgCl.

Additional phase and conductivity studies on the AgI + Ag$_2$S system[25] indicate that the conductance is primarily ionic in nature. AgI + Ag$_2$Te and AgI + Ag$_2$Se mixtures were reported to provide no compounds, based on phase diagrams,[77] in contrast to the stable Ag$_3$SI mentioned previously for AgI + Ag$_2$S molten mixtures.

Activities of AgCl and AgBr determined by mass spectrometry in molten Ag$_2$S were described by Chang and Toguri.[21] Positive deviations from ideality were noted, whereas Bell *et al.* found negative deviations due to interfering junction potentials in the emf concentration cell they used.[3] G^E, H^E, and S^E values for AgCl + Ag$_2$S and AgBr + Ag$_2$S were reported, with the H^E predictions being based on the assumption that pure molten

Ag_2S was a nonpolar liquid.[21] The H^E values predicted[21] did not agree with those measured,[5] suggesting that the assumption is incorrect and is not in accord with the high degree of electronic conduction in Ag_2S being due to an electron-hopping or exchange mechanism involving ionic species, as described in Section 5, page 375, for molten sulfide-rich mixtures.

The results of the Canadian cryoscropic, activity, and molar volume studies on $AgCl + Ag_2S$ melts, together with the cryoscopic studies by Blachnik and Kahleyss, indicate that dilute (up to 15 mol %) solutions of Ag_2S in $AgCl$ are extensively ionized and form almost ideal solutions. The study to be described in Section 6 indicates that a drastic change in melt structure occurs between the two terminal $AgCl$ and Ag_2S dilute solution regions.

Electrical conductivities[4] of $AgCl + Ag_2S$ (Figure 6) show that pure $AgCl$ has a very low activation energy for conductance and that it conducts ionically. As the Ag_2S concentration rises to 20 mol % the specific conductance decreases by a third to ~ 2.7 ohm^{-1} cm^{-1} at 420°C, and the activation energy for ionic conduction rises (Figure 7). Above 20 mol % Ag_2S the conductivity increases rapidly with increasing temperature and with increasing sulfide concentration, reaching magnitudes of 200 ohm^{-1} cm^{-1}, suggesting an electronic conductivity mechanism. Over the 20–36 mol % Ag_2S region the apparent activation energy for presumably electronic conduction is composition independent, which is typical of semiconductor behavior. The significance of conclusions drawn from apparent activation energies of conductance when there is a rapidly varying mixed mode of conductance as the temperature and composition change is, however, debatable. Above 50 mol % Ag_2S there is an extremely rapid increase in conductivity with sulfide concentration, due to structural changes in the melt

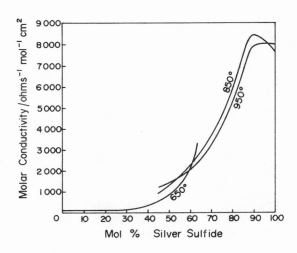

FIGURE 6. Molar conductivity of $AgCl + Ag_2S$ melts.

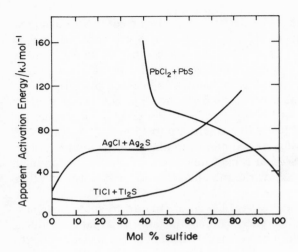

FIGURE 7. Activation energy
for conductance.

markedly influencing the electron mobility. The temperature dependence becomes negative from about 60 mol % Ag_2S between 700 and 950°C. Pure heavy metal sulfides are known to be electronic conductors even in the liquid state and hence electrowinning of such metals from their pure melts is not possible. However, studies of solutions of such sulfides in chloride melts[2-4, 6, 8, 29] show them to be, in general, ionic conductors over substantial composition and temperature ranges.

Phase studies[24] by dta for $AgCl + Ag_2Se$, $AgCl + Ag_2Te$, $AgBr + Ag_2Se$, and $AgBr + Ag_2Te$ show the peritectic compounds Ag_3TeBr (mp 446°C) and Ag_5Te_2Cl (mp 517°C). The results suggest that the self-association of Ag_2S molecules, as found in the sulfide-rich molten sulfide + halide mixtures,[2, 23] decreases on going to the corresponding selenide mixtures. Replacing Cl^- by Br^- apparently has no effect in this regard on the chalcogen-rich side. (cf. Sn and Pb systems in Section 3). It seems clear that compound formation increases in the order S < Se < Te in $Ag_2Y + AgX$ melts (Y = chalcogen, X = Cl, Br), and there is a corresponding increase in self-association of Ag_2Y molecules in the order Te < Se < S. In halide-rich (AgCl, AgBr) mixtures the available evidence points to dissociation of the chalcogenide solute to ions.[2, 24]

Thompson and Flengas[42] used a sulfur vapor electrode in $Ag_2S + AgCl$ solutions from 460 to 860°C with up to 12 mol % Ag_2S and from emf's of the formation cell

$$\ominus Ag_{(s)} | Ag_2S_{(l)} + AgCl_{(l)} | graphite, S_{(g)} \oplus$$

found Nernstian behavior only in very dilute silver sulfide solutions. The activities of Ag_2S at 825°C (the mp) showed positive deviations from ideality as a function of composition, consistent with a tendency of halide to associate with halide and presumably sulfide with sulfide, i.e., a tendency towards self-association and immiscibility. As mentioned above, the previously suspected incongruently melting compound in the $AgCl + Ag_2S$ system was not confirmed. It is apparent that, in general, immiscibility is as often the rule as the exception in metal chloride + metal sulfide mixtures, particularly in reciprocal systems.

5. Copper and Iron Systems: Comparison with Other Systems

Garbee and Flengas[8] determined the phase diagrams (Figure 1), electrical conductivity, and densities of $Cu_2S + CuCl$, and $FeS + FeCl_2$ melts up to 16 mol % FeS. It was found that for $Cu_2S + CuCl$ solutions up to approximately 30 mol % Cu_2S and below 1000°C the conductance is essentially ionic. For solutions richer in sulfide electronic (semiconductive) conductance becomes appreciable, increasing with more sulfide. In $FeS + FeCl_2$ melts similar conductance behavior pertains, with ionic conductance prevailing below 10 mol % sulfide.

In dilute sulfide solutions in $CuCl_2$ the presence of $(Cu_2S)_2$ dimers was proposed in contrast to the Tl and Ag cases (Section 6). At higher sulfide compositions a continuous sulfide network was proposed[8] which can act as a suitable path for electron transfer. Excitation of electrons from the valence band of sulfides to the conductance band of metal ions is facilitated by the presence of sulfide polymers. As the concentration of ionically conducting metal chloride increases in a binary MS + MCl mixture the sulfide polymer dissociates and the energy gap between valence and conduction bands is presumed to thereby increase, thus lowering the conductivity, the solutions eventually becoming ionic conductors.

As previously noted, since the conductivity of SnS, PbS, Ag_2S, Tl_2S, Cu_2S, and FeS is predominantly electronic, electrodeposition of metal from the sulfide-rich melts is impossible. However, in the ionically conducting regions metal should be obtainable by electrolysis and this has been confirmed in the case of Cu metal extraction with high current efficiency from $Cu_2S + CuCl$ melts.[8] In addition, FeS was extracted from $FeS + Cu_2S$ by solvent extraction with molten CuCl. The extraction was attributed to the exchange reaction

$$FeS_{(s)} + 2CuCl_{(l)} \rightarrow FeCl_{2(l)} + Cu_2S_{(l)}$$

which has $\Delta G° = -41,000$ kJ/mol at 1100°C and which thus produced the soluble ionic $FeCl_2$ melt. By operating at about 660°C the ratio of Fe to Cu in the liquid mixture was 1.5 times that in the original solid matte, thus suggesting that molten salt solvent extraction together with electrolysis could provide a basis for extracting copper from Cu + Fe mattes.[8]

The phase studies by A. Rabenau et al.[69] of CuX + Y (Y = Se, Te) systems showed CuClSe$_2$ (mp 319°C), CuBrSe$_3$ (mp 338°C), and CuISe$_3$ (mp 394°C) to be stable components of the pseudobinary systems. In the tellurium systems two series of compounds existed: CuXTe and CuXTe$_2$ with CuClTe$_2$ and CuBrTe melting congruently.

Differences in structural behavior in these systems are also indicated[31] by the excess enthalpies of mixing for CuCl + Se and CuBr + Se mixtures being endothermic (up to 2.3 kJ mol^{-1}), whereas CuCl + Te and CuBr + Te mixtures exhibit exothermic enthalpies (up to -0.6 kJ mol^{-1}) at similar temperatures (818 K). Molten CuCl and probably CuBr can be regarded structurally as consisting of close-packed halide ions with Cu$^+$ ions distributed in tetrahedral holes in the halide lattice. Molten Se consists of long chains just above the melting point which shorten as the temperature rises, accompanied by ring formation. In Te melts chains, layers, and finally a network grid structure pertain as the temperature rises from the melting point.[78, 79] Se and Te species can be considered as being incorporated in the octahedral holes in the CuCl melt, the initial endothermic process being the breaking of Se–Se or Te–Te covalent bonds. The tendency of molten Te to exhibit higher coordination and stronger interaction than Se in these mixtures, as found for pure Te, but not pure molten Se, can be considered to result in a net release of energy, resulting in the exothermic mixing process with CuX melts as found experimentally. A tendency towards ideal behavior as the temperature rises would thus be expected, and was found.[31]

Although considerable work on halide + chalcogenide systems has been published it is clear that large areas of endeavor have still to be unraveled. The upsurge in interest, in sulfide melts in particular, has gained momentum with the so-called energy crisis and considerable effort has gone into development of a sulfide battery program.[46] The fact that further phase studies are necessary has been exemplified by the unusual set of compounds, such as K$_6$LiFe$_{24}$S$_{26}$Cl, which were found to occur in the LiCl + KCl eutectic used as part of the sulfide battery program at Argonne National Laboratory.[76] The negative electrode in the LiCl + KCl electrolyte is a Li–Al alloy, the positive one is FeS, and phases such as that above have been found on the anodes, thus seriously limiting battery life. This phase has also been prepared externally from a battery by reaction of Li$_2$FeS$_2$ with Fe in LiCl + KCl eutectic at 400–450°C.

Comparison of Tl, Ag, Pb, and Cu Results

As TlCl is added to Tl$_2$S melts the specific conductivity decreases overall and the per cent ionic contribution increases as shown in Figure 8 for thallium mixtures.[29] The conductivity at 500°C is seen (Figure 8) to

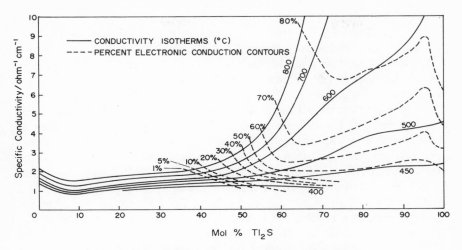

FIGURE 8. Specific conductivity of TlCl + Tl$_2$S melts.

increase rapidly as the Tl$_2$S concentration rises above ~ 60 mol % and similarly for AgCl + Ag$_2$S solutions at ~ 50 mol % Ag$_2$S (Figure 6). A similar sudden increase in (specific) conductivity at ~ 60 mol % PbS in PbCl$_2$ + PbS solutions[7] was found at all temperatures, with almost negligible increase in conductivity from 0 to ~ 60 mol % PbS. It seems as though the conductance mechanism change is not a gradual one, but in such binary melts there appears to be a sudden change over comparatively small composition regions. The initial dependence of conductance in the chloride-rich solutions in binary melts of Ag and Tl is of the type shown in Figure 6, for the Ag mixtures.[3] It is noteworthy, however, that the specific conductance for pure molten Ag$_2$S is practically temperature independent changing from 202.6 ohm^{-1} cm^{-1} at 840°C to 201.5 ohm^{-1} cm^{-1} at 960°C. In marked contrast are the results for pure molten Tl$_2$S where the specific conductance is much smaller and increases rapidly from 2.5 ohm^{-1} cm^{-1} at 455°C to 10.9 ohm^{-1} cm^{-1} at 601°C.[29] This has been independently confirmed.[64]

As the sulfide concentration rises the conductance mechanism changes in an undefined manner from that of ionic conductance in halide-rich mixtures to an electronic conductance mechanism in chalcogenide (sulfide)-rich melts. A typical apparent activation energy of conductance isotherm is shown for lead, silver, and thallium mixtures in Figure 7. Although detailed analysis of such apparent activation energy curves is not warranted as explained previously, it can be concluded that in general a large increase in activation energy for (electronic) conduction occurs as sulfide concentration rises for the Ag and Tl systems. The marked contrast between the Ag and Tl systems on the one hand and Pb on the other is noteworthy. The PbS-rich melts appear to be less covalent than the Tl_2S and Ag_2S rich melts as evidenced by the high activation energy for $PbS + PbCl_2$ mixtures at 40 mol % sulfide. The dissociation of PbS in such mixtures is also confirmed by phase diagram studies, as described elsewhere in this chapter.

A comparison of partial molar volumes for metal sulfide and metal chloride, respectively, in sulfide-rich mixtures at, say, 90 mol % sulfide at 700°C for the Cu (Reference 8) and Ag (Figures 4 and 5) systems shows very different molar volumes of about $32/26$ cm^3 mol^{-1} and $37/31, 23$ cm^3 mol^{-1}, compared to the Pb case of $35/59$ cm^3 mol^{-1}.[6] Clearly the usual band theory of electronic conductance is not applicable in the liquid state owing to the lack of long-range interactions. The high sulfide partial molar volumes suggest a continuous sulfide network, the degree of polymerization being dependent on the chloride concentration, the metal chlorides having a structure-breaking effect on the sulfide network with the corresponding rapid decrease in electronic conduction. As the sulfide concentration rises the electronic contribution to conduction increases and the total conductance rises. The electronic conductance in chalcogenide-rich melts is probably a random charge transfer between sulfide/sulfur species in the melt, the electrons coming from sulfur species such as polysulfide ions or the continuous sulfur network. The resultant conductance will thus depend on the mobility and availability of electrons, i.e., on the temperature as well as the chloride concentration and valence of the metal ion, resulting in the complex activation energy dependence found experimentally (Figure 7). In the Ag melts, as the chalcogenide changes from Ag_2S to Ag_2Te the electronic conductance rises sharply, from 202 ohm^{-1} cm^{-1} for Ag_2S to 255 ohm^{-1} cm^{-1} for Ag_2Se to 500 ohm^{-1} cm^{-1} for Ag_2Te at temperatures just above the melting points,[4, 24] indicating an even greater degree of association and ease of availability of electrons as the anion size increases. The differing molar volumes from 20–60 mol % sulfide shown for $AgCl + Ag_2S$ in Figures 4 and 5 remains to be explained. The deviations from additivity in Figure 4 do, however, agree with the thermodynamic analysis for the $AgCl + Ag_2S$ system to be described in Section 6. In the $PbCl_2 + PbS$ melts slight positive deviations from additivity were found for the molar volumes,[6] again indicating the uniqueness of the Pb system.

The excess enthalpies for $AgX + Ag_2S$ (X = Cl, Br, I) mixtures (Figure 2) indicated that these melts were not regular solutions, and relatively high negative excess entropies were reported, viz., of the order of -3 J K^{-1} mol^{-1}, in agreement with the proposed ordered structures.

In the $Cu_2S + CuCl$ solutions molar volume data indicated that Cu_2S exists as an associated structure into which added CuCl can initially be accepted. As the chloride concentration rose the sulfide structures collapsed and the dissociated Cu_2S polymers could be incorporated in the CuCl melt structure. Garbee and Flengas[8] suggested $(Cu_2S)_2$ dimers existed in the sulfide-rich melt. Molar volumes for Cu_2S + CuCl show a distinct transition at ~ 30 mol % Cu_2S at 1000°C, where the molar volumes of the solutions change rapidly from decreasing with increasing sulfide concentration to increasing as the sulfide concentration increases. Similar trends occur at other temperatures. The molar conductivity at 1000°C for Cu_2S + CuCl (Reference 8) shows a rapid increase as the Cu_2S component increases above ~ 30 mol %. Again it appears as though a sudden change in melt structure is occurring over a small composition range.

6. Analysis of $AgCl + Ag_2S$ and $TlCl + Tl_2S$ Molten Mixtures

The foregoing comparisons have prompted an analysis of available data for Ag and Tl binary chloride + sulfide molten salt solutions.

6.1. Silver Chloride + Silver Sulfide Solutions

Darken has found, for binary[63] and ternary[71] metallic solutions, that the excess free energy can be represented as a quadratic formulation. In particular, for binary solutions of components 1 and 2 it is found that in a region for which the activity coefficient of the solvent γ_1 may be represented as

$$\log \gamma_1 = \alpha N_2^2$$

that for the solute must be represented as

$$\log \gamma_2 = \alpha N_1^2 + I$$

where I is the integration constant and N_i is number of moles of i. In general I is not zero unless the above two equations are valid over the entire composition range. In effect, the constant allows for a change of standard state from pure component 1 in the 1-rich melts to (hypothetical) pure 1 with the structure of pure component 2. The phase diagram of the $TlCl + Tl_2S$ system has been determined by the method of cooling curves (Figure 1)[2] and that for $AgCl + Ag_2S$ by cooling curve and dta methods (Figure 3).[3, 23, 70] Recently, enthalpies of mixing for the $AgCl + Ag_2S$ molten system have been determined calorimetrically[5] which result in conclusions that are compatible with data obtained for the $TlCl + Tl_2S$ system,[2] whereas enthalpies of mixing determined earlier for $AgCl + Ag_2S$ appear in error.[20] In addition the new ΔH_{fusion}° value for Ag_2S[20] (7.87 kJ mol^{-1})

differs considerably from that used by Bell and Flengas[3] (11.3 kJ mol^{-1}) in their analysis of the AgCl + Ag$_2$S system. Accordingly a reanalysis of the AgCl + Ag$_2$S phase diagram using the new ΔH^{mix} values will be given.

Table 1 shows g_i^E, the partial molar excess free energies for the solvent i, calculated on the liquidus and at 800 K using the van't Hoff equation and published heats of fusion and specific heats.[20, 72–74]

Near $X_i = 1$, where X_i is the mole fraction of AgCl or Ag$_2$S, the one foreign particle limiting liquidus slope can be calculated from the limiting van't Hoff cryoscopic equation

$$X_i = \frac{-(\Delta H_f^\circ)_i}{R(T_f)_i^2}(\Delta T)$$

where $\Delta T = T_{liquidus} - T_f$, T_f being the freezing point of pure i, ΔH_f the enthalpy of fusion. The limiting slopes were found to agree with the experimental data.[2]

Figure 9 shows plots of g_{AgCl}^E (800 K) vs. $X_{Ag_2S}^2$ and $g_{Ag_2S}^E$ (800 K) vs. X_{AgCl}^2. The limiting region at the AgCl-rich end in Figure 9 was calculated using the Gibbs–Duhem relation. For these molten AgCl + Ag$_2$S salt mixtures three distinct composition regions are found as evidenced by Figure

TABLE 1. Thermodynamic Data for AgCl + Ag$_2$S Mixtures

$10^{-2}X_{Ag_2S}$, mol %	$T_{liquidus}$, K	g_i^E on liquidus, J mol^{-1}	Δh_i, J mol^{-1}	$S_i^E = \dfrac{\Delta h - g_i^E}{T_{liq}}$, J K^{-1} mol^{-1}	$(g_i^E)_{800}$ from $\Delta h_i - 800 S_i^E$, J mol^{-1}
i = AgCl					
0.0	726	0	0	0	0
0.1	696	63	55	−0.02	54
0.2	672	264	200	−0.09	276
0.3	658	711	404	−0.46	774
(0.4) extrapolated	647	1305	639	−1.05	1477
i = Ag$_2$S					
(0.3) extrapolated	623	2594	1046	−2.47	3021
0.4	681	1987	602	−2.05	2243
0.5	773	1950	310	−2.13	2017
0.6	837	1536	130	−1.67	1469
0.7	869	908	38	−1.00	841
0.8	921	410	0	−0.46	368
0.9	984	13	0	0.00	0
1.0	1103	0	0	0	0

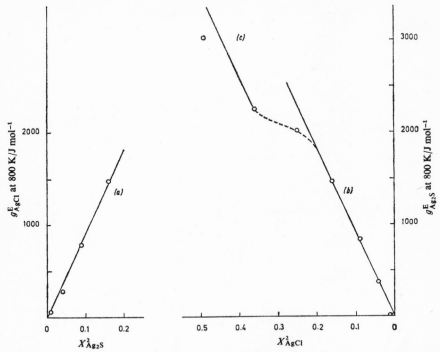

FIGURE 9. g^E for AgCl + Ag$_2$S mixtures at 800 K showing terminal regions where $g^E_{AgCl} = 9100 X^2_{Ag_2S}$ J mol^{-1} [curve (a)], $g^E_{Ag_2S} = 9260 X^2_{AgCl}$ J mol^{-1} [curve (b)], and $g^E_{Ag_2S} = 9100 X^2_{AgCl} - 1033$ J mol^{-1} [curve (c), calculated from Gibbs–Duhem relation and curve (a)]. The transition region is dotted.

9: (i) A terminal solution region rich in AgCl which behaves regularly with regular solution parameter $a = 9.100$ kJ, (ii) solutions rich in Ag$_2$S which behave regularly with regular solution parameter $b = 9.260$ kJ, and (iii) a transition region connecting the two terminal regions where nonregular solution behavior pertains.

The formalism of Darken described above is clearly fitted by the AgCl + Ag$_2$S melts, with S^E_i exhibiting similar behavior to g^E_i, as shown in Figure 10. The structural interpretation of the quadratic formalism is that in the AgCl-rich solutions the structure of AgCl is maintained, while in Ag$_2$S-rich solutions the structure of Ag$_2$S is maintained. There is a narrow transition region between $X_{AgCl} = 50$ mol % and $X_{AgCl} = 60$ mol % in which the structure changes. The narrowness of the transition region in this system is particularly noteworthy and the thermodynamics in the transition region is undoubtedly more complex than in the relatively simple terminal regions.

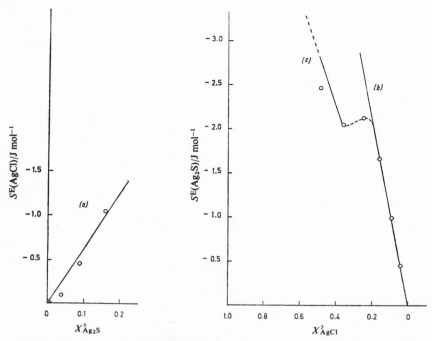

FIGURE 10. S^E for AgCl + Ag$_2$S mixtures showing limiting terminal regions were
$S^E(\text{AgCl}) = 6.11\ X^2_{\text{Ag}_2\text{S}}$ J mol^{-1} [curve (a)], $S^E(\text{Ag}_2\text{S}) = 10.46\ X^2_{\text{AgCl}}$ J mol^{-1} [curve (b)], and
$S^E(\text{Ag}_2\text{S}) = 6.11\ X^2_{\text{AgCl}} - 0.17$ J mol^{-1} [curve (c), calculated from Gibbs–Duhem relationship
and curve (a)]. The transition region is dotted.

6.2. *Thallous Chloride + Thallous Sulfide Mixtures*

The limiting one-foreign-particle liquidi for Tl$_2$S + TlCl agree with the
experimental data in the dilute solution limit, as was found for the silver
salts.[2] In both systems, in infinitely dilute solution, each M$_2$S (M = Ag or
Tl) species adds one foreign particle to the chloride solvent (e.g., S^{2-} ion,
M$_2$S molecule ...) and vice versa for the sulfide limit. This behavior contrasts with that of Cu$_2$S in CuCl melts, where (Cu$_2$S)$_2$ dimers are indicated
(Section 5), and with Bell and Flengas' earlier analysis,[3] in which they
found two foreign particles for dilute solutions of AgCl in Ag$_2$S calculated
from their limiting slopes using the older ΔH_f values.

It seems reasonable to assume that, in the sulfide solvents, the chlorides are dissociated. Hence, the foreign particle in the sulfide solvents is the
Cl$^-$ ion. It then follows that the Tl$^+$ (or Ag$^+$) cations are incorporated into
the Tl$_2$S (or Ag$_2$S) structures, and are not "foreign."

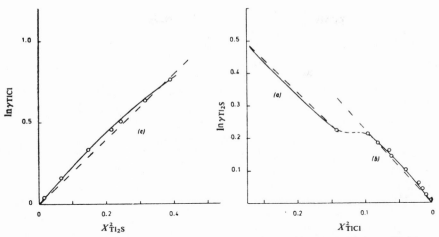

FIGURE 11. Nonisothermal plots of lnγ for TlCl + Tl$_2$S melts calculated from the phase diagram (Figure 1), showing limiting terminal regions where the dashed lines correspond to lnγ_{Tl_2S} = 2.0 X^2_{TlCl} − 0.06 [curve (a), calculated from Gibbs–Duhem relation], lnγ_{Tl_2S} = 2.3 X^2_{TlCl} [curve (b)], and lnγ_{TlCl} = 2.0 $X^2_{Tl_2S}$ in the TlCl-limiting region [curve (c)]. The transition region is dotted. Curves (a) and (b) for 648 < T < 719 K. Curve (c) for 648 < T < 681 K.

Plots of ln γ_{Tl_2S} vs. X^2_{TlCl} and ln γ_{TlCl} vs. $X^2_{Tl_2S}$ are shown in Figure 11 where ln $\gamma_i = g^E_i/RT$ was calculated from the cryoscopic data. Using the Gibbs–Duhem integration two nearly linear limiting regions are found together with a very narrow transition region, in the range 60 mol % < X_{Tl_2S} < 70 mol %. This has also been found to be the region where the specific conductivity increases extremely rapidly with sulfide concentration (Figure 8). Clearly, as for the silver system, and perhaps also the copper system (see Section 5), there appears to be a drastic and sudden change in melt structure in the region between two well-behaved limiting regions. The values of ln γ_i in Figure 11 are not isothermal of course, although the temperature ranges are not large at 648 < T < 681 K for ln γ_{TlCl} and 648 < T < 719 K for ln γ_{Tl_2S}. Furthermore, because the values are nonisothermal, the Gibbs–Duhem integration is not strictly applicable although the small temperature ranges involved permit the ln γ values to be treated as essentially isothermal. Although no enthalpy data exist for TlCl + Tl$_2$S it is reasonable to suppose S^E_i and Δh_i for the thallium system would be similar to that for silver. We conclude that in the region X_{Tl_2S} > 66 mol % (approx.) the solution appears to remain very ordered, and added TlCl is not breaking up the ordered Tl$_2$S structure, and furthermore the added TlCl is entering this structure in some ordered fashion. For

X_{Tl_2S} less than about 66 mol % the activity of Tl_2S drops drastically. It seems that in the transition region there is a sudden catastrophic collapse of the ordered Tl_2S structure on adding $TlCl$. This structural explanation of the transition region, between the Tl_2S-rich and $TlCl$-rich terminal zones, which was found thermodynamically, is also proposed for the $AgCl + Ag_2S$ system. The evidence seems to suggest that the structure of the major component in each terminal region is maintained. The negative S_i^E values (Table 1) for the $AgCl + Ag_2S$ indicating an ordered structure are significant in this respect. For the chlorides the ionic nature of the bonding in e.g., $AgCl$, $PbCl_2$, $CuCl$, and $TlCl$, is dominant; and in Ag_2S, Tl_2S, PbS, and Cu_2S, as well as a number of selenides and tellurides, the polymeric or network structure with covalent interaction appears prevalent, based on the evidence described previously. Clearly, as the composition varies a transition must occur, and the experimental facts indicate that this transition region is not broad enough to obliterate the well-behaved terminal regions. For the silver and thallium sulfide + chloride binaries it is apparent that the transition regions are, in fact, rather narrow, being 50–60 mol % for silver and 55–70 mol % for thallium. The significant change in melt structure in, for example, $AgCl + Ag_2S$ and $TlCl + Tl_2S$ melts, over a narrow composition range is not paralleled by a *correspondingly sudden* change in densities for $AgCl + Ag_2S$,[3, 26] (but see Figures 4 and 5) nor for $TlCl + Tl_2S$ melts,[29] as a function of composition. Predictions about structure and bonding in melts cannot be made based on density data alone in much the same way as it has been shown that predictions about the existence or otherwise of complexes in melts cannot be made based on conductivity data.[30]

Further experimental data are needed before similar structural conclusions can be drawn concerning the other halide + chalcogenide systems described in this chapter. For example, viscosity and spectroscopic data for all the systems described, of which almost none exists, could prove valuable.

ACKNOWLEDGMENTS

Supporting grants from the University of Natal Research Fund, the South African Council for Scientific & Industrial Research, and AECI Trust Fund are acknowledged.

References

1. *Trends in Electrochemistry*, Eds. J. O'M. Bockris, D. A. J. Rand, and B. J. Welch, Plenum, New York (1976), p. 267.
2. H. C. Brookes, H. J. Coppens, and A. D. Pelton, *J. Chem. Soc. Faraday Trans. 1*, **74**, 2193 (1978).

3. M. C. Bell and S. N. Flengas, *J. Electrochem. Soc.* **111**, 569, 1440 (1964), and personal communication.
4. M. C. Bell and S. N. Flengas, *J. Electrochem. Soc.* **111**, 575 (1964).
5. R. Blachnik and Hoppe, *J. Chem. Thermodyn.* **8**, 631 (1976).
6. M. C. Bell and S. N. Flengas, *J. Electrochem. Soc.* **113**, 27 (1966).
7. M. C. Bell and S. N. Flengas, *J. Electrochem. Soc.* **113**, 31 (1966).
8. A. K. Garbee and S. N. Flengas, *J. Electrochem. Soc.* **119**, 631 (1972).
9. L. Yang, G. M. Pound, and G. Derge, *Trans. AIME, J. Met.* **206**, 783 (1956).
10. P. Hagenmuller, J. Rouxel, J. David, A. Colin, and B. Le Neidr, *Z. Anorg. Allg. Chem.* **1**, 323 (1963).
11. P. Palvadeau and J. Rouxel, *Bull. Soc. Chim. Fr.*, 2698 (1967).
12. J. Rouxel and P. Palvadeau, *Bull. Soc. Chim. Fr.*, 2044 (1967).
13. G. Handfield, M. D'Amboise, and M. Bourgon, *Can. J. Chem.* **44**, 853 (1966).
14. V. M. Glazov and N. M. Makhmudova, *Izvest. Akad. Nauk. SSSR Neorg. Mater.* **6**, 1409 (1970).
15. F. G. Bodewig and J. A. Plambeck, *J. Electrochem. Soc.* **117**, 618 (1970).
16. J. Robinson, B. Gilbert, and R. A. Osteryoung, *Inorg. Chem.* **16**, 3040 (1977).
17. F. G. Bodewig and J. A. Plambeck, *J. Electrochem. Soc.* **116**, 607 (1969).
18. J. Greenberg, B. R. Sundheim, and D. M. Gruen, *J. Chem. Phys.* **29**, 461 (1958).
19. G. Delarue, *Bull. Soc. Chim. Fr.*, 906, 1654 (1960).
20. W. T. Thompson and S. N. Flengas, *Can. J. Chem.* **49**, 1550 (1971).
21. S. K. Chang and J. M. Toguri, *J. Chem. Thermodyn.* **7**, 423 (1975).
22. S. Djierle, *Acta Chem. Scand.* **12**, 1427 (1958).
23. R. Blachnik and H. Kahleyss, *Thermochim. Acta* **3**, 145 (1971).
24. R. Blachnik and G. Kudermann, *Z. Naturforsch.* **286**, 1 (1973).
25. T. Takahashi, O. Yamamoto, and H. Honi, *Denki Kagaku* **35**, 181 (1967).
26. R. Blachnik and J. E. Alberts, *Z. Naturforsch.* **316**, 163 (1976), and personal communication.
27. D. N. Gruen, R. L. McBeth, M. S. Foster, and C. E. Crouthamel, *J. Phys. Chem.* **70**, 472 (1966).
28. L. G. Boxall, H. L. Jones, and R. A. Osteryoung, *J. Electrochem. Soc.* **121**, 212 (1974).
29. H. C. Brookes and H. J. Coppens, previously unpublished work on Tl_2S + TlCl mixtures.
30. D. Inman, A. D. Graves, and A. A. Nobile, *Specialist Periodical Reports, Electrochemistry* Vol. 2, The Chemical Society, London (1972), p. 77.
31. R. Blachnik and J. Schöneich, *Z. Anorg. Allg. Chem.* **429**, 131 (1977).
32. *Characterization of Solutes in Non-aqueous Solvents*, Ed. G. Mamantov, Plenum, New York (1978).
33. L. M. Toth and B. F. Hitch, *Inorg. Chem.* **17**, 2207 (1978).
34. C. E. Bamberger, J. P. Young, and R. G. Ross, *J. Inorg. Nucl. Chem.* **36**, 1158 (1974).
35. K. Paulsen and R. A. Osteryoung, *J. Am. Chem. Soc.* **98**, 6866 (1976).
36. D. M. Gruen, R. L. McBeth, and A. J. Zielen, *J. Am. Chem. Soc.* **93**, 6691 (1971).
37. W. Giggenbach, *Inorg. Chem.* **10**, 1308 (1971).
38. J.-P. Bernard, A. de Haan, and H. van der Poorten, *C.R. Acad. Sci. Ser. C* **276**, 587 (1973).
39. B. Cleaver, A. J. Davies, and D. H. Schiffrin, *Electrochim. Acta* **18**, 747 (1973).
40. C. H. Liu, A. J. Zielen, and D. M. Gruen, *J. Electrochem. Soc.* **120**, 67 (1973).
41. P. L. King and B. J. Welch, *Proc. Australas. Inst. Min. Metall.*, **246**, 7 (1975).
42. W. T. Thompson and S. N. Flengas, *Can. J. Chem.* **46**, 1611 (1968).
43. F. G. Bodewig and J. A. Plambeck, *J. Electrochem. Soc.* **117**, 904 (1970).
44. J. H. Kennedy and F. Adams, *J. Electrochem. Soc.* **119**, 1518 (1972).
45. *Proceedings of the International Symposium on Molten Salts*, Ed. J. P. Pemsler *et al.*, Electrochemical Society, New Jersey (1976).
46. C. A. Melendres, C. C. Sy, and B. Tani, *J. Electrochem. Soc.* **124**, 1060 (1977).
47. B. Tani, *Am. Mineral* **62**, 819 (1977).

48. A. N. Belous, T. A. Kusnitsyna, and A. A. Velikanov, *Elektrokhim. Termodin. Svoistva Ionnykh Rasplavov*, 53 (1977).
49. A. D. Pelton and S. N. Flengas, *Can. J. Chem.* **48**, 2016 (1970).
50. M. L. Sholokhovich, D. S. Lesnyk, and G. A. Buchalova, *Dokl. Akad. Nauk. S.S.S.R.* **103**, 261 (1955).
51. B. Reuter and K. Hardel, *Z. Anorg. Chem.* **340**, 158 (1965).
52. R. Blachnik and F. W. Kasper, *Z. Naturforsch.* **296**, 159 (1974).
53. A. V. Novoselova, I. N. Odin, V. A. Trifonov, and B. A. Popovkin, *Izv. Akad. Nauk. S.S.S.R. Neorg. Mater.* **3**, 2101 (1967).
54. A. V. Novoselova, I. N. Odin, N. R. Valitova, and B. A. Popovkin, *Izv. Akad. Nauk S.S.S.R. Neorg. Mater.* **4**, 777 (1968).
55. A. V. Novoselova, I. N. Odin, and B. A. Popovkin, *Izv. Akad. Nauk S.S.S.R. Neorg. Mater.* **2**, 1397 (1966).
56. A. V. Novoselova, I. N. Odin, and B. A. Popovkin, *Izv. Akad. Nauk S.S.S.R. Neorg. Mater.* **6**, 381 (1970).
57. A. V. Novoselova, I. N. Odin, I. N. Fedoseeva, and B. A. Popovkin, *Izv. Akad. Nauk S.S.S.R. Neorg. Mater.* **6**, 135 (1970).
58. A. V. Novoselova, I. N. Odin, and B. A. Popovkin, *Izv. Akad. Nauk S.S.S.R. Neorg. Mater.* **6**, 257 (1970).
59. I. S. Morozov and C. F. Li, *Zh. Neorg. Khim.* **8**, 1688 (1962).
60. A. V. Novoselova, M. K. Todriya, I. N. Odin, and B. A. Popovkin, *Izv. Akad. Nauk S.S.S.R. Neorg. Mater.* **7**, 500 (1971).
61. A. V. Novoselova, M. K. Todriya, I. N. Odin, and B. A. Popovkin, *Izv. Akad. Nauk Neorg. Mater.* **7**, 1266 (1971).
62. A. V. Novoselova, I. N. Odin, and B. A. Popovkin, *Zh. Neorg. Khim.*, 2659 (1969).
63. L. S. Darken, *Trans. AIME* **239**, 80 (1967).
64. Y. Nakamura, personal communication.
65. D. K. Belaschenko, I. A. Magidson, and F. L. Konopelko, *Ukrain. Fiz. Z.* **12**, 66 (1967).
66. A. A. Velikanov and V. F. Zinchenko, *Electrokhim.* **11**, 1862 (1975).
67. A. Rabenau and H. Rau, *Z. Anorg. Allg. Chem.* **369**, 295 (1969).
68. A. V. Novoselova, I. N. Odin, and B. A. Popovkin, *Zh. Neorg. Khim.* **14**, 1402 (1969).
69. A. Rabenau, H. Rau, and G. Rosenstein, *Z. Anorg. Allg. Chem.* **374**, 43 (1970).
70. R. Blachnik, personal communication.
71. L. S. Darken, *Trans. AIME* **239**, 90 (1967).
72. J. Lumsden, *Thermodynamics of Molten Salt Mixtures*, Academic, New York (1966).
73. G. J. Janz, *Molten Salts Handbook*, Academic, New York (1967).
74. *Handbook of Chemistry and Physics*, Ed. R. C. Weast, Chemical Rubber, Cleveland (1971–1972), B244.
75. *Molten Salts: Characterization and Analysis*, Ed. G. Mamantov, Marcel Dekker, New York (1969).
76. F. R. Mrazek and J. E. Battles, *J. Electrochem. Soc.* **124**, 1556 (1977).
77. T. Takahashi, K. Kuwabara, O. Yamamoto, and S. Watanabe, *Denki Kagaku*, **37**, 717 (1969).
78. A. E. Ioffe and A. R. Regel, *Progr. Semicond.* **4**, 239 (1960).
79. A. Epstein and H. Fritzsche, *Phys. Rev.* **93**, 922 (1954).

18

Recent Aspects of Sulfur and Sulfide Electrochemistry in Ionic Liquids

V. Plichon, A. de Guibert, and M. N. Moscardo-Levelut

1. General Review

The most important part of these electrochemical investigations on sulfur and sulfide compounds in ionic liquids has been performed in two melts: LiCl–KCl and chloroaluminates. Therefore, we shall first focus attention on these two media.

1.1. LiCl–KCl Eutectic

The behavior of sulfur species in LiCl–KCl eutectic at 420°C has been critically reviewed in a recent detailed monograph on the sulfur electrode in nonaqueous media.[1] We shall present only a short summary of the results obtained in this melt.

1.1.1. Solubility of Sulfide Ion and Sulfur

Sulfide ion has a low solubility in purified melts. The solubility of Li_2S is about 1.5×10^{-3} (ion fraction unit).[2, 3] Na_2S has been found insoluble,[4] or of solubility limited by precipitation of Li_2S.[5]

Sulfur seems to be insoluble. Solubility is increased by the formation of polysulfide ions when both S^{2-} and sulfur are present. Na_2S_2 is easily dissolved but higher degree polysulfides are not stable and evolve sulfur immediately.[4]

V. Plichon, A. de Guibert, M. N. Moscardo-Levelut • Laboratoire de Chimie Analytique des milieux réactionnels, ESPCI, 10, rue Vauquelin, F75231 Paris Cedex 05, France.

1.1.2. Electrochemical Behavior

Owing to the low solubility of Li_2S and Na_2S in the eutectic (Cleaver et al.[4] noticed no voltammetric oxidation peak for Na_2S), most studies have been performed with polysulfide[4] or sulfur[6, 7] as starting material.

1.1.2.1. *Reduction of Sulfur.* A two-step reduction of sulfur now seems probable[4-7] (but see comment by Tischer and Ludwig[1] on Reference 6).

The first step leads to a blue solution of an anion radical, and the second one to a colorless polysulfide, most probably S_2^{2-} (References 1, 4, and 7)

$$S \xrightarrow{+n_1e} \text{blue species} \xrightarrow{+n_2e^-} S_2^{2-}$$

The blue species is observed during either voltammetric experiments (at a gold electrode) or long-term electrolysis.[7] The reverse process, that is, oxidation of polysulfide to the blue species, then back to sulfur, has been performed with both methods.[4-7]

Some discrepancies exist when sulfide is the starting material. Weaver and Inman,[5] during their chronopotentiometric experiments on the oxidation of Na_2S added to the melt, suggest a polymerization reaction prior to sulfur film formation. At a graphite electrode Birk and Steunenberg[8] obtained the two expected voltammetric peaks for the oxidation of Li_2S leading to sulfur, but during the backward scan, three reduction peaks appeared. A final reduction to $S(-II)$ was claimed.

Possible mechanisms have been discussed by Tischer and Ludwig[1] and by Chivers,[14] who tried to reconcile these discrepancies; they speculated on the nature of the polysulfides.

1.1.2.2. *Oxidation of Sulfur.* Sulfur is oxidized in a single step.[6] Delarue[9] and Plambeck[6] suggest that S_2Cl_2 is the product formed on the electrode. However, attempts to isolate and identify it were always unsuccessful.

1.1.3. Spectrophotometry

An absorption peak of $\lambda_{max} = 618$ nm characterizes the blue color of "dissolved sulfur" in LiCl–KCl. It has been ascribed to several species: firstly to neutral molecules of S_2,[10] and later to polysulfide ions.[11-13] The exact nature of the polysulfide ion is still under discussion; contenders include $S_2^{\bullet-}$ and $S_3^{\bullet-}$, with the possibility of an equilibrium between these species and their dimers. In a recent review on chalcogenide anions, arising from a comparison with organic solvents, Chivers[14] concludes that the blue color is undoubtedly that of $S_3^{\bullet-}$ in all media. Previous authors had been somewhat less definitive.

1.2. AlCl₃–NaCl Melts

The oxidation of sulfide ions and the existence and nature of cationic species of sulfur have been the subject of rather intensive investigations recently, especially in relation to the properties of secondary batteries for energy storage. $AlCl_3$–NaCl is also an attractive melt because of its well-characterized acidic and basic properties near the equimolar composition.[15, 16]

1.2.1. Solubility of Sulfide Ion and Sulfur

The solubility of sulfur (as S) in $AlCl_3$–$NaCl_{sat}$ melt at 175°C is less than 2.1×10^{-2} m.[17] Sulfur solutions have the same pale yellow color in the presence and absence of sulfide ion, thus precluding the formation of species such as $S_3^{\bullet-}$ or $S_2^{\bullet-}$, which are responsible for the blue and green colors in LiCl–KCl.

Solubility increases with temperature and the acidity of the solvent.[18] In the $AlCl_3$–NaCl 63–37 mol % melt at 175°C, the electrochemical results indicate that the sulfur solubility is higher than 0.5 F (as monomeric sulfur).[19] Raman spectra of sulfur dissolved in pure Al_2Cl_6 or in $AlCl_3$–NaCl melt show that S_8 is the predominant entity.[18] This result is inconsistent with the experiments of Paulsen and Osteryoung, who deduced from potentiometric measurements that sulfur is dissolved as S_2.[20]

Marassi and Mamantov have indicated that Na_2S does not seem very soluble in the basic melt[17] and dissolves only very slowly. They have not published more quantitative data.

1.2.2. Acid–Base Chemistry of Sulfide Ion

An acid–base equilibrium exists between S^{2-} ions and the melt.[20, 21] In acidic media, the sulfide ion appears to behave as a trichloro- base, according to

$$3Al_2Cl_7^- + S^{2-} \rightleftharpoons AlSCl + 5AlCl_4^- \tag{1}$$

while in more basic solutions, the equilibrium is

$$2Al_2Cl_7^- + S^{2-} \rightleftharpoons AlSCl_2^- + 3AlCl_4^- \tag{2}$$

AlSCl and $AlSCl_2^-$ are probably solvated[21] as $Al_2SCl_5^-$ and $Al_2SCl_6^{2-}$; no evidence has been offered for this hypothesis. The equilibrium constants for equations (1) and (2) have not been calculated, owing to the lack of precision of titrations, especially in basic media.[21]

The different basicity of sulfide ion in acidic or basic media is invoked to explain the increase of solubility in acidic medium.

1.2.3. Electrochemical Studies

The most important features are the absence of polysulfides in this melt and the oxidation of S into cationic species S^{n+}. The general redox sequence is a two-step one. For instance, for sulfide oxidation

$$S(-II) - 2e^- \rightarrow S \tag{3}$$

$$S - ne^- \rightarrow S^{n+} \tag{4}$$

$$S^{n+} \rightarrow \dots \text{ chemical step}$$

Electrochemical processes are slow, but reverse reactions such as the reduction of S^{n+} or S occur and experiments performed on sulfide or sulfur solution give the same results.

1.2.3.1. Basic Melt (Saturated with NaCl). (a) *First step.* At 175°C equation (3) is consistent with data of pulse, cyclic voltammetry, and chronoamperometry.[17, 20] The voltammograms obtained from current–time curves for the sulfide solutions are similar to those obtained for elemental sulfur, except for the shift of the zero current line; in the same way, cyclic voltammograms show the same peaks except for the initial scan. Experimental results have been obtained on tungsten, glassy carbon, and platinum electrodes, but an extra reduction peak dependent on the history of the electrode is evident on platinum.

Both cathodic and anodic peaks are drawn out at low temperature; the sulfur–sulfide system becomes more reversible[17] at 250°C.

(b) *Second step.* Mamantov and co-workers[17] have published a reaction scheme for the oxidation of Na_2S at 175°C, based on pulse, cyclic voltammetric, or chronoamperometric data:

$$S(-II) \rightarrow S + 2e^- \tag{3a}$$

$$S \rightarrow \tfrac{1}{2}S_2^{2+} + e^- \tag{5}$$

A chemical reaction following equation (5) and leading to S_2Cl_2 is suggested.[17, 20] Such a mechanism certainly improves our understanding of sulfur cations, as well as aiding the interpretation of cyclic voltammetric behavior. Moreover, the same voltammetric peaks are obtained when experiments are carried out with an S_2Cl_2-containing melt.

The electrochemical behavior changes at higher temperature (250°C). The second step becomes more dominant. The modified scheme can be written

$$S(-II) \rightarrow S + 2e^- \tag{3a}$$

$$S \rightarrow S(+II) + 2e^- \tag{6}$$

By comparison with the reduction wave for sulfur, this oxidation wave increases rapidly with temperature: the ratio of anodic to cathodic peak changes from 0.5 to 1.0 between 175 and 250°C. A change in mechanism from a one-electron to a two-electron process may be deduced and is consistent with the fact that higher oxidation states are more stable in more basic melts (the basicity is increased at 250°C because of the higher solubility of NaCl).

1.2.3.2. Acidic Melts. The general shape of curves is the same as in basic melts. The waves obtained by pulse polarography are poorly defined. The stoichiometry of the sulfur reduction has been deduced from coulometry[20]: an exchange of 2.00 ± 0.01 electrons per mole of sulfur is found.

No mechanism has been suggested for the electrochemical oxidation of sulfur in slightly acidic media (51 to 53 mol % $AlCl_3$): the cyclic voltammogram of a sulfur solution exhibits, for this step, one oxidation and two reduction peaks which have not been interpreted.[22]

In very acidic media (67 mol % $AlCl_3$, at 150°C), spectrophotometric measurements coupled with theoretical calculations of spectra have led to the conclusion that four cationic species of sulfur are produced[23] during electro-oxidation. S($+IV$) and S($+II$) oxidation states were identified with certainty. For the two others, S_2^{2+} and S_4^{2+} would be the most likely species obtained.

An electrochemical study of sulfur oxidation in $AlCl_3$–NaCl (63–37 mol %) confirms the production of four cationic species: Marassi *et al.*[19] have proved that the oxidation of sulfur proceeds from S_8 to SCl_3^+ and involves intermediates S_8^+ (S_{16}^{2+}), S_8^{2+}, and S($+I$), this later one being stable above 200°C.

1.3. Miscellaneous

1.3.1. NaCl–KCl

Rempel and Malkova[24] give a half-wave potential of -0.7 V vs. chlorine for the oxidation of 0.01 N sulfide ion in equimolar NaCl–KCl at 700°C.

Solubilities of Pb^{2+}, Bi^{3+}, Sb^{3+}, and Cu^+ sulfides have been reported in NaCl–KCl.[25, 26]

1.3.2. Molten Thiocyanates

Together with LiCl–KCl, the sulfur–sulfide system is reviewed for this melt in Reference 1. The whole range of polysulfides from S_2^{2-} to S_6^{2-} can exist in molten thiocyanate.[4, 27] Polysulfides can be formed by anodic oxidation of the solvent.

Considerable dissociation of polysulfides into radical anions can occur and produces the blue and green colors characteristic of dilute solutions.

Polysulfide ions can be oxidized to sulfur.[4] A characteristic stripping peak is obtained for the redissolution of sulfur.

2. Sulfide Ion Oxidation in Lead Chloride

2.1. Preparative Electrolysis of Lead

Lead preparation by electrolysis of $PbCl_2$–PbS in the molten state has been attractive for a long time; the first patent on the subject is more than half a century old.[28] Production has been developed to pilot plant scale on several occasions,[29-31] in Wales (Halkyn), Canada, and Australia. The last one was at Port Pirie (Australia) and produced 1800 tons a year.

Few details are available concerning the mode of operation of the plants, especially the electrode processes. Liquid lead is recovered at the cathode without problem; sulfur is obtained at the anode, where most difficulties arise. A scum forms on the baths and losses of lead chloride occur during its removal.

2.1.1. Improvement of the Cathodic Yield

Laboratory scale experiments have been undertaken in the fifties, with the principal aim of improving electrolysis yields.[32-35] The influence on cathodic yields of melt composition, temperature, and current density was investigated. A more systematic study of these parameters was performed recently[36-38]; the following results were obtained:

(1) A diminution of PbS concentration always improves the cathodic yield.[37]

(2) At a fixed lead sulfide concentration, yield is higher when the current density is increased.[37]

(3) The influence of temperature and melt composition has not been studied independently: an addition of alkali chloride is necessary to lower the melting point of mixtures. Yield becomes higher when KCl or NaCl is added to the melt.[38]

(4) Different electrode arrangements have shown that the cathode efficiency is improved when diaphragms are employed.[37]

2.1.2. Behavior at the Anodic Side

Owing to experimental problems of sulfur recovery anodic yields have not been measured in a quantitative way.[37] Attempts to understand the anodic process were only partly successful. Welch and co-workers have suggested that dissolution of sulfur in the baths is the factor which limits electrolytic yields.[38] The following experimental results agree with their suggestion:

(1) A higher temperature—which increases the rates of dissolution—lowers the electrolysis yield.

(2) Losses of lead are proportional to times of electrolysis; higher current densities which shorten times of electrolysis produce an increase of the cathodic yield.

2.2. Mechanism of the Electrochemical Oxidation

Unfortunately, the mechanisms of these electrode reactions could not be determined by preparative electrolysis. Two investigations in this area have been carried out recently, using $PbCl_2$–$NaCl$ (Reference 39) and $PbCl_2$–KCl (References 40–43).

2.2.1. Chronopotentiometric Study of PbS Oxidation in Molten PbCl₂–NaCl (Reference 39)

Skyllas and Welch have investigated PbS oxidation by current reversal chronopotentiometry at 440°C on a glassy carbon electrode. The reaction is diffusion controlled at PbS concentrations below 0.8 M. The interpretation of chronopotentiometric curves fits well with the simple mechanism

$$S^{2-} \underset{k_b}{\overset{k_f}{\rightleftharpoons}} S + 2e^-$$

The electron transfer is slow. Sulfur has a limited solubility in the melt. The electrochemical reaction is followed by the dissolution of sulfur. The authors have discussed the possibility of polysulfide formation reactions. By comparison with the theoretical chronopotentiograms calculated for these two reactions, they concluded that such a reaction takes place. With the assumption that the rate of dissolution is first order with respect to S^{2-}, it can be written

$$S + S^{2-} \overset{k}{\rightarrow} S_2^{2-}$$

$k = 8$ sec^{-1}, the first-order rate constant, has been deduced from a comparison with theoretical chronopotentiometric curves.

2.2.2. PbS Oxidation in Molten PbCl₂–KCl (References 40–43)

We have investigated the PbS oxidation in melts of composition 77 mol % $PbCl_2$ and 23 mol % KCl at 440°C. Our experiments have been performed using transient techniques (double-step chronoamperometry[41] and ac impedance measurements[42]) as well as by controlled potential electrolysis coupled with visible absorption spectrophotometry[43]; these methods provide information about the behavior of species on different time scales.

PbS oxidizes in two steps (Figure 1). The first step leads to the formation of sulfur, in agreement with the oxidation of sulfide ion in $PbCl_2$–NaCl; the second one produces an unstable cationic species of sulfur.

2.2.2.1. First Oxidation Step. (a) *Reaction at the electrode.* Both chronoamperometric and ac impedance measurements show that this first step is a rapid two-electron transfer limited by the diffusion of sulfide ion.

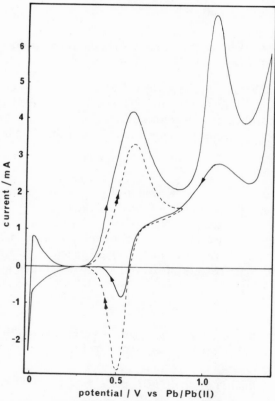

FIGURE 1. Cyclic voltammogram for PbS solution in $PbCl_2$–KCl 77–23 (mol %) at 440°C; glassy carbon electrode; scan rate 0.1 V sec⁻¹.

The reaction is reversible and can be written

$$S^{2-} \rightleftharpoons S + 2e^-$$

The diffusion coefficient of sulfide ion has been calculated from chronoamperometric curves and found to be[41] $D = 0.34 \times 10^{-5}$ cm^2 sec^{-1}

The stripping cathodic peak obtained by cyclic voltammetry and the unusual shape of cathodic currents in the chronoamperometric curves suggest the existence of a sulfur deposit on the electrode after the anodic process. Impedance measurements have produced evidence that the deposit does not passivate the electrode.

It has been shown that the amount of sulfur on the electrode rapidly reaches a limit[41] and, from this result, that sulfur dissolves in the bath.

(b) *Dissolution of sulfur.* The dissolution of sulfur by a polysulfide formation reaction has been followed by spectroelectrochemistry. Controlled potential oxidation of sulfide solutions coupled with spectrophotometry presents two particular features attributed to the formation of a polysulfide ion in the melt:

(i) The current does not follow an exponential decay, showing that the electron transfer involves subsequent reactions.

(ii) An intermediate compound absorbing in the same wavelength range as the sulfide ion [i.e., below 500 nm (Reference 43); see Figure 2] appears during electrolysis.

The nature of the polysulfide has not been determined owing to the lack of a definite absorption maximum for these ions.

2.2.2.2. Second Oxidation Step. Sulfur is also the final product in this step, although a different mechanism seems to be involved. (a) *Reaction at the electrode.* The reaction is non-diffusion-controlled and corresponds to the oxidation of sulfur to a cationic species, with an apparent exchange of less than one electron per atom of sulfur.[41] Cyclic volammetry has shown that this step is totally irreversible (Figure 1).

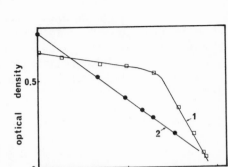

FIGURE 2. Typical decrease in the optical density during electrolysis: (1) at the first wave, (2) at the second wave, at $\lambda = 500$ and, 510 nm, respectively.

(b) *Subsequent reaction.* The production of a cationic species of sulfur disagrees with our electrolysis results, which always show the exchange of only 2 F per mole of sulfide in this step as in the first one.[43] Sulfur is the final product in all cases.

We suggest the following explanations for the experimental results. The production of a cationic species is only an intermediate step; reaction with the chloride ions of the melt produces a sulfur chloride entity. A rapid decomposition of the sulfur chloride is then most probable. No polysulfide formation is involved during this decomposition, as confirmed by the linear decrease of the visible absorption with time of electrolysis.

3. Sulfur Compounds in Sodium Hydroxide–Water Mixtures

3.1. General Review

Sodium hydroxide–water "ionic" mixtures will be quite arbitrarily defined as media containing more than 0.2NaOH in mole fraction (12 M). The general properties of these solvents have been discussed by Tremillon et al.[44] (see also chapters of this book dealing with water in melts).

Published thermochemical data for more dilute media are of only peripheral interest in knowing the range of existence of sulfur compounds in our media: extrapolation for basic solutions of potential–pH diagrams established[45] at 25°C leads to the prediction that only sulfide and sulfate ions are thermodynamically stable. However, several metastable species have been observed. Kinetic parameters then become important and must be taken into consideration.

3.1.1. Sulfide Ion

The only experiments on the electrochemical oxidation of sulfide ion have been carried out in dilute aqueous hydroxide solutions, i.e., beyond the scope of the present chapter. Nevertheless, it is worth noting that the products of oxidation of sodium sulfide solutions in 1–4 M NaOH are polysulfides,[46, 47] whilst the ultimate product of prolonged electrolysis is sulfate. Whether thiosulfate is formed as an intermediate is not of concern in this context.

3.1.2. Polysulfide Ions

Most results concerning stability,[48] electrochemical behavior,[49] and spectrophotometry[50, 51] of polysulfide solutions have been obtained in melts of low basicity (NaOH < 2 M).

In high pH ranges (OH^- concentration up to 18 M), Giggenbach[52] has calculated the absorption spectra of individual polysulfide ions S_x^{2-} and the equilibrium distribution of the various species at 25°C as a function of OH^- concentration. His results must be treated with caution because in the calculations he uses a second dissociation constant[53] of H_2S equal to 17 instead of the usual value of 13.[54] The absorption maxima and corresponding molar absorbancies are reported in Table 1 and compared with the values of Schwarzenbach et al.[50] at pH 10. Giggenbach ascribes the discrepancy for pentasulfide ion to the use by Schwarzenbach of equimolar Na_2S_4–Na_2S_5 instead of pure sodium pentasulfide.

An increase of sodium hydroxide concentration displaces equilibria towards smaller polysulfide ions as observed in aqueous solutions.[48]

3.1.3. Sulfur

Ugorets et al.[55] have studied the disproportionation of sulfur in NaOH 4–15 M at temperature up to 200°C. The reaction can be written

$$2(x + 1)S + 6OH^- \rightarrow S_2O_3^{2-} + 2S_x^{2-} + 3H_2O$$

with an average value for x of 2.5–4, there being no sulfite formation.

In more concentrated hydroxide solutions,[56, 57] the reaction of sulfur with the melt yields sulfite and polysulfide ions, according to

$$(2x + 1)S + 6OH^- \rightarrow SO_3^{2-} + 2S_x^{2-} + 3H_2O$$

In this last case, sodium sulfite crystallizes out. Only small amounts of thiosulfate remain in the melt.

TABLE 1. Spectrophotometric Characteristics of Sulfide and Polysulfide Ions Reported by W. Giggenbach[52] and G. Schwarzenbach and A. Fischer[50] *

Ion	λ_1/nm	ε_1/M cm^{-1}	λ_2/nm	ε_2/M cm^{-1}
S^{2-}	—	—	—	—
S_2^{2-}	358	850		
S_3^{2-}	303	2280	416	190
S_4^{2-}	303	3420	367	960
	300*	3800*	370*	1000*
S_5^{2-}	299	8000	374	2560
	300*	5200*	375*	1550*

3.2. Sulfur Compounds in Equimolar NaOH–H$_2$O Melt

This section describes the results of our experiments on sulfur compounds in equimolar NaOH–H$_2$O, at 100°C. The working temperature seems a convenient one with regard to melting point (65°C),[58] viscosity,[59] conductivity[60] of the melt, and corrosion (at 100°C Pyrex glass is free from attack). A home-made spectroelectrochemical device[61] using quartz fiber optics and a modular spectrophotometer adapted from the cell described[62] for studies in molten SbCl$_3$ allows us to record uv spectra of sulfur compounds inside the electrolytic cell.

3.2.1. Sulfide Solutions

Sodium sulfide solutions kept under an oxygen-free atmosphere are stable for several hours. The solutions are colorless and no absorption is noticed between 300 and 500 nm. When oxygen is present yellow polysulfide solutions are obtained.

Sulfide oxidation has been investigated by means of steady state voltammetry at a rotating disk electrode. Although sulfide ion is electrooxidizable on gold, glassy carbon, or platinum electrodes, the gold electrode, which gives more reliable results, has been exclusively used here.

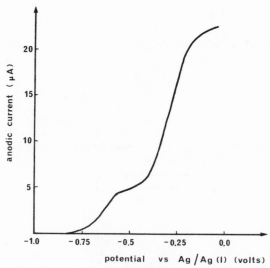

FIGURE 3. Steady state voltammetric curve for Na$_2$S oxidation in NaOH–H$_2$O (1–1) at 100°C; $C_{Na_2S} = 8.3 \times 10^{-3}$ M; gold electrode area $= 2.10^{-3}$ cm^2; rotation speed, 1000 r.p.m.; scan rate, 5 mV sec^{-1}.

The voltammetric curve (Figure 3) shows two main oxidation waves which are proportional to the concentration of sulfide ion and follow the Levich equation. Comparison with the height of a standard wave [reduction of Cr(VI) to Cr(III) in the same melt] indicates a transfer of two electrons per mole of Na_2S for the first wave and 7 ± 1 electrons for the second wave, assuming the same diffusion coefficient for all species.

The first wave appears to be the oxidation of sulfide ion to sulfur according to

$$S^{2-} \rightarrow S + 2e^-$$

Nevertheless, we have not yet succeeded in observing the reduction of sulfur to sulfide ion by cyclic voltammetry, as occurs in $PbCl_2$–KCl.

The second oxidation wave leads to an oxygenated compound of sulfur, either thiosulfate, sulfite, or sulfate. The nature of this species is now under investigation.

3.2.2. Polysulfide Solutions

3.2.2.1. Stability, Absorption Spectra, and Electrochemical Behavior. We have succeeded in observing the two polysulfides S_2^{2-} and S_3^{2-}. 10^{-2} M disulfide solutions are stable for several hours, whereas the trisulfide ion slowly disproportionates.

The absorption spectra of sulfur–sodium sulfide mixtures [S_2^{2-} predominant species; see part (b) below] and dissolved sulfur in the absence of Na_2S [S_3^{2-} predominant species; see part (a) below] are shown in Figure 4. They present the same broadband with $\lambda_{max} = 365$ nm, in agreement with the formation of disulfide ion (Table 1) as the predominant (curve a) or nonpredominant (curve b) species. Another maximum around 300 nm ascribed by Giggenbach[52] to the absorption of trisulfide ion must merge into the absorption edge arising from the solvent.

The electrochemical properties of the two polysulfides are summarized in Table 2. They are both reduced to sulfide at the same potential and can be oxidized at the same potential as the second oxidation step of sulfide. Only the disulfide exhibits the first oxidation step. This essential difference allows amperometric titration of disulfide ion at the first oxidation wave (in absence of sulfide ion) and of the trisulfide ion (for known disulfide ion concentrations) at the reduction wave.

3.2.2.2. Preparation of Polysulfide Solutions. We have used the following methods to prepare polysulfide solutions[61]: disproportionation of sulfur, additions of sulfur to sulfide solutions and controlled potential oxidation of Na_2S.

(a) *Disproportionation of sulfur.* Dissolution of sulfur in the absence of sulfide ion in the melt leads, initially, to S_3^{2-}. This result is evident from the

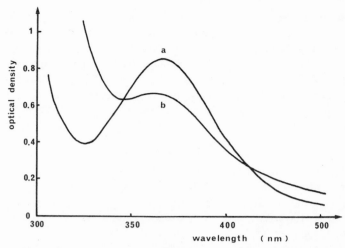

FIGURE 4. Absorption spectra of polysulfide solutions in NaOH–H_2O (1–1) at 100°C, obtained (a) by dissolution of 6×10^{-3} M sulfur in a 1.5×10^{-2} M sulfide solution, (b) by disproportionation of 1.5×10^{-2} M sulfur. Optical pathlength = 0.315 cm.

absence of the first oxidation wave on the voltammetric curves after dissolution of sulfur, whilst the reduction wave and the absorption spectrum support the presence of at least one polysulfide.

Electrolysis on the plateau of the reduction wave[61] confirms the formation of trisulfide ion and thiosulfate or sulfite according to one of the following reactions:

$$S_8 + 6OH^- \rightarrow 2S_3^{2-} + S_2O_3^{2-} + 3H_2O$$

or

$$\tfrac{7}{8}S_8 + 6OH^- \rightarrow 2S_3^{2-} + SO_3^{2-} + 3H_2O$$

At longer times (i.e., 15 min), trisulfide ion disproportionates; this is accompanied by an increase in the first oxidation wave and the optical density at 365 nm together with a decrease in the reduction wave.

(b) *Additions of sulfur to sulfide solutions.* Sulfur added to an Na_2S solution reacts with sulfide ion. A mixture of S_2^{2-} and S_3^{2-} is obtained, the concentration of each polysulfide depending on the amount of sulfur added, relative to the initial concentration of Na_2S.

Absorption measurements (at 365 nm) and amperometry at the reduction wave[61] have shown that, at low sulfur/sulfide ratio, the polysulfide formation reaction is the predominant one compared to the disproportionation of sulfur. Disulfide ion predominates. However, no matter what

TABLE 2. Half-Wave Potential $E_{1/2}$ on Gold Microelectrode of Sulfide, Disulfide, and Trisulfide in Equimolar NaOH–H_2O Melt at 100°C [Ag/Ag(I) Reference Electrode]

Ions	Stability[a]	$E_{\text{I, ox}}$, V	$E_{\text{II, ox}}$, V	E_{red}, V
S^{2-}	Yes	−0.63 V	−0.26	—
S_2^{2-}	Yes	−0.63 V	−0.26	−0.98
S_3^{2-}	No	—	−0.26	−0.98

[a] Stability measured after 3 hr under N_2 atmosphere.

the excess of S^{2-}, no pure S_2^{2-} solutions have been prepared and the extinction coefficient of disulfide ion has not been determined using this method.

At higher S/S^{2-} ratio, the competition between the disproportionation of sulfur and the polysulfide formation reaction favors the formation of the trisulfide ion.

(c) *Controlled potential oxidation of Na$_2$S.* At the first oxidation wave of Na_2S, the electrochemical reaction

$$S^{2-} \rightarrow S + 2e^-$$

is followed by a chemical reaction in the bulk of the melt. The yellow color of polysulfides developed during oxidation provides obvious evidence of this. Appearance of polysulfide can be followed in Figure 5, where the optical density at $\lambda = 365$ nm during electrolysis is plotted. The curve can be divided into three parts:

1. In part AB, sulfide ion is firstly oxidized to disulfide, which absorbs at $\lambda = 365$ nm (Table 1).

2. Beyond the consumption of one faraday (part BC) S_2^{2-} is oxidized into S_3^{2-} at the same potential. The reaction leads to a decrease of optical density.

3. In part CD, for a consumption of more than the 1.3 F required for the oxidation of S_2^{2-} into S_3^{2-}, the optical density should be constant and the oxidation current should fall nearly to zero, since S_3^{2-} cannot be oxidized at this potential. In practice, a slow decrease in the absorption is observed owing to the slow disproportionation of S_3^{2-} into sulfide (or disulfide). The oxidation current of S^{2-} (or S_2^{2-}) is limited by the rate of the disproportionation reaction and reaches zero when the most stable sulfur species is obtained, i.e., above a current consumption of more than 4 F per mole of sulfide ion.

The simultaneous disappearance of the second oxidation wave provides strong supporting evidence for such a reaction scheme.

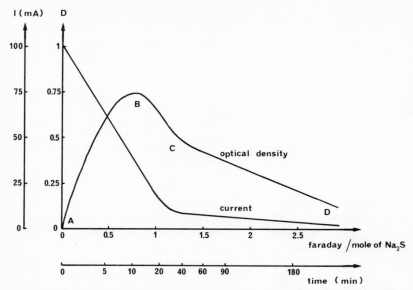

FIGURE 5. Variations of the optical density at 365 nm and of the electrolysis current (a) versus the number of faradays per mole of sodium sulfide consumed, and (b) versus time. Optical pathlength = 0.315 cm; $C_{Na_2S} = 1.14 \times 10^{-2}$ M.

We can sum up the first step of Na_2S oxidation by the following scheme: At short times (i.e., $t < 15$ min)

$$S^{2-} - 2e^- \rightarrow S$$

followed by

$$S^{2-} + S \rightarrow S_2^{2-}$$

$$3S_2^{2-} - 2e^- \rightarrow 2S_3^{2-}$$

At longer times (i.e., $t > 15$ min), disproportionation of S_3^{2-} is slow.

5. Comparison between Melts

5.1. Lewis Basicity of Melts

Most experimental studies on polysulfides in ionic liquids have been performed in five melts which cover a large range of Lewis basicity. Three of them are very basic: LiCl–KCl, thiocyanates, and $NaOH–H_2O$. The order of Lewis basicity must be

$$OH^- > Cl^- > SCN^-$$

TABLE 3. Stability of Polysulfides S_x^{2-} or Radical Anions $S_{x/2}^{\bullet-}$ in Ionic Melts

Lewis basicity	Melt	θ, °C	Stability	x
Basic	NaOH–H$_2$O	100	Very stable	$x \leq 3$
	LiCl–KCl	450	Very stable	$x/2 \geq 2$
	MSCN	210	Very stable	$2 \leq x \leq 6$
Medium	PbCl$_2$–MCl	440	Slightly stable	$x \leq 2$
Low	AlCl$_3$–NaCl	175	Not stable	—

On the other hand, AlCl$_3$ is a strong Lewis acid. In equimolar AlCl$_3$–NaCl, the corresponding anion AlCl$_4^-$ has a very weak Lewis basicity.

The PbCl$_2$–MCl melts $(M = Na^+$ or $K^+)$ lie in between these two classes. PbCl$_2$ is a Lewis acid, but, though the existence of PbCl$_x^{(x-2)-}$ anions has been proved, their stability is certainly low and it is difficult[40] to obtain an accurate value for x. The following order of basicity of the melts can thus be established:

$$NaOH–H_2O > LiCl–KCl > MSCN \gg PbCl_2–MCl \gg AlCl_3–NaCl$$

5.2. Stability of Polysulfides

The stability of polysulfide ions is summarized in Table 3. A strong correlation exists between Lewis basicity and the stabilization of polysulfides. This correlation is not new: a cursory inspection of molecular solvents shows that polysulfides and radical anions are stabilized in moderately or strongly basic media (e.g., DMSO, DMF, HMPA,* NH$_3$) and cationic species as S_x^{n+} in strongly acidic media (e.g., oleum, superacids).

5.3. Electrochemical Properties

The electrochemical oxidation of sulfide ion seems to correspond to the general scheme

$$S^{2-} - 2e^- \rightarrow S(0) \tag{3b}$$

$$S(0) + (x-1)S^{2-} \overset{k}{\rightarrow} S_x^{2-} \text{ (and/or } S_{x/2}^{\bullet-}) \tag{7}$$

---

* HMPA, hexamethyl phosphoramide in current usage in the U.K. and the U.S., is also known as HMPT, hexamethyl phosphoric triamide.

The reversibility of the electrochemical step (3b) depends on the nature and the temperature of the melt: the reaction is slow in chloroaluminates and has been found more reversible in LiCl–KCl or $PbCl_2$–KCl.

Cyclic voltammetry provides information about the rate of the chemical step (7): if the reaction is infinitely slow, i.e., if S_x^{2-} is unstable, a cathodic peak for the reduction of S(0) should be observed. Experimentally, in the less basic melt $AlCl_3$–NaCl where reaction (7) does not occur, this peak is easily observed. Basicity of melts and stability of S_x^{2-} increase from $PbCl_2$–MCl ($k = 8$ sec^{-1} with $M = Na^+$)[39] and MSCN where a cathodic stripping peak is still observed, through to $NaOH$–H_2O where no sulfur deposit could be observed.

In most case, the electrochemical reduction of sulfur has been studied only by indirect methods (cyclic voltammetry), because of the low solubility of sulfur (in the absence of sulfide).

ACKNOWLEDGMENT

We would like to express our thanks to "Société Nationale Elf Aquitaine" for supporting part of this work in $NaOH$–H_2O melt.

References

1. R. P. Tischer and F. A. Ludwig, in *Advances in Electrochemistry and Electrochemical Engineering*, Vol. 10, Eds. C. W. Tobias and H. Gerisher, Interscience, New York (1977).
2. C. H. Liu, A. J. Zielen, and D. M. Gruen, *J. Electrochem. Soc.* **120**, 67 (1973).
3. M. L. Saboungi, J. J. Marr, and M. Blander, *J. Electrochem. Soc.* **125**, 1567 (1978).
4. B. Cleaver, A. J. Davies, and D. J. Schiffrin, *Electrochim. Acta* **18**, 747 (1973).
5. M. J. Weaver and D. Inman, *Electrochim. Acta* **20**, 929 (1975).
6. F. G. Bodewig and J. A. Plambeck, *J. Electrochem. Soc.* **116**, 607 (1969).
7. J. H. Kennedy and F. Adamo, *J. Electrochem. Soc.* **119**, 1518 (1972).
8. J. R. Birk and R. K. Steunenberg, *Adv. Chem. Ser.* **140**, 18 (1978).
9. G. Delarue, *Bull. Soc. Chim. Fr.*, 906 (1960); *Chim. Anal.* **44**, 91 (1962).
10. J. Greenberg, B. R. Sundheim, and D. M. Gruen, *J. Chem. Phys.* **29**, 461 (1958).
11. F. G. Bodewig and J. A. Plambeck, *J. Electrochem. Soc.* **117**, 904 (1970).
12. W. Giggenbach, *Inorg. Chem.* **10**, 1308 (1971).
13. D. M. Gruen, R. L. McBeth, and A. J. Zielen, *J. Am. Chem. Soc.* **93**, 6691 (1971).
14. T. Chivers, in *Homoatomic Ring Chains and Macromolecules of Main Group Elements*, Ed. A. Rheingold, Elsevier, Amsterdam (1977), Chap. 22.
15. G. Torsi and G. Mamantov, *Inorg. Chem.* **10**, 1900 (1971).
16. B. Tremillon and J. P. Duchange, *J. Electroanal. Chem.* **44**, 395 (1973).
17. R. Marassi, G. Mamantov, and J. Q. Chambers, *J. Electrochem. Soc.* **123**, 1128 (1976).
18. R. Huglen, F. W. Poulsen, G. Mamantov, R. Marassi, and G. M. Begun, *Inorg. Nucl. Chem. Lett.* **14**, 167 (1978).
19. R. Marassi, G. Mamantov, M. Matsunaga, S. E. Springer, and J. P. Wiaux, *J. Electrochem. Soc.* **126**, 231 (1979).
20. K. A. Paulsen and R. A. Osteryoung, *J. Am. Chem. Soc.* **98**, 6866 (1976).

21. J. Robinson, B. Gilbert, and R. A. Osteryoung, *Inorg. Chem.* **16**, 3040 (1977).
22. R. Marassi, G. Mamantov, and J. Q. Chambers, *Inorg. Nucl. Chem. Lett.* **11**, 245 (1975).
23. R. Fehrmann, N. J. Bjerrum, and F. W. Poulsen, *Inorg. Chem.* **17**, 1195 (1978).
24. S. I. Rempel and E. M. Malkova, *Zh. Prikl. Khim.* **25**, 558 (1951).
25. O. P. Mohapatra, C. B. Alcock, and K. T. Jacob, *Met. Trans.* **4**, 1755 (1973).
26. D. H. Kerridge, in *The Chemistry of Non-Aqueous Solvents*, Vol. VB, Ed. J. J. Lagowski, Academic, New York (1978), pp. 269–329.
27. P. Cescon, F. Pucciarelli, V. Bartocci, and R. Marassi, *Talanta* **21**, 783 (1974).
28. C. P. Townsend, Br. Pat. 182478 (1906).
29. A. R. Gibson and S. Robson, Br. Pat. 448328 (1934).
30. J. B. Richardson, *Trans. Inst. Min. Met.* **46**, 339 (1936); *Bull. Inst. Mining Met.* **387**, 1 (1936); **389**, 35 (1937); **390**, 27 (1937); **395**, 7 (1937).
31. A. P. Newall, *Trans. Inst. Chem. Eng.* **16**, 77 (1938).
32. G. Angel and E. Garnum, *Tek. Tid.* **75**, 279 (1945).
33. H. Sawamoto and T. Saito, *J. Mining Inst. Japan* **68**, 555 (1952).
34. H. Winterhager and R. Kammel, *Z. Erzbergbau Metallhuetenwes.* **9**, 97 (1956).
35. T. Yanagase, Y. Suginohara, and I. Kyono, *Denki Kagaku* **36**, 129 (1968).
36. P. L. King and B. J. Welch, *J. Appl. Electrochem.* **2**, 23 (1972).
37. B. J. Welch, P. L. King, and R. A. Jenkins, *Scand. J. Metall.* **1**, 49 (1972).
38. P. L. King and B. J. Welch, *Proc. Australas. Inst. Min. Metall.* **246**, 7 (1973).
39. M. Skyllas and B. J. Welch, *Electrochim. Acta* **23**, 1157 (1978).
40. A. de Guibert and V. Plichon, *J. Electroanal. Chem.* **90**, 399 (1978).
41. A. de Guibert, V. Plichon, and J. Badoz-Lambling, *J. Electrochem. Soc.* **126**, 1902 (1979).
42. D. Lelievre, A. de Guibert, and V. Plichon, *Electrochim. Acta* **24**, 1243 (1979).
43. A. de Guibert, V. Plichon, and J. Badoz-Lambling, *J. Electroanal. Chem.* **105**, 143 (1979).
44. B. Tremillon and R. Doisneau, *J. Chim. Phys.* **71**, 1445 (1974); R. Doisneau, Thèse Paris, No. CNRS A.O. 9017 (1973).
45. G. Valensi, Centre Belge d'Etudes de Corrosion, Rapport Technique 207-1, 207-22 (March 1973).
46. W. R. Fetzer, *J. Phys. Chem.*, 1787 (1928).
47. H. Gerischer, *Ber. Bunsenges Phys. Chem.* **54**, 540 (1950).
48. L. Gustafsson and A. Teder, *Svensk Papperstidn.* **72**, 249 (1969).
49. P. L. Allen and A. Hickling, *Chem. Ind. (London)*, 1558 (1954); *Trans. Faraday Soc.* **63**, 1626 (1957).
50. G. Schwarzenbach and A. Fischer, *Helv. Chim. Acta* **43**, 1365 (1960).
51. A. Teder, *Svensk Papperstidn.* **70**, 197 (1967).
52. W. Giggenbach, *Inorg. Chem.* **11**, 1201 (1972).
53. W. Giggenbach, *Inorg. Chem.* **10**, 1333 (1971).
54. L. G. Sillén, *Stability Constants of Metal Ion Complexes*, Vol. 17, The Chemical Society Burlington House, London (1964); Vol. 25 (1971); *I.U.P.A.C., Dissociation Constants of Inorganic Acids and Bases in Aqueous Solution*, Butterworths, London (1969); S. Ramachandra Rao and L. G. Hepler, *Hydrometallurgy* **2**, 293 (1977).
55. M. Z. Ugorets, T. Rustembekov, K. M. Akmetov, and E. A. Buketov, *Tr. Khim. Met. Inst. Akad. Nauk. Kaz. S.S.R.* **17**, 77 (1972).
56. E. R. Bertozzi, U.S. Pat. 2,796,325 (June 18, 1957).
57. M. B. Berenbaum, in *Encyclopedia of Chemical Technology*, Eds. R. E. Kirk and D. F. Othmer, Vol. 16, Wiley, New York, p. 255.
58. R. Cohen-Adad, *Rev. Chim. Min.* **1**, 451 (1964).
59. H. Gmelin, *Handbuch der Anorg. Chem.*, Vol. 21–2 (1965).
60. *Landolt-Börnstein Tabällen*, Springer-Verlag (1969).
61. M. N. Moscardo-Levelut and V. Plichon, to be published.
62. D. Bauer, C. Colin, and M. Caude, *Bull. Soc. Chim. Fr.*, 943 (1973).

Chemical and Electrochemical Studies in Room Temperature Aluminum-Halide-Containing Melts

Helena Li Chum and R. A. Osteryoung

1. Introduction

Molten salts have been used in industrial processes, such as for the production of aluminum, for many years. Applications in areas of molten salt battery systems, molten salt nuclear reactors, extraction from metal ores, electrorefining, etc., have prompted researchers to study the fundamental properties of molten salt systems for the past 30 years.[1] Recently, owing partly to the relatively low working temperatures (150–200°C), the molten aluminum halides have been widely studied, e.g., AlX_3/MX, where M^+ is an alkali metal cation.[1] Very recently, work on "room temperature" AlX_3-based systems has been initiated.

The primary scope of this review is the room temperature molten salt systems based on aluminum halides; however, it is pertinent to review some of the properties of the higher-temperature aluminum halide–alkali metal halide melts as background for the discussion of the low-temperature systems.

Several aspects of the chemistry of chloroaluminate melts have been reviewed. Boston[2] reviewed the molten salt chemistry of haloaluminates up to 1971 covering physical properties, solvent properties, and some aspects of electrochemical and spectroscopic studies. Fung and Mamantov[3]

_____
Helena Li Chum and R. A. Osteryoung • Department of Chemistry, State University of New York, Buffalo, New York. Dr. Chum's present address: Solar Energy Research Institute, Golden, Colorado 80401.

detailed the electroanalytical chemistry of molten salts, particularly of molten haloaluminates, emphasizing the electrochemical behavior of the inorganic solutes that had been studied with emphasis on transition metal ions. Jones and Osteryoung,[4] in 1975, reviewed organic reactions in molten tetrachloroaluminate solvents of the following types: condensation–addition, dehydrogenation–addition (Scholl reactions), and rearrangement–isomerizations. Plambeck[5] summarized the various inorganic systems studied in these melts, hydrogen, elements of the Groups IIIA, IVA, VA, VIA, VIIA, as well as transition metal elements, including actinides. A summary of electromotive force series in the 50 : 50 mol % $AlCl_3$: $NaCl$ at 175°C is presented. Mamantov and Osteryoung[1] described the acid–base-dependent redox chemistry in molten chloroaluminates, basically from the point of view of the authors' own research.

With this brief introduction it is now appropriate to consider aspects of the chemistry which makes the chloroaluminate systems unique.

2. Aluminum Halide–Alkali Metal Halide Molten Salt Systems

2.1. Acid–Base Chemistry

The aluminum halide–alkali metal halide melts display wide acidity changes with varying melt compositions. The acid–base properties can determine the redox and coordination chemistry in these media, and several research groups have investigated this particular aspect over the past ten years.[6–10] The acid–base chemistry of the $AlCl_3$/$NaCl$ molten salt system, in the range 49–70 mol % of $AlCl_3$, can be represented[9,10] by the following three equilibria (175°C):

$$2AlCl_3(l) \rightleftarrows Al_2Cl_6(l) \qquad (K_0 = 2.86 \times 10^7) \qquad (1)$$

$$AlCl_4^- + AlCl_3 \rightleftarrows Al_2Cl_7^- \qquad (K_2 = 2.4 \times 10^4) \qquad (2)$$

$$2AlCl_4^- \rightleftarrows Al_2Cl_7^- + Cl^- \qquad (K_m = 1.06 \times 10^{-7}) \qquad (3)$$

where the K values are on the mole fraction scale. The dominant acid–base equilibrium is described by equation (3), where $Al_2Cl_7^-$ is the Lewis acid and the Cl^- is the base. This equilibrium is an autosolvolysis reaction, analogous to water autoprotolysis. The pCl^- ($-\log[Cl^-]$), on the mole fraction scale, of a neutral melt (1 : 1 $AlCl_3$: $NaCl$) is 3.5 at 175°C. At this composition, the system can be considered essentially as $Na^+AlCl_4^-$. This melt can be made more acid (excess $AlCl_3$) by anodizing an Al wire[10] or by addition of $AlCl_3$. The melt can be made more basic (excess $NaCl$) by cathodizing an Al wire or by addition of $NaCl$. An Al electrode is employed to monitor the melt's acidity.[9] At 175°C, the pCl of the most basic melt saturated in NaCl is 1.1. The major acidity (pCl) changes occur in the vicinity of the 1 : 1 $AlCl_3$: NaCl mole ratio.

2.2. Species Characterization

The nature of the species present in these equilibria has been studied by Raman spectroscopy.[11-15] Mamantov *et al.*[11] have reported Raman spectra of sodium chloroaluminate melts and have reported evidence for species $AlCl_4^-$, $Al_2Cl_7^-$, Al_2Cl_6, and possibly $Al_3Cl_{10}^-$.[11] For the $AlCl_4^-$ species, the major Raman vibrational frequencies and their respective polarizations were 351 (*P*), 121 (*D*), 490 (*D*), and 186 (*D*) cm^{-1}, in order of decreasing intensity. For pure molten Al_2Cl_6, the major Raman-active frequencies were 341 (*P*), 218 (*P*), 119 (*D*), and 104 (*P*) cm^{-1}. By assuming that in the range from 50/50 to 58/42 mol % $AlCl_3$/NaCl, the only Raman-active species are $AlCl_4^-$ and $Al_2Cl_7^-$, the following vibrational frequencies were assigned to $Al_2Cl_7^-$: 313 (*P*), 432 (*P*), ~ 100 (*D*), and ~ 165 (*D*) cm^{-1}. Owing to overlap of broadbands of $AlCl_4^-$ at 121 and 186 cm^{-1}, the frequencies at 100 and 165 cm^{-1} were only estimated. Since the spectra in the high-acidity range (> 66 mol % $AlCl_3$) displayed other bands which could not be explained solely on the basis of $AlCl_4^-$, $Al_2Cl_7^-$, and Al_2Cl_6, species such as $Al_3Cl_{10}^-$ were postulated.[11] Raman spectra of the corresponding bromide and iodide melts have also been reported.[13]

Raman spectroscopic data for the potassium chloroaluminate melts have been reported by Oeye and co-workers.[12, 14] Spectral data for $AlCl_4^-$ agree well with those found in the sodium chloroaluminate melt.[11] The four absorption bands of $AlCl_4^-$ verify T_d symmetry. Better low-frequency spectroscopic data for $Al_2Cl_7^-$ were also obtained[12, 14]: 99 (*D*), 164 (*D*), 312 (*P*), and 435 (*D*) cm^{-1}. An earlier report[15] of Raman spectra of the 50/50 mol % $AlCl_3$/KCl indicated a much larger number of bands (9) than observed for $AlCl_3$/NaCl (4). This melt was related to a distorted $AlCl_4^-$ (C_{2v}).[15] According to Maroni and Cairns,[16] a number of the reported bands are weak and poorly defined and may be spectral artifacts.

A normal coordinate analysis for $Al_2Cl_7^-$ assuming a D_{3d} model with a linear Al–Cl–Al bridge does not reproduce the experimental frequencies.[12] Structural determinations[17] by X-ray crystallography on the analogous $Al_2Br_7^-$ ion either as a potassium or ammonium salt show that the anion consists of two $AlBr_4^-$ tetrahedra sharing one corner in a staggered arrangement with a bent Al–Br–Al bridge (109.3° and 107.7°, respectively, for the K^+ and NH_4^+ salts). The symmetry of the $Al_2Br_7^-$ anion is close to C_{2v}. Structural determinations by the nuclear quadrupole resonance (NQR) double-resonance technique confirmed the structure of $Al_2Br_7^-$ and led to a reasonable qualitative bond model for this ion.[18] The structure of the compound $Te_4(Al_2Cl_7)_2$[19] was determined and the $Al_2Cl_7^-$ anion also consists of two tetrahedra sharing one corner in staggered arrangement with a bent Al–Cl–Al bridge (110.8°). Normal coordinate calculations assuming similar symmetry for the $Al_2Cl_7^-$ anion might yield better agreement between calculated and experimental frequencies, but is clearly more difficult to do.

2.3. Electrochemistry of the Chalcogenide Ions and Chalcogens

Mamantov et al.[20, 21] and Paulsen and Osteryoung[22] have studied the electrochemistry of sulfur and sulfide. Sulfur has been quanitatively reduced in both acid and basic melts at 175°C. Nernst plots for the sulfur/sulfide redox couple have a slope corresponding to a four-electron process, described as

$$S_2 + 4e^- \rightleftarrows 2S^{2-}$$

The electrochemistry of oxide and chalcogenide ions in these molten salt systems was more quantitatively investigated by Osteryoung and co-workers.[23–26] By titrating O^{2-}, S^{2-}, Se^{2-}, or Te^{2-} species in the melt with electrogenerated $Al_2Cl_7^-$ ions, it was shown that in the basic-to-neutral melts these ions behave as dibases as described by the equilibrium

$$X^{2-} + 2Al_2Cl_7^- \rightleftarrows AlXCl_2^- + 3AlCl_4^- \tag{4}$$

where X^{2-} may be O^{2-}, S^{2-}, Se^{2-}, or Te^{2-}, whereas in neutral-to-acidic melts they became tribases, the equilibrium then becoming

$$X^{2-} + 3Al_2Cl_7^- \rightleftarrows AlXCl + 5AlCl_4^- \tag{5}$$

The base strength of the ions in the acid melt was found to be in the order $O^{2-} > S^{2-} > Se^{2-} \approx Te^{2-}$.[24]

Oxide does not appear to be electroactive[23] in this molten salt system (cf. Reference 1) but the electrochemistry of sulfur,[22] selenium,[26] tellurium,[25] and their compounds has been thoroughly investigated using a variety of electrochemical methods. Sulfur was shown to oxidize very close to the anodic limit of the melt and it is therefore uncertain what the nature of the oxidation product is though in basic melts it appears to be S_2^{2+}, whereas in acidic melts some higher oxidation state species appears likely. Spectrophotometric, potentiometric, and Raman data[27] indicate that the oxidation product is a S(IV) species. Addition of $SCl_3^+ AlCl_4^-$ (Reference 28) to the melts reproduces the electrochemical and spectral behavior[1, 27] observed in the electrochemical oxidation of sulfur.

The electrochemical behavior of selenium[26] is similar; it is reduced to selenide, more reversibly than sulfur is to sulfide, and can be oxidized to Se(IV). In basic melts this oxidation goes via a stable Se(II) intermediate, whereas in acidic melts the oxidation is directly to Se(IV). The behavior of tellurium[25] is rather different as the metal is insoluble in the melt. In the most basic melts at temperatures above 250°C, Te(IV) is initially reduced to a soluble Te(II) species, and then to the metal. In more acidic melts and at lower temperatures the Te(IV) is reduced directly to Te. Te metal rapidly reacts with Te(IV) to form Te_4^{2+}.

The nature of the chalcogen species involved in chloroaluminate melts, as determined by an interesting combination of potentiometric and spectrophotometric studies by Bjerrum and co-workers has been reviewed recently.[27] For instance, entities such as Te_2^{2+}, Te_4^{2+}, Te_6^{2+}, and Te_8^{2+}, stabilized by the large anions, $AlCl_4^-$ and $Al_2Cl_7^-$, are present in the melt as well as Te^{2+}, which is stabilized by Cl^-.[27, 29, 30] Bjerrum and co-workers[31, 32] have also studied the chloro- complexes of the Te^{4+} cation in $AlCl_3/KCl$ melts and found the complexes $TeCl_6^{2-}$, $TeCl_5^{2-}$, $TeCl_4$, and $TeCl_3^+$, which are linked by melt acidity-dependent equilibria. $Te_4(AlCl_4)_2$ and $Te_4(Al_2Cl_7)_2$, where formally the tellurium is in a $\frac{1}{2}$ oxidation state, have been isolated and characterized.[19, 33] This agrees with earlier data by Bjerrum and Smith.[34]

Similarly, Bjerrum and co-workers[27] have presented evidence for ions: Se_4^{2+}, Se_8^{2+}, Se_2^{2+}, Se_{16}^{2+} as the most likely other low oxidation states formed by reduction of Se(IV) with elemental selenium. The Se(IV) species is $SeCl_3^+$ in most of the basic and most of the acidic range as deduced from spectroscopic and potentiometric data. For sulfur[35] the most consistent interpretation of potentiometric and spectrophotometric data was S_2^{2+}, S_4^{2+}, S_8^{2+}, S_{16}^{2+}, together with S(II) and S(IV) most likely as SCl_3^+.

2.4. Stability of Radical Ions

From a chemical point of view, another interesting finding is the stability of radical ions in these melts. Several aromatic amine radical cations were shown to be remarkably stable in these melts[36] at 175°C, in comparison to the poor stability found in aprotic organic solvents.[4] It has been suggested that this stability is a result of the "totally" andydrous nature of these melts, whereas most organic solvents contain varying residual amounts of water depending on the purification procedure. Triphenylamine,[36] N,N-dimethylaniline,[36, 37] N,N,N',N'-tetramethylbenzidine,[38] quinone-hydroquinone systems,[39] sulfur heterocycles,[40] were studied and showed the interesting interplay of acid–base chemistry and redox processes with radical cation stabilization.

3. Room Temperature Molten Salt Systems—Introductory Remarks

All the electrochemical and chemical reactions mentioned so far involve temperatures greater than 150°C. It is very clear (1) that the operating temperatures were above the boiling point of many organic solutes whose electrochemistry could be of interest and (2) that rapid homogeneous reactions could take place between melts and organic compounds.[4, 41] For most other applications of the haloaluminates, a melt system with similar solvent properties operating at lower temperatures, e.g., room temperature, would be advantageous.

In 1951 a melt composed of 67 mol % aluminum chloride and 33 mol % of the quaternary organic salt ethylpyridinium bromide (a 2 : 1 molar ratio), which is molten at room temperature, was reported by Hurley and Wier.[42] Osteryoung and co-workers utilized this solvent for electrochemical and photochemical investigations.[43–47] The electrochemical oxidation of iron diimine complexes was shown to be a reversible process. Upon irradiation with low-energy photons, the iron(II) diimine complexes were shown to be converted to the corresponding stable iron(III) compounds, with electron transfer to the acceptor ethylpyridinium cations.[46]

An interesting feature of the 2 : 1 $AlCl_3$: ethylpyridinium bromide melt is its miscibility with cosolvents, e.g., benzene,[42] toluene, etc. Hurley and Wier[42] had observed that this mixed solvent system was better than the pure melt for electroplating of aluminum. Upon dilution with benzene, the conductivity of the solutions increases compared to the pure melt, while the viscosity decreases. This solvent system, melt plus benzene, proved very suitable to the study of organic[43, 44] and organometallic[45] compounds. An advantage of the mixed solvent is that a number of organic and organometallic compounds, insoluble in the pure melt, are soluble in the melt plus benzene.[45] In the 50% (v/v) melt and benzene solution hexamethylbenzene (HMB) can be anodically oxidized to pentamethylbenzene and diphenylmethane.[44] This was the first reasonable synthesis in a molten salt system, where HMB can be selectively and quantitatively demethylated to mesitylene and diphenylmethane in a Friedel–Crafts-type reaction.[44] The same solvent system has been employed to study the electrochemical oxidation of six metal carbonyls.[45] Chromium hexacarbonyl and iron pentacarbonyl were found to be reversibly oxidized to the corresponding 17 electron cations.[45] In acetonitrile, only the former compound was found to undergo reversible one-electron oxidation.[48] However, this melt is not very suitable for investigating acid–base-dependent reactions, since decreasing the acidity by increasing the ethylpyridinium bromide content raises the melting point sharply at the 50 : 50 mol % composition.

Similar systems based on toluene plus $AlBr_3$ and KBr in a 2 : 1 mole ratio have been employed by Gileadi and co-workers[49–53] primarily with interest in electroplating of aluminum.

4. Aluminum Chloride–N-(n-butylpyridinium) Chloride (BuPyCl) Molten Salt Systems

Recently, the $AlCl_3$–N-(n-butylpyridinium) chloride system (1 : 1 mole ratio) was found to melt at 27°C, and the 2 : 1 mole ratio exists as a liquid at temperatures well below ambient. In fact, a mole ratio range of 0.75 : 1 to 2 : 1 can be employed at temperatures only slightly greater than 27°C, very close to room temperature![47, 54]

4.1. Species Characterization[54]

Raman spectroscopy has been used to study the species composition of the molten salt system $AlCl_3$: BuPyCl over the 0.75 : 1 to 2.1 : 1 mole ratio range[54] at 30°C. Table 1 summarizes the experimental frequency shifts (cm^{-1}) for the vibrational peaks. To characterize these spectra, the peak positions were compared to the assignments for $AlCl_4^-$ and $Al_2Cl_7^-$ (see above) in the $AlCl_3$/MCl (M = Na^+, K^+)[11, 12, 14] melts, and the spectra of Al_2Cl_6 (Reference 11) (see above) and n-butylpyridinium cation (major frequencies in order of decreasing intensity in cm^{-1}: 1024, 125, 88–103, 644, 774). There is a straightforward correlation between the frequencies 351 (P), 126 (D), 484 (D), and 184 (D) cm^{-1} found in the low-$AlCl_3$-content melts (0.75 : 1–1.0 : 1.0) and the frequencies assigned to $AlCl_4^-$ anion.[11, 12, 14] As the $AlCl_3$ content increases, new vibrational peaks at 102 (shoulder) (D), 163 (D), 315 (P), and 434 (P) cm^{-1} become dominant, and the bands assigned to $AlCl_4^-$ do not contribute to the observed

TABLE 1. Summary of Experimental Raman Frequency Shifts (cm^{-1}) for Aluminum Chloride: n-Butylpyridinium Chloride Mixtures at Room Temperatures from Reference 54[a]

Molar ratio [AlCl₃] : [RCl]				
0.75 : 1.0	1.0 : 1.0	1.5 : 1.0	1.75 : 1.0	2.0 : 1.0
			−95 sh	−102 sh
−126 sh (0.9)	−127 sh (1.3)	−126 sh		
		−159 (1.2)	−163 (1.5)	−163 (2.3)
−186 (2.1)	−184 (2.0)	−183 (2.8)	−183 (1.6)	−182 (0.9)
−298 (0.9)	−295 (0.7)	−296 sh (1.4)	−295 sh (1.6)	−295 sh (1.1)
		−314 (8.8)	−316 (10.0)	−315 (10.0)
−351 (10.0)	−352 (10.0)	−351 (10.0)	−352 (5.0)	
−433 (0.1)	−433 (0.1)	−434 (2.7)	−433 (3.3)	−434 (2.2)
−485 (0.5)	−483 (0.4)			
−653 (2.8)	−650 (2.4)	−651 (3.2)	−647 (4.7)	−651 (2.7)
−775 (0.7)	−770 (0.5)	−774 (0.2)	−772 (0.2)	−769 (1.0)

[a] sh = shoulder.

spectra within the experimental error. These frequencies compare very well with the previously assigned frequencies for the $Al_2Cl_7^-$ anion, observed by Mamantov et al.[11] and Oeye et al.[12, 14] (vide supra). Owing to the overlap of the strong peak at 315 cm^{-1} in the high aluminum chloride content region or the strong peak at 351 cm^{-1} in the basic side, and the major Al_2Cl_6 Raman-active frequency at 341 cm^{-1}, Al_2Cl_6 cannot be detected if present in small concentrations.

The molten salt system $AlCl_3$: BuPyCl can be described as follows as a function of composition.

2 : 1 Mole Ratio of AlCl$_3$: BuPyCl. At this mole ratio, the molten salt is best described as *n*-butylpyridinium dialuminum heptachloride, $BuPy^+Al_2Cl_7^-$. There is no evidence from the Raman spectra for the presence of $AlCl_4^-$ anions, contrary to what was observed in the $AlCl_3$–alkali metal halides.[11, 12, 14] These facts suggest that at this composition the association

$$2AlCl_4^- + Al_2Cl_6 \rightleftarrows 2Al_2Cl_7^- \tag{6}$$

must be virtually complete.

1 : 1 Mole Ratio of AlCl$_3$: BuPyCl. The major aluminum-containing component anion in this composition is the $AlCl_4^-$ anion. A small amount of the $Al_2Cl_7^-$ anion may be present at this composition (cf. band intensities at 433 cm^{-1}). Therefore, at this composition the melt contains $AlCl_4^-$, $Al_2Cl_7^-$, Cl^-, and possibly traces of Al_2Cl_6. Thus, the equilibrium

$$2AlCl_4^- \rightleftarrows Al_2Cl_7^- + Cl^- \tag{7}$$

identical to the dominant equilibrium in the $AlCl_3$/NaCl systems [equation (3)], may be considered to exist in the room temperature molten salt.

At intermediate compositions it is possible to estimate the ratio of the mole fractions of $Al_2Cl_7^-$ and $AlCl_4^-$ species by assuming that these mole fractions are proportional to the integrated intensity of the strongest bands for each species.[12, 54] The ratios of 1.2 : 1 and 3.3 : 1 were obtained, respectively, for the 1.5 : 1.0 and 1.75 : 1.0 molar ratios of $AlCl_3$: BuPyCl. Therefore, as the melt becomes $AlCl_3$ rich, the molar excess of $AlCl_3$ is converted into $Al_2Cl_7^-$. At the limit, the 2 : 1 mole ratio $AlCl_3$: BuPyCl, this species is the major and possibly the only aluminum-containing species in the melt.

The large organic pyridinium cation does not influence the absorption frequencies of the anionic species, $AlCl_4^-$ or $Al_2Cl_7^-$. This finding agrees well with previous data by Rytter and Oeye[58] on the influence of the alkali metal cation size on the absorption frequencies of $AlCl_4^-$ anion. The smaller cation, Li^+, affects strongly asymmetric vibrations, whereas practically identical frequencies are obtained for Na^+ and K^+ within experimental error.

The major effect of the organic cation is to promote more extensive $Al_2Cl_7^-$ formation at the $2:1$ mole ratio than is found in the higher-temperature haloaluminate–alkali metal halide molten salt systems.

4.2. Acid–Base Chemistry[55]

In order to quantify[55] the equilibrium described by equation (7) in the basic region, potentiometric titrations were carried out, by using an aluminum reference electrode, which behaves as a reversible chloride activity indicator electrode. From the Nernst equation

$$E = E_0 + (RT/3F)\ln(a_{Al^{3+}}/a_{Al(0)}) \tag{8}$$

incorporating the association constants for reactions (7) and formation of $AlCl_4^-$:

$$Al^{3+} + 4Cl^- \rightleftarrows AlCl_4^- \tag{9}$$

one obtains the following relation:

$$E = (RT/3F)\ln(a^{\circ}_{AlCl_4^-}/a^{i}_{AlCl_4^-}) + (4RT/3F)\ln(a^{i}_{Cl^-}/a^{\circ}_{Cl^-}) \tag{10}$$

where superscripts $\circ$ and i refer to reference and variable states, respectively.

The description of the electrochemical cell can be given as follows:

$$Al \mid BuPyCl, AlCl_3(ref) \mid fritted\ disk \mid AlCl_3(c), BuPyCl \mid Al$$

where BuPyCl is *n*-butylpyridinium chloride. Figure 1 shows some results of potentiometric titrations in the $2.0:1.0$ to $0.6:1.0$ molar ratio of $AlCl_3 : BuPyCl$ at 30, 60, and 120°C. The sigmoidal curves in Figure 1 differ considerably from those obtained for the $AlCl_3/NaCl$ system.[9, 10] First, the potential difference between the two plateau regions is 2–3 times larger than that observed in the high-temperature melts, indicating a greater pCl change in the low-temperature melt. Second, a change in the slope of the curve in the acid region is observed in the low temperature but not in the $AlCl_3/NaCl$ system.[9] Third, in the basic portion of the curve, potentiometric data are less reproducible than those in the acid region, owing to the appearance of a blue-colored species at the Al electrode, indicating that secondary reactions are taking place.

From the potentiometric data it is possible to assess an approximate conditional equilibrium constant for equation (7) by using equation (10):

$$K = \frac{[Al_2Cl_7^-][Cl^-]}{[AlCl_4^-]^2} = 3.8 \times 10^{-13} \quad (30°C)$$

The solid line in Figure 1 is a potential function represented by equation (10) with this equilibrium constant. The model is inappropriate in the acid region above $2:1$ ratio.

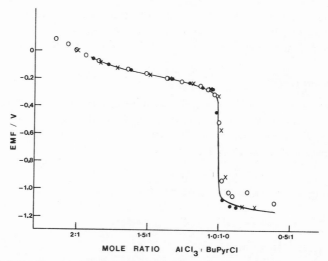

FIGURE 1. Potentiometric data for $AlCl_3$: 1-butylpyridinium chloride versus $Al(0)/2$: 1 mole ratio mixture reference; ×, 30°C; ○, 60°C; ●, 120°C; —, theory (30°C). After Gale and Osteryoung.[55]

In the basic region mixed potentials are measured owing to the spontaneous reduction of $BuPy^+$ cations by elemental aluminum. Therefore, the constant K, determined by calculations over the entire acid and basic range, is a lower limit because of the positive shift in the measured potential due to the $BuPy^+$ reduction.

A plot of the species distribution in this system from the 0.6 : 1.0 to the 1.8 : 1.0 mole ratio of $AlCl_3$: BuPyCl is shown in Figure 2. It is interesting to note that the chloride ion mole fraction almost spans the range found in the proton/hydroxyl equilibrium in water.

The potentiometric curves at higher temperatures up to 175°C can also be characterized solely on the basis of equation (7). At 175°C, $K < 1.19 \times 10^{-8}$, similar, but less than 1.06×10^{-7},[9] the constant for this equilibrium in the $AlCl_3$/NaCl system. These findings are consistent with the data of Torsi and Mamantov[59] for the $AlCl_3$/alkali metal halide systems which indicate that the greater the polarizing power of the cation, the larger the autosolvolysis constant at a given temperature.

When comparing spectroscopic and potentiometric data for the $AlCl_3$/alkali metal halide and $AlCl_3$/BuPyCl systems, one can observe that whereas in the former Al_2Cl_6 is a species present throughout the studied composition and temperature ranges, being particularly important on the acid side, this Lewis acid does not seem to be a component of the room

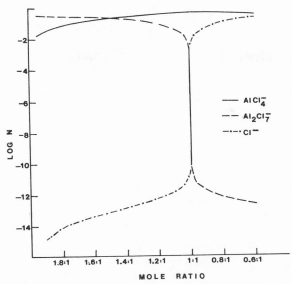

FIGURE 2. Mole fraction, N, of the major species in the melt at 30°C as a function of the net AlCl$_3$: RCl mole ratio. After Gale and Osteryoung.[55]

temperature molten salt. Even at temperatures as high as 175°C, little aluminum chloride vapor is detected over the 2 : 1 AlCl$_3$: BuPyCl melt. For the AlCl$_3$: BuPyCl system in the molar ratios of 0.75 : 1 to 2 : 1, the acid–base chemistry is described by equation (7), whereas for the AlCl$_3$/alkali metal halides two additional equilibria involving AlCl$_3$ and Al$_2$Cl$_6$ [equations (1) and (2)] must be invoked.

4.3. Electrochemical and Spectroscopic Studies of Aromatic Hydrocarbons[56]

The electrochemical oxidation of 12 aromatic hydrocarbons: mesitylene, biphenyl, naphthalene, durene, phenanthrene, pentamethylbenzene, hexamethylbenzene, pyrene, anthracene, 9,10-diphenylanthracene, benzo(a)pyrene, and benzo(e)pyrene, was studied in the AlCl$_3$: BuPyCl systems in the 0.75 : 1 to 2 : 1 molar ratio, at 40°C.[56] All the hydrocarbons studied were shown to undergo a one-electron oxidation to the radical cation at a potential that is independent of the melt composition. The radical cations in these molten salt systems were found to be more stable than in acetonitrile.[60–63]

Figure 3 illustrates a linear correlation between the half-peak potential (proportional to the half-wave potential) and the first ionization potential

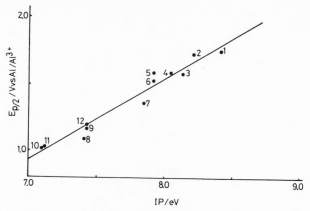

FIGURE 3. The correlation between $E_{p/2}$ values for the oxidation of aromatic hydrocarbons in the AlCl$_3$: n-butylpyridinium chloride molten salt and their photoionization potentials. 1, mesitylene; 2, biphenyl; 3, napthalene; 4, durene; 5, phenanthrene; 6, pentamethylbenzene; 7, hexamethylbenzene; 8, pyrene; 9, anthracene; 10, 9,10-diphenylanthracene; 11, benzo-(a)pyrene; and 12, benzo(e)pyrene. After Robinson and Osteryoung.[56]

(I.P.) of the parent hydrocarbon determined from photoelectron spectroscopy.[64–65] The equation

$$E_{p/2} = 0.61 \text{ I.P.} - 3.34 \qquad (11)$$

was found to fit the experimental data with a correlation coefficient (CC) of 0.97. An analogous correlation with similar slope and intercept was obtained using the data from acetonitrile experiments[60–63] (CC = 0.96). These findings suggest that there is no strong interaction between the acid species in the melt and the hydrocarbons.

After referring all $E_{p/2}$ potentials to that of the ferrocene/ferricinium ion couple in the melt and in acetonitrile, it has been observed that the oxidation potentials in the melt are about 0.1 V positive to those in aceto-nitrile, implying that it is slightly more difficult to oxidize these compounds in the melt than in acetonitrile.

Though the electrochemical data and correlations with I.P. seem to suggest that the hydrocarbons do not interact with acid species in the melt, it was observed that some of the hydrocarbon solutions change color in acid melts suggesting that the hydrocarbons interacts with the solvent. When these solutions are made basic by addition of BuPyCl, they can be reversibly decolorized.

There are several reports in the literature of the existence of solid state σ complexes between AlCl$_3$ and hydrocarbons.[66–69] Table 2 compares the uv–visible spectra of solutions of some of the hydrocarbons studied in the

TABLE 2. Spectroscopic Data for Aromatic Hydrocarbons in a 1.2 : 1 Aluminum Chloride: *n*-Butylpyridinium Chloride Molten Salt[56] and for Solid State Complexes with $AlCl_3$[66, 67, 69]

Aromatic hydrocarbon	λ_{max}/nm melt[56]	λ_{max}/nm solid state complex
Mesitylene	367	365[67]
Durene	382	380[67]
Pentamethylbenzene	380	385[67]
Hexamethylbenzene	398	400[67]
Naphthalene	393	410,[67] 377[69]
Phenanthrene	417	425,[67] 653[69]
Pyrene	451, 465, 475, 492	520,[67] 480[69]
Anthracene	417	418,[69] 560,[67] 420[66]

1.2 : 1.0 $AlCl_3$: BuPyCl melt and the solid state σ complexes.[66-69] It has been found that in melts more basic than 1.0 : 1.0, the spectra obtained correspond to those of the parent hydrocarbons. Though extinction coefficients were not reported, it has been observed that as the acidity increases, the intensity of the colored species increased. This behavior indicates that an acid–base type of equilibrium between acid species in the melt, $Al_2Cl_7^-$, and the hydrocarbons is occurring. It has been suggested than an equilibrium of the type

similar to that involved in the formation of carbonium ions[69] may exist, as in the example shown with anthracene. From the absence of any acid–base effects in the electrochemical oxidations of these hydrocarbons in the $AlCl_3$/BuPyCl melts it was concluded that either the equilibrium of equation (12) is very fast and the hydrocarbon is more readily oxidized than the σ complex, or alternatively the σ complex and the hydrocarbon oxidize at the same potential.

An interesting but so far puzzling phenomenon was observed in acid melts for hydrocarbons with a $E_{p/2}$ of less than 1.4 V at a glassy carbon electrode versus Al reference electrode. In the more acid melts, these hydrocarbons [hexamethylbenzene, pyrene, anthracene, 9,10-diphenylanthracene, benzo(*a*)pyrene and benzo(*e*)pyrene] are slowly but spontaneously oxidized to the cation radicals by the molten salt itself. The nature of the oxidizing agent or its reduction product is unknown. The spontaneous oxidation ceases when the acidity of the melt is lowered.

4.4. Electrochemical Study of the Fe(III)/Fe(II) System

Hussey et al.[57] have studied the Fe(III)/Fe(II) electrode reaction in the 2 : 1 $AlCl_3$: BuPyCl melt at 30°C and in the 1 : 1 $AlCl_3$: BuPyCl melt at 35°C, employing a rotating disk electrode, cyclic voltammetry, and potentiometry.

At a glassy carbon or tungsten working electrode in the acid melt the reduction $Fe(III) + e^- \rightarrow Fe(II)$ was found to be moderately rapid, with a standard heterogeneous electron transfer rate constant of 1.3×10^{-4} cm sec^{-1} and $\alpha = 0.4$. The heat of activation for this process is ~ 60 kJ mol^{-1}. From Nernst plots the standard electrode potential ($E°$) for the Fe(III)/Fe(II) couple at 30°C is 2.01 V, with a $dE°/dt = 9.5 \times 10^{-4}$ mV/°C. From rotating disk data, the diffusion coefficient of the electroactive species was found to be 4.4×10^{-7} cm^2 sec^{-1}, with an activation energy for diffusion of 21 kJ mol^{-1}.

The electrochemical reduction of Fe(III) in the basic melt is a one-electron reversible process. The standard electrode potential is 0.125 V (35°C) with $dE°/dt = 4.7 \times 10^{-4}$ mV/°C. The diffusion coefficient of the electroactive species in this medium is 4.6×10^{-7} cm^2 sec^{-1} (35°C).

The marked increase in the rate of the Fe(III)/Fe(II) electrode reaction in going from the 2 : 1 to the 1 : 1 $AlCl_3$: BuPyCl melts, and also the almost 2-V difference in standard potentials coincides with the marked difference in anionic species observed in these melts, reported above. That the Fe(III) species present in these two media are different solvates and/or complexes is to be expected, and is further supported by the different colors exhibited by the ion in these media: orange (acid) to canary yellow (basic). A more complete conclusion about the nature of the species present in these media must await further investigations. The electrochemical behavior of Fe(III) ions illustrates the influence of the melt composition, therefore acidity, on the redox and coordination chemistry of this transition metal ion in these media.

TABLE 3. Physical Properties of Some Alkyl Pyridinium Halide–Aluminum Halide Melts from Reference 47

Melt	Temp/°C	Density/g cm^{-3}	Viscosity/cP	Electrochemical window/V[a]
2 : 1 $AlCl_3$: BuPyCl	25	1.33	27.0	0.0 to 2.1
1 : 1 $AlCl_3$: BuPyCl	40	1.24	20.9	−0.4 to 2.0
2 : 1 $AlCl_3$: EtPyBr	25	1.52	25.0	−0.2 to 1.8
2 : 1 $AlBr_3$: EtPyBr	25	2.20	~ 50	0.2 to 1.6

[a] At a glassy carbon electrode versus an Al wire reference electrode in the same melt.

5. Conclusions

The newly developed molten salt systems based on aluminum halides and alkylpyridinium halides, principally the $AlCl_3$: *n*-butylpyridinium chloride system, appear to be very suitable media for the study of chemical and electrochemical acid-dependent reactions at a very convenient temperature range close to room temperature (see Table 3). It is worth emphasizing that these new ionic liquids offer new possibilities for structural studies over a wide range of temperatures. Miscibility of these molten salt systems with aromatic cosolvents such as benzene, toluene, etc., suggests the possibility of expanding the number of organic and organometallic compounds which may be investigated, in addition to improving the physical properties of the solvent system by decreasing the viscosity and increasing the electrical conductivity.

From a more practical point of view, it appears that these systems may also be useful battery electrolytes. Research of possible electrode systems is therefore another branch for future research.

ACKNOWLEDGMENTS

Support from the Air Force Office of Scientific Research and the Office of Naval Research is acknowledged. Helpful discussions with Dr. J. H. Christie and Dr. R. I. Gale are appreciated.

References

1. G. Mamantov and R. A. Osteryoung, in *Characterization of Solutes in Nonaqueous Solvents*, Ed. G. Mamantov, Plenum, New York (1976), pp. 223–249.
2. C. R. Boston, in *Advances in Molten Salt Chemistry*, Vol. 1, Eds. J. Braunstein, G. Mamantov, and G. P. Smith, Plenum, New York (1971), pp. 129–163.
3. K. W. Fung and G. Mamantov, in *Advances in Molten Salt Chemistry*, Vol. 2, Eds. J. Braunstein, G. Mamantov, and G. P. Smith, Plenum, New York (1973), pp. 199–254.
4. H. L. Jones and R. A. Osteryoung, in *Advances in Molten Salt Chemistry*, Vol. 3, Eds. J. Braunstein, G. Mamantov, and G. P. Smith, Plenum, New York (1975), pp. 121–176.
5. J. A. Plambeck, *Fused Salt Systems*, Vol. X of *Encyclopedia of Electrochemistry of the Elements*, Ed. A. J. Bard, Marcel Dekker, New York (1976), pp. 233–254.
6. B. Tremillon and G. Letisse, *J. Electroanal. Chem.* 17, 37 (1968).
7. G. Torsi and G. Mamantov, *Inorg. Chem.* 10, 1900 (1971).
8. A. A. Fannin, L. A. King, and D. W. Seegmiller, *J. Electrochem. Soc.* 119, 801 (1972).
9. L. G. Boxall, H. L. Jones, and R. A. Osteryoung, *J. Electrochem. Soc.* 120, 223 (1973).
10. L. G. Boxall, H. L. Jones, and R. A. Osteryoung, *J. Electrochem. Soc.* 121, 212 (1974).
11. G. Torsi, G. Mamantov, and G. M. Begun, *Inorg. Nucl. Chem. Lett.* 6, 553 (1970).
12. S. J. Cyvin, P. Klaeboe, E. Rytter, and H. A. Oeye, *J. Chem. Phys.* 52, 2776 (1970); H. A. Oeye, E. Rytter, P. Klaeboe, and C. J. Cyvin, *Acta Chem. Scand.* 25, 559 (1971).
13. G. M. Begun, C. R. Boston, G. Torsi, and G. Mamantov, *Inorg. Chem.* 10, 886 (1971).

14. E. Rytter, H. A. Oeye, S. J. Cyvin, B. N. Cyvin, and P. Klaeboe, *J. Inorg. Nucl. Chem.* **35**, 1185 (1973).
15. K. Balasubrahmanyam and L. Nanis, *J. Chem. Phys.* **42**, 676 (1965).
16. V. A. Maroni and E. J. Cairns, in *Molten Salts—Characterization and Analysis*, Ed. G. Mamantov, Marcel Dekker, New York (1969), p. 245.
17. E. Rytter, B. E. D. Rytter, H. A. Øye, and J. Krogh-Moe, *Acta Crystallogr.* **29B**, 1541 (1973); **31B**, 2177 (1975).
18. N. Weiden and A. Weiss, *J. Magn. Reson.* **30**, 403 (1978).
19. T. W. Couch, D. A. Lokken, and J. D. Corbett, *Inorg. Chem.* **11**, 357 (1972).
20. R. Marassi, G. Mamantov, and J. Q. Chambers, *Inorg. Nucl. Chem. Lett.* **11**, 245 (1975).
21. R. Marassi, G. Mamantov, and J. Q. Chambers, *J. Electrochem. Soc.* **123**, 1128 (1976).
22. K. A. Paulsen and R. A. Osteryoung, *J. Am. Chem. Soc.* **98**, 6866 (1976).
23. B. Gilbert and R. A. Osteryoung, *J. Am. Chem. Soc.* **100**, 2725 (1978).
24. J. Robinson, B. Gilbert, and R. A. Osteryoung, *Inorg. Chem.* **16**, 3040 (1977).
25. J. Robinson and R. A. Osteryoung, *J. Electrochem. Soc.* **125**, 1784 (1978).
26. J. Robinson and R. A. Osteryoung, *J. Electrochem. Soc.* **125**, 1454 (1978).
27. N. Bjerrum, in *Characterization of Solutes in Nonaqueous Solvents*, Ed. G. Mamantov, Plenum, New York (1976), pp. 251–271.
28. H. E. Doorenbos, J. C. Evans, and R. U. Kagel, *J. Phys. Chem.* **74**, 3385 (1970).
29. N. J. Bjerrum, *Inorg. Chem.* **9**, 1965 (1970); **10**, 2578 (1971); **11**, 2648 (1972).
30. R. Fehrmann, N. J. Bjerrum, and M. A. Andreasen, *Inorg. Chem.* **15**, 2187 (1976).
31. J. M. von Barnes, N. J. Bjerrum, and D. R. Nielsen, *Inorg. Chem.* **13**, 1708 (1974).
32. F. W. Paulsen, N. J. Bjerrum, and D. R. Nielsen, *Inorg. Chem.* **13**, 2693 (1974).
33. D. J. Prince, J. D. Corbett, and B. Garbisch, *Inorg. Chem.* **9**, 2731 (1970).
34. N. J. Bjerrum and G. P. Smith, *J. Am. Chem. Soc.* **90**, 4472 (1968).
35. R. Fehrmann, N. J. Bjerrum, and F. W. Paulsen, *Inorg. Chem.* **17**, 1195 (1978).
36. H. L. Jones, L. G. Boxall, and R. A. Osteryoung, *J. Electroanal. Chem.* **38**, 476 (1972).
37. H. L. Jones and R. A. Osteryoung, *J. Electroanal. Chem.* **49**, 281 (1974).
38. D. E. Bartak and R. A. Osteryoung, *J. Electrochem. Soc.* **122**, 600 (1975).
39. D. E. Bartak and R. A. Osteryoung, *J. Electroanal. Chem.* **74**, 68 (1976).
40. K. W. Fung, J. Q. Chambers, and G. Mamantov, *J. Electroanal. Chem.* **47**, 81 (1973).
41. V. R. Koch, L. L. Miller, and R. A. Osteryoung, *J. Org. Chem.* **39**, 2416 (1974).
42. F. H. Hurley and J. P. Wier, *J. Electrochem. Soc.* **98**, 203 (1951).
43. H. L. Chum, V. R. Koch, L. L. Miller, and R. A. Osteryoung, *J. Am. Chem. Soc.* **97**, 3264 (1975).
44. V. R. Koch, L. L. Miller, and R. A. Osteryoung, *J. Am. Chem. Soc.* **98**, 5277 (1976).
45. H. L. Chum, D. Koran, and R. A. Osteryoung, *J. Organomet. Chem.* **140**, 349 (1977).
46. H. L. Chum, D. Koran, and R. A. Osteryoung, *J. Am. Chem. Soc.* **100**, 310 (1978).
47. J. Robinson, R. C. Bugle, H. L. Chum, D. Koran, and R. A. Osteryoung, *J. Am. Chem. Soc.* **101**, 3776 (1979).
48. C. J. Pickett and D. Pletcher, *J. Chem. Soc. Chem. Commun.*, 660 (1974).
49. E. Peled and E. Gileadi, *Plating* **62**, 342 (1975); *J. Electrochem. Soc.* **123**, 15 (1976).
50. A. Reger, E. Peled, and E. Gileadi, *J. Electrochem. Soc.* **123**, 639 (1976).
51. E. Peled, A. Mitavsky, A. Reger, and E. Gileadi, *J. Electroanal. Chem. Int. Electrochem.* **75**, 677 (1977).
52. E. Peled, A. Mitavsky, and E. Gileadi, *Z. Phys. Chem.* **96**, 111 (1976).
53. S. Ziegel, E. Peled, and E. Gileadi, *Electrochim. Acta* **23**, 363 (1978).
54. R. J. Gale, B. Gilbert, and R. A. Osteryoung, *Inorg. Chem.* **17**, 2728 (1978).
55. R. J. Gale and R. A. Osteryoung, *Inorg. Chem.* **18**, 1603 (1979).
56. J. Robinson and R. A. Osteryoung, *J. Am. Chem. Soc.* **101**, 323 (1979).
57. C. L. Hussey, L. A. King, and J. S. Wilkes, *J. Electroanal. Chem.*, **102**, 321 (1979).
58. E. Rytter and H. A. Oeye, *J. Inorg. Nucl. Chem.* **35**, 4311 (1973).

59. G. Torsi and G. Mamantov, *Inorg. Chem.* **11**, 1439 (1972).
60. H. Lund, *Acta Chem. Scand.* **11**, 1232 (1957).
61. W. C. Niekam and M. M. Desmond, *J. Am. Chem. Soc.* **86**, 4811 (1964).
62. E. S. Pysch and M. C. Yang, *J. Am. Chem. Soc.* **85**, 2124 (1963).
63. R. E. Sioda, *J. Phys. Chem.* **72**, 2322 (1968).
64. W. C. Herndon, *J. Phys. Chem.* **98**, 887 (1976).
65. F. I. Vilesov, *Zn. Fiz. Khim.* **35**, 2010 (1961).
66. M. Sato and Y. Aoyama, *Bull. Chem. Soc. Jpn.* **46**, 631 (1973).
67. S. J. Costanzo and W. B. Jurinski, *Tetrahedron* **23**, 2571 (1967).
68. M. M. Perkampus and Th. Kranz, *Z. Phys. Chem. Neue Folge* **34**, 213 (1962).
69. M. M. Perkampus and Th. Kranz, *Z. Phys. Chem. Neue Folge* **38**, 295 (1963).

Polarization Energy in Ionic Melts

J. Lumsden

1. Introduction

The main objective of this chapter is to emphasize the importance of polarization energy in determining the thermodynamic properties of molten salt mixtures. Firstly, consideration is given to those typical ionic compounds, such as alkali metal halides, in which each anion is symmetrically surrounded by cations, so that the anions are unpolarized in the pure salts; it is shown that the main term in the enthalpy of mixing of two molten alkali metal halides with a common anion is the polarization energy due to the dipole generated in the polarizable anion because of the different sizes of its neighboring cations. Secondly, consideration is given to those compounds, such as magnesium chloride, that are generally regarded as not being typically ionic, because they form layer lattices, in which the three cations with which each anion is in contact are all congregated around one hemisphere; it is shown that, because of the ratio of the polarizability of the anion to the cube of the $Mg^{2+}-Cl^-$ distance, electrostatic asymmetry is to be expected in both solid and liquid magnesium chloride, and the unsymmetrical enthalpy of mixing in a liquid mixture of an atypical and a typical ionic compound is exemplified quantitatively for the $MgCl_2-CaCl_2$ system.

In, for example, the liquid LiCl–KCl system, the polarization of the chloride ion is the result of unequal opposing electrostatic forces exerted by the Li^+ and K^+ cations and can therefore be regarded as competitive polarization. On the other hand, the polarization of the chloride anion in magnesium chloride is a cooperative effect. Another term in the enthalpy of mixing is ascribable to van der Waals energy; this is often a small term, but

J. Lumsden • Imperial Smelting Processes Limited, Avonmouth, Bristol, BS11 9HP, England.

is sometimes important, particularly with cations that are not of inert-gas structure. Even without estimating the parameters involved, recognizing the relative importance of these three terms is useful for explaining the form of the variation with composition of the enthalpy of mixing in binary system with a common ion; for competitive polarization it is approximately a symmetrical function of composition expressed in equivalent fractions, while for van der Waals energy it is a symmetrical function of composition expressed in volume fractions; insofar as these two types of interaction are concerned, the dissolution of B in A is a simple converse of the dissolution of A in B. When cooperative polarization is involved, however, there is no such reciprocal relation between solvent and solute; dissolving calcium chloride in magnesium chloride is a qualitatively different process from dissolving magnesium chloride in calcium chloride.

When sodium is dissolved in liquid sodium chloride the solute electron becomes localized and thus can be regarded as behaving as an anion. The Na–NaCl phase diagram has a closed miscibility gap that is approximately a symmetrical function of composition. Thermodynamic analysis of the phase diagrams of other mixtures of alkali metals with their halides shows that the excess free energy is a symmetrical function of composition expressed in volume fractions. This indicates that, from the thermodynamic viewpoint, there is no important qualitative change of structure as the composition changes from pure halide to pure metal; the important term is the positive enthalpy of mixing arising from some effect analogous with van der Waals energy.

Van der Waals energy is electrostatic in origin, arising from the fluctuating polarization of an ion owing to the fluctuating electric field generated by the quantum-dictated motion of electrons in a neighboring ion. Alternatively, van der Waals energy can be ascribed to the tendency of electrons to keep in phase with each other while they traverse their quantum-dictated courses. This definition, applied to electrons in metals, relates to the difference between the actual electrostatic energy of a metal and that calculated on the assumption that the electrons are anchored at lattice points. Similarly, the fluctuating polarization between a localized electron and a polarizable halide anion can be calculated. It is shown that the van der Waals energies, thus defined, account for the experimental heat of solution of sodium in its halides.

2. Competitive Polarization

Førland[1] pointed out that the exothermic enthalpy of mixing of two alkali metal halides with a common anion can be explained by the decrease of electrostatic energy due to the difference between the sizes of the cations, this decrease being partly coulombic and partly ascribable to polarization of the anion. The essential feature of Førland's model is that in a mixture of X^+Z^- and Y^+Z^- the cation–anion separations remain at the same respective values, r_{XZ} and r_{YZ}, as in the pure salts. The local arrangement is that of the NaCl structure. In pure XZ all the Z^- anions are in contact with diametrically opposed pairs of X^+ cations, and to each of these $X^+Z^-X^+$ groups can be assigned an electrostatic energy:

$$E_{XZX} = -2e^2/r_{XZ} + e^2/2r_{XZ} \qquad (1)$$

Similarly, in pure YZ each $Y^+Z^-Y^+$ group can be assigned an energy:

$$E_{YZY} = -2e^2/r_{YZ} + e^2/2r_{YZ} \qquad (2)$$

In a mixture of XZ and YZ there are some $X^+Z^-Y^+$ groups, the electrostatic energy of which can be written

$$E_{XZY} = -e^2/r_{XZ} - e^2/r_{YZ} + e^2/(r_{XZ} + r_{YZ}) \qquad (3)$$

From equations (1)–(3) the change of electrostatic energy in forming this $X^+Y^-Z^+$ chain is

$$\Delta E_c = -e^2(r_{XZ} - r_{YZ})^2/4r_{XZ}r_{YZ}(r_{XZ} + r_{YZ}) \qquad (4)$$

If α is the polarizability of the anion, the polarization energy of this $X^+Z^-Y^+$ group is

$$\Delta E_p = -\alpha e^2(r_{XZ} - r_{YZ})^2(r_{XZ} + r_{YZ})^2/2(r_{XZ}r_{YZ})^4 \qquad (5)$$

The Førland model, with the pattern of the ionic structure and the cation–anion distance remaining unchanged in the solution, if applied to a solid solution, is obviously incorrect, as a solid solution of alkali metal halides with a common anion is formed endothermically; some thermochemically significant readjustment of the ionic structure is necessary. Some readjustment, less drastic than for the solid, must be necessary to accommodate the ion of different size; this readjustment could be regarded as thermochemically insignificant only if, in accordance with the Førland model, the linear dimension of the solution were an additive function of composition, this implying a decrease of volume on mixing. Thus, the linear dimension of an equimolar mixture of XZ and YZ would be

$$r = (r_{XZ} + r_{YZ})/2 \qquad (6)$$

so that the volume would be proportional to

$$r^3 = (r_{XZ} + r_{YZ})^3/8 \qquad (7)$$

In fact, the volumes of mixtures of alkali metal salts are substantially additive, so that

$$r^3 = (r_{XZ}^3 + r_{YZ}^3)/2 \qquad (8)$$

which is greater than estimate (7) by

$$3(r_{XZ} - r_{YZ})^2(r_{XZ} + r_{YZ})/8 \qquad (9)$$

The relative increase in volume compared with estimate (7) is

$$3(r_{XZ} - r_{YZ})^2/(r_{YZ} + r_{YZ})^2 \qquad (10)$$

The relative increase in linear dimensions is therefore

$$(r_{XZ} - r_{YZ})^2/(r_{XZ} + r_{YZ})^2 \qquad (11)$$

This relative linear expansion introduces an equal relative decrease in the magnitude of the Madelung coulombic energy, and thus introduces a positive electrostatic energy term that is of the same form as the coulombic energy term (4):

$$\Delta E_c = Ae^2(r_{XZ} - r_{YZ})^2/2r_{XZ}r_{YZ}(r_{XZ} + r_{YZ}) \qquad (12)$$

where A is the Madelung constant.

If the multiplying factor for expression (4) is estimated as twice the Madelung constant, the two coulombic terms [expressions (4) and (12)] would be numerically equal and would cancel each other. As this multiplying factor is difficult to estimate, no deductive conclusion can be confidently drawn concerning the sign, let alone the magnitude, of the enthalpy of mixing attributable to coulombic energy, without a detailed specification of the pattern of ionic arrangement. No similar difficulty arises with the deductive approach to the polarization energy. Changing the assumed interionic distances by 1% would change the estimated polarization energy by 4%, but no uncertainty arises concerning the order of magnitude or the sign of the polarization energy.

There is already abundant evidence[2] that, for mixtures of alkali metal halides, the polarization energy is of overwhelming importance; the experimental interaction parameters, after allowing for the effect of interionic van der Waals energy, is proportional to the polarization expression (5). A simple deductive calculation of the polarization energy can be made only when the coordination number is known. There is good evidence[3] that the coordination number is approximately four in some alkali metal halides in which the cation is considerably smaller than the anion. In such a salt XZ each X^+ cation is surrounded at a distance r_{XZ} by four Z^- anions, each of which is in contact also with three other X^+ cations also at a distance r_{XZ}. In a dilute solution in XZ of a similar salt YZ each Y^+ cation is surrounded at a distance r_{YZ} by four Z^- anions, each of which is in contact with three X^+ cations sited at a distance r_{XZ}. In each of the four anions surrounding the cation there is generated a polarization energy given by equation (5), so that the polarization energy per solute anion is

$$E_p = -2\alpha e^2(r_{XZ} - r_{YZ})^2(r_{XZ} + r_{YZ})^2/(r_{XZ}r_{YZ})^4 \qquad (13)$$

In view of the fact that the factors in expression (13) include the square of the primary dipole and the inverse sixth power of mean interionic distances, the polarization energy in more distant anions will be relatively small.

Four salts for which the coordination number is almost exactly four are the fluorides and chlorides of lithium and sodium. Apart from measurements by X-ray and neutron diffraction, the 20%–24% increase of volume on melting is consistent with a decrease of coordination number from six to four. With the zinc blende structure assumed, the cation–anion distance can be calculated from the molar volume. Hence it is calculated that the ratio of the cation–anion distance in the liquids at their melting points to that in the solids at ambient temperature are as follows: LiF, 0.99; LiCl, 0.97; NaF, 0.98; NaCl, 0.97. Since the relative differences between the cation–anion distances are known accurately for the solids and are likely to remain almost constant, calculations are made of the polarization energy on the assumption that in the liquids these cation–anion distances are all 0.98 times those for the solids. These contributions of polarization energy for dilute solutions in the LiF–NaF and LiCl–NaCl systems are, respectively, -8.3 and -5.1 kJ mol^{-1}. There is a small contribution from intercationic van der Waals energy,[4] 0.3 kJ for solution of NaF in LiF, 0.2 kJ for LiF in NaF, and 0.1 kJ for NaCl in LiCl or LiCl in NaCl. The calculated sums of polarization energy and van der Waals energy are then as follows (with the corresponding experimental measurements[5, 6] in parentheses): NaF in LiF -8.0 (-8.2); LiF in NaF -8.1 (-7.9); NaCl in LiCl and LiCl in NaCl -5.0 (-4.7); all in kJ mol^{-1}.

In these two systems the contribution of coulombic energy to the enthalpy of mixing is approximately zero. This deductive calculation of the polarization energy can be carried out with confidence only when both components have a coordination number of four. For other systems the total interionic electrostatic energy (coulombic and polarization) can be calculated from the experimental enthalpy of solution and the estimated van der Waals energy. Thus, for the LiCl–KCl system,[6] this electrostatic energy is calculated as $-17.9-0.8 = -18.7$ kJ for KCl in LiCl and $-17.6-0.5 = -18.1$ kJ for LiCl in KCl. A calculation assuming fourfold coordination would be -21.5 kJ for the polarization energy of solution.

While the only cases for which quantitative calculations can be made are LiCl–NaCl and LiF–NaF, it seems likely that polarization energy is the main electrostatic term in the enthalpy of mixing of alkali metal halides; the difference between the coordination numbers would tend to make the competitive polarization less efficient. Even without any quantitative explanation of this loss of efficiency, it is of interest to point out that in the LiCl–KCl system the contributions of electrostatic energy to the enthalpies of the two dilute solutions differ by only 3%; although the ratio of interionic distances is 1.224, the corresponding volume ratio is 1.83. This is in sharp contrast to the relation, readily explicable when simple additive van der Waals forces are operating, stating that in mixtures of nonpolar liquids the two enthalpies of dilute solution are proportional to the molar volumes of the solutes. A qualitative explanation of the very different relation in ionic solution is that when one component has, like lithium chloride, a strong preference for a coordination number of four, it constrains the other component to adopt this same coordination number locally.

Solutions in which the enthalpy of mixing is a symmetrical function of composition can be defined as regular solutions. Unless the enthalpy of mixing is large enough to cause appreciable departures from randomness, the symmetrical function is approximately parabolic. Førland[7] pointed out that in ionic mixtures the enthalpy of mixing is a parabolic function of composition expressed in equivalent fractions.

When the ratio of interionic distances becomes much greater than the 1.224 for the LiCl–KCl system, the two enthalpies of dilute solution are no longer approximately equal; thus, in the LiF–KF system, for which the ratio of interionic distances is 1.33, the heats of dilute solution are −21.3 and −15.7 kJ, respectively, for KF and LiF as solutes; allowing for van der Waals energy, the calculated electrostatic energy terms are −23.4 for KF in LiF and −16.7 for LiF in KF.

In general, the electrostatic energy term in the enthalpy of mixing is approximately a parabolic function of composition so long as the ratio of interionic distances is less than 1.22. Exceptions, however, are found in the LiBr–KBr and LiI–KI systems,[6, 8] in which the enthalpy of dilute solution is more negative for the potassium halide than for the lithium halide. This can be ascribed to the small size of the lithium cation, which is therefore free to site itself away from the center of its tetrahedral cavity. This reduces the electrostatic energy to a certain extent in the pure salt, and to a greater extent by the ion siting itself nearer anions in contact with larger cations.

Attention has been concentrated on mixtures of alkali metal halides with a common anion, because these are the mixtures with large enthalpies of mixing. In fact, the simplest argument for the relative unimportance of the coulombic energy is the very small enthalpy of mixing of some mixtures with a common cation,[9] such as LiF–LiCl, for which the coulombic factor [equation (4)] is greater than that for LiCl–KCl.

As has been pointed out,[10] the polarization of the chloride anions surrounding each lithium cation for a dilute solution of lithium chloride in potassium chloride means that the outer electrons of the anion are displaced toward the lithium cation, so that the Li–Cl bond can be regarded as partially covalent. Such an argument gives a qualitative explanation of why the mobility of the lithium cation is so much less in a solution of lithium chloride in potassium chloride than in pure lithium chloride. For quantitative calculation, however, it seems preferable to regard the octet of electrons as still belonging to the chloride ion.

In some liquid mixtures the enthalpy of mixing can be approximately equated to the excess free energy of mixing, calculated on the assumption that there is no term due to vibrational entropy. Usually the vibrational entropy change in a reaction is of the same sign as the enthalpy. As an example of ionic interchange between two salts where only cation–anion interactions are involved:

$$LiF + KCl = LiCl + KF$$

$$\Delta H = 74 \text{ kJ}, \qquad \Delta S = 10 \text{ J K}^{-1}, \qquad \Delta S / \Delta H = 1.3 \times 10^{-4} \text{ K}^{-1} \quad (14)$$

This $\Delta S / \Delta H$ ratio varies, but it remains of this order of magnitude for reciprocal pairs of alkali metal halides.

An example where the enthalpy[6] change is attributable mostly to competitive polarization of an anion is provided by dissolution of potassium chloride in lithium chloride:

$$KCl = KCl(LiCl)$$

$$\Delta \bar{H} = 18 N_{LiCl}^2 \text{ kJ}, \qquad \Delta \bar{S}^{xs} = 5.9 N_{LiCl}^2 \text{ J K}^{-1} \quad (15)$$

$$\Delta \bar{S}^{xs} / \Delta \bar{H} = 3.3 \times 10^{-4} \text{ K}^{-1}$$

This value is typical of such anion–anion interactions and is noticeably higher than when, as in equation (14), the enthalpy is attributable to coulombic energy.

3. Cooperative Polarization

The indubitably ionic alkali metal halides form crystals in which each ion is symmetrically surrounded by ions of opposite sign. Electrostatic symmetry is likewise attained in dihalides that form CaF_2-type lattices. The other dihalide lattices lack complete electrostatic symmetry; in some of these, like calcium chloride, there is approximately electrostic symmetry, but others, like magnesium chloride, form layer lattices with a grossly unsymmetrical grouping of cations round each anion. To explain these layer lattices, it is not necessary to invoke covalent bonds.

In some ionic assemblages the lowest electrostatic energy can be attained with an arrangement less symmetrical than is geometrically possible, provided that, for a given cation–anion distance l, the polarizability α of the anion rises above a certain value. As shown by Debye,[11] for an ionic molecule $X_2^+ Y^{2-}$, the minimum electrostatic energy is attained with a linear arrangement only when $l^3/\alpha > 8$; when $l^3/\alpha < 8$ the stable molecule is nonlinear. It has been pointed out[12, 13] that this can explain why some dihalides (CaF_2, SrF_2, $SrCl_2$, BaF_2, BaI_2) form angular molecules in the gaseous phase while others (BeF_2, $BeCl_2$, MgF_2, $SrBr_2$, SrI_2) form linear molecules.

In Table 1 the l^3/α ratio is calculated for the solid halides of magnesium, calcium and strontium. For the eight salts with their ratio greater than 6 the arrangement of cations round anions is at least approximately symmetrical. The other four, with this ratio less than 6, crystallize in layer lattices, with the three neighboring cations all grouped round one hemisphere of each anion; this is consistent with the stable structure being determined by minimization of electrostatic energy, with the polarization energy becoming increasingly important when the l^3/α ratio becomes smaller, the critical value lying between 5.9 and 6.8.

It may further be noted that the dihalides that crystallize in lattices with approximate electrostatic symmetry form Førland-type regular solutions with each other, but not with dihalides that crystallize in layer lattices. Thus, the two molar enthalpies of dilute solution[14] are almost equal in the $CaCl_2$–$SrCl_2$ system at -1.82 and -1.84 kJ, in the $SrCl_2$–$BaCl_2$ at -2.00 and -2.03 kJ, and in the $CaCl_2$–$BaCl_2$ system at -8.7 and -8.5 kJ, respectively. On the other hand, in the $MgCl_2$–$CaCl_2$ system the heats of dilute solution are 0.0 and $+10.0$ kJ, respectively, for $MgCl_2$ and $CaCl_2$ as solutes, and in the $MgCl_2$–$SrCl_2$ system are -5.3 and $+11.3$ kJ, respectively, for $MgCl_2$ and $SrCl_2$ as solutes.

It is reasonable to presume that, in the liquid state also, there is some unsymmetrical arrangement of cations round each anion in salts such as magnesium chloride, but that, as in lithium chloride, the cations occupy tetrahedral rather than octahedral interstices between the anions. A local grouping very favorable to polarization would consist of two anions each in contact with the same pair of cations. In order to attain efficient polarization of all four anions with which it is in contact, each cation must enter

TABLE 1. The Ratio l^3/α in Dihalides[a]

MgF_2	$MgCl_2^{* b}$	$MgBr_2^*$	MgI_2^*
11.7	5.9	5.2	4.3
CaF_2	$CaCl_2$	$CaBr_2$	CaI_2^*
18.3	8.5	7.2	5.9
SrF_2	$SrCl_2$	$SrBr_2$	SrI_2
21.8	9.9	8.4	6.8

[a] l, cation–anion distance; α, polarizability of anion.
[b] *, indicates layer lattice.

into two polarizing partnerships, with one suitably sited cation to polarize two of the anions and with another cation, situated in the opposite direction, to polarize the other two anions. Because of this catenated cooperation, effective polarization of this type is greatly weakened locally if one of the magnesium cations is replaced by a larger cation such as strontium, in the vicinity of which electrostatic energy is minimized only if its neighboring cations are arranged symmetrically. Consequently, in a dilute solution of strontium chloride in magnesium chloride, each strontium ion engenders negative electrostatic energy by competitively polarizing the surrounding anions, but also engenders positive electrostatic energy by disrupting the cooperative polarization by the surrounding magnesium cations.

Similarly, in a very dilute solution of magnesium chloride in strontium chloride, the only negative electrostatic energy term is the competitive polarization of the anions. Even in moderately dilute solutions, where there is the geometrical possibility of two neighboring magnesium ions siting themselves so as to effect cooperative polarization of two anions, the overriding requirement by the strontium chloride solvent, that all cations shall be arranged symmetrically, prohibits such asymmetry. A polarized pattern is possible only when several magnesium cations are grouped together so as to form a polarized domain where sufficient negative cooperative energy is engendered to outweigh the increase of electrostatic energy in the boundary layer where strontium ions are present. A similar conclusion would apply to solutions of magnesium chloride in calcium chloride, but the domains would be smaller.

The qualitative conclusion is thus reached that, in dilute solution in calcium chloride or strontium chloride, magnesium chloride behaves as a normal ionic compound, and cooperative polarization by groups of magnesium ions does not occur to any appreciable extent until there is a considerable concentration of magnesium chloride. Over such a range the enthalpy of mixing should be a parabolic function of composition:

$$x\text{MgCl}_2 + (1 - x)\text{XCl}_2: \qquad \Delta H = bx(1 - x) + cx \qquad (16)$$

where b is the interaction parameter between the salt XCl_2 and MgCl_2 in the unpolarized state, and c is the enthalpy of conversion of 1 mol of liquid magnesium chloride to the unpolarized state. The integral heat of mixing per mole of magnesium chloride is

$$\Delta H/x = b(1 - x) + c \qquad (17)$$

In Figure 1 this function[14] is plotted for magnesium chloride in solution in the chlorides of calcium and strontium. As expected, they fit a straight line over a considerable range, and, extrapolated to the MgCl_2 composition, give 5.0 kJ for the heat of conversion of 1 mol of magnesium chloride to its unpolarized form.

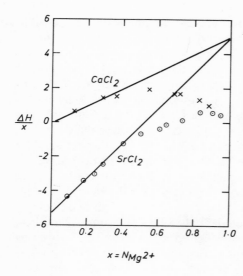

FIGURE 1. Heat of mixing (ΔH). $\times$, $x MgCl_2 + (1 - x)CaCl_2$; $\odot$, $x MgCl_2 + (1 - x)SrCl_2$.

At the higher concentrations of magnesium chloride the difference between these extrapolated lines and the experimental points indicates the energy decrease caused by cooperative polarization. These differences are plotted in Figure 2, which illustrates the essential feature of a cooperative effect, namely, that it increases more than proportionately with the concentration of the active species. The simplest algebraic formulation for such a relationship is to represent the cooperative polarization energy (E_p) as some power of the mole fraction of magnesium chloride:

$$E_p = -cx^n \tag{18}$$

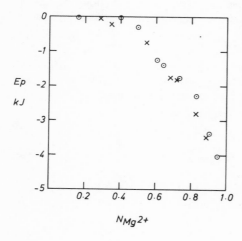

FIGURE 2. Cooperative polarization energy E_p in $MgCl_2 - CaCl_2$ and $MgCl_2 - SrCl_2$ melts per mole of $MgCl_2$. $\times$, $CaCl_2$; $\odot$, $SrCl_2$.

This form of equation can be justified by postulating that a chloride ion becomes cooperatively polarized only when its n nearest cations are all magnesium. It implies that, in a dilute solution of the unpolarized chloride in magnesium chloride, each solute cation cancels the cooperative polarization energy of $n - 1$ molecules of magnesium chloride.

For the $CaCl_2$–$MgCl_2$ system it is found that equation (18) applies with $n = 4$:

$$xMgCl_2(1 - x)CaCl_2: \qquad E_p = -cx^4 \qquad (19)$$

Adding this to equation (16) gives

$$xMgCl_2 + (1 - x)CaCl_2: \qquad \Delta H = bx(1 - x) + cx(1 - x^3) \qquad (20)$$

and

$$\Delta H/x(1 - x) = (b + c) + c(x + x^2) \qquad (21)$$

Figure 3 shows that the experimental enthalpies of mixing agree well with equation (21).

In Figure 4 the same method of plotting is used for the enthalpies of mixing[15] of calcium chloride with the dichlorides of cobalt, iron, and manganese. The parameters are shown in Table 2.

The c parameter is determined by simple electrostatic energy and becomes larger as the cation becomes smaller; if b were also determined by electrostatic energy, it would become more negative as the cation becomes smaller. $CoCl_2$ has the least negative value of b and the largest value of c because of its high van der Waals energy.

When, as in the $MgCl_2$–$CaCl_2$ system, one of the components is cooperatively polarized, the enthalpy of conversion of $MgCl_2$ from the unpo-

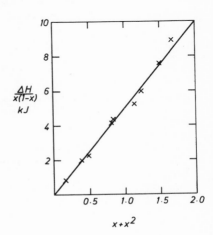

FIGURE 3. Heat of mixing (ΔH). $xMgCl_2 + (1 - x)CaCl_2$.

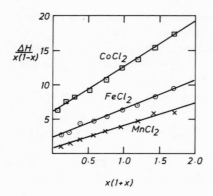

$$\frac{\Delta H}{x(1-x)}$$

$x(1+x)$

FIGURE 4. Heat of mixing (ΔH). $xXCl_2 + (1-x)CaCl_2 : X = Co, Fe, Mn.$

larized to the polarized structure represents the difference between the negative polarization energy and a positive coulombic energy arising from the unsymmetrical arrangement of the cations around the anions. In view of the fact that the ratio of the excess entropy to the enthalpy is higher for polarization than for coulombic energy, the possibility arises that this ratio would become still higher for cooperative polarization.

The most accurate information on the free energy of mixing in the $MgCl_2$–$CaCl_2$ system is derivable from the phase diagram, particularly the $MgCl_2$ branch of the liquidus.[16] The partial excess free energy $(RT \ln \gamma)$ of magnesium chloride is shown in Figure 5, where it is compared with the partial heat of solution of magnesium chloride, calculated from the integral heat of solution from Table 2:

$$MgCl_2(CaCl_2): \quad \Delta\bar{H} = -5.0(1-x)^2 + 5.0(1-4x^3+3x^4) \quad (22)$$

where x is the mole fraction of magnesium chloride and the two terms represent the contributions from competitive and cooperative polarization. Hence is calculated, from each liquidus measurement, the excess partial entropy of magnesium chloride:

$$\Delta\bar{S}^{xs} = (\Delta\bar{H} - RT \ln \gamma)/T \quad (23)$$

TABLE 2. Enthalpy of Mixing
$xXCl_2 + (1-x)CaCl_2:$
$\Delta H = bx(1-x) + cx(1-x^3) \text{ kJ}$

XCl_2	$MgCl_2$	$FeCl_2$	$MnCl_2$	$CoCl_2$
b	−5.0	−1.7	−2.3	−0.4
c	5.0	4.1	3.2	6.5

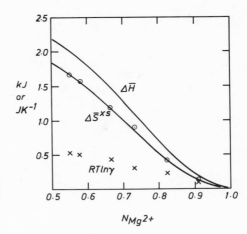

FIGURE 5. The partial excess free energy $(RT\ln\gamma)$ of $MgCl_2$ in solution in $CaCl_2$ at liquidus temperature. $RT\ln\gamma = \Delta\bar{H} - T\Delta\bar{S}^{xs}$.

The relation between the excess entropy and the enthalpy in Figure 5 could well be represented by the simple relationship

$$\Delta\bar{S}^{xs}/\Delta\bar{H} = 8.0 \times 10^{-4}\ \text{K}^{-1} \tag{24}$$

No generalized physical significance can, however, be attached to this ratio. The enthalpy of mixing is the sum of the two terms of equation (22), each of which can be expected to have its characteristic entropy/enthalpy ratio. It is found, in fact, that by using the ratio of equation (15) for the competitive-polarization term and double this ratio for the cooperative-polarization term, good agreement is attained, as can be seen from Figure 4, where the line represents the equation

$$MgCl_2(CaCl_2): \qquad \Delta S^{xs} = -1.65(1-x)^2 + 3.3(1-4x^3+3x^4) \tag{25}$$

Equation (20) appears to have some general significance, since, with the exponent of equation (18) fixed at 4, it gives an adequate representation of the enthalpy of mixing of calcium chloride with four dihalides that crystallize in layer lattices and therefore are cooperatively polarized, while the relative values of the two parameters are reasonably consistent with the interionic separations and the van der Waals intercationic energies. The generality of equation (20) is, however, limited, because with the cation larger than calcium there would be more resistance to cooperative polarization and the exponent of equation (18) would be increased. It has been found that, whatever exponent is chosen, the enthalpy of mixing of magnesium chloride with strontium chloride cannot represent the experimental results within their apparent limits of error, while with barium chloride as the other component there is not even approximate agreement. The reason is clear from Figure 6, in which the function $\Delta H/x(1-x)$ is plotted for

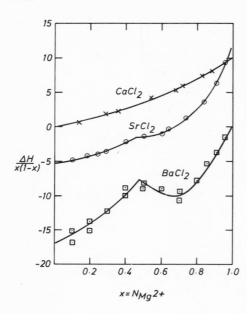

FIGURE 6. Enthalpy of mixing (ΔH).
$x = N_{Mg^{2+}}$
$x\text{MgCl}_2 + (1 - x)\text{XCl}_2 : \text{X} = \text{Ca, Sr, Ba.}$

mixtures of magnesium chloride with alkaline earth chlorides. The MgCl_2–CaCl_2 curve is smooth, but there is a sharp change in the slope of the MgCl_2–BaCl_2 curve at about the equimolar composition and the indication of a similar, less pronounced, break in the MgCl_2–SrCl_2 curve.

It must be concluded that in the MgCl_2–SrCl_2 and MgCl_2–BaCl_2 systems the liquid changes its structure at around the equimolar composition, although there is no corresponding change in the MgCl_2–CaCl_2 system. The difficulty of making a definitive calculation of the thermodynamics of this apparent change of structure arises from the fact that, as shown in Figure 2, the points for the MgCl_2–SrCl_2 system never differ greatly from those for the MgCl_2–CaCl_2 system. The following suggested explanation must therefore be regarded as provisional.

In halides of the postulated structure the local arrangement of anions resembles a face-centered cubic lattice with, in one unit cell, four anions and eight sites available for two cations. With one cation on the $\frac{1}{4}\frac{1}{4}\frac{1}{4}$ site, in a symmetrical arrangement the other cation would be at $\frac{3}{4}\frac{3}{4}\frac{3}{4}$. In the unsymmetrical arrangement postulated for magnesium chloride the other cation would be at $\frac{1}{4}\frac{1}{4}\frac{3}{4}$, giving rise to strong polarization. A weaker polarization could arise, however, if the other cation were at $\frac{1}{4}\frac{3}{4}\frac{3}{4}$.

Calcium chloride has the unpolarized structure, and in dilute solution in calcium chloride, magnesium chloride conforms to this structure. At higher MgCl_2 concentrations, domains of strongly polarized chloride ions are formed; around these domains there will be a boundary layer of weakly polarized chloride anions. This weakly polarized state, apart from being a compromise to form a buffer between the strongly polarized and unpolarized zones, may become the stable structure of the bulk solution over a range of Ca^{2+}/Mg^{2+} ratios. From Figure 6 it is clear that this structural transition must progress smoothly for the $MgCl_2$–$CaCl_2$ system, but with a slight break in the $MgCl_2$–$SrCl_2$ system and a pronounced break in the $MgCl_2$–$SrCl_2$ system and a pronounced break in the $MgCl_2$–$BaCl_2$ system.

The properties of magnesium chloride, and of other compounds that form layer lattices, have generally been taken as a symptom of partial covalent bonding. If partial covalent bonding means some tendency to share electrons, this can be reconciled with the present approach, for the postulated cooperative polarization means that the electrons belonging to the chloride ion are no longer symmetrically grouped around the chlorine nucleus but their electrostatic center has moved in the direction of the nearest cations, which thus can be said to be tending to share the electrons that in a purely ionic model should all belong to the chlorine. This is a possible definition of a partial covalent bond, but it does not seem a very useful definition, for it would imply that, in a dilute solution of lithium chloride in potassium chloride, the Li–Cl bonds are partially covalent.

The principal tenet of the present chapter is that many compounds that can be represented as composed of inert-gas ions can usefully be regarded as so constituted, any apparent electron sharing being ascribed to polarization. This model is obviously preferable when meaningful calculations can be made of thermodynamic properties, as for mixtures of alkali metal halides. This same polarization model explains both the crystalline structure of magnesium chloride and the thermodynamic properties of liquid mixtures of magnesium chloride with alkaline earth chlorides, and it can obviously be extended to mixtures of magnesium chloride with alkali metal chlorides. In liquid solutions the question of cooperative polarization becomes important only for the $MgCl_2$-rich compositions. When magnesium chloride is the minor component in solution in calcium chloride or potassium chloride, the Mg^{2+} cation polarizes the surrounding chloride anions, the polarization energy being much larger with the very unequal competition between the small doubly charged Mg^{2+} and the large singly charged K^+. As was pointed out by Flood,[17] for most of the composition range between KCl and K_2MgCl_4 the solution behaves as an ideal ionic mixture with the two anions Cl^- and $MgCl_4^{2-}$, and the common cation K^+. It has been pointed out[18] that this formation of the $MgCl_4^{2-}$ anion merely means that the electrostatic energy is minimized with an arrangement in which no chloride ion is shared between two magnesiums, so that the number of competitively polarized anions is maintained at the maximum of four for each Mg^{2+} cation.

The polarizability of the chloride anion, in conjunction with the smallness of the magnesium cation, explains why $MgCl_4^{2-}$ anions are formed in ionic melts and why the Mg^{2+} cations are unsymmetrically arranged around each chloride anion in crystalline $MgCl_2$. Analogously, the properties of some other XY_2 compounds can be similarly explained. In one example, SiO_2, the mere qualitative fact that one crystalline form is optically active suggests that asymmetry is an essential feature of the controlling factor.

4. Fluctuating Polarization

Apart from competitive and cooperative polarization, a further possibility arises from the fluctuating polarization of an anion by the vibrational motions relative to the surrounding cations; this effect is irrelevant for the enthalpy of mixing, and the corresponding interaction with the nearest anions is, in general, small. When, however, the anion is a localized electron, its amplitude of vibration is sufficient to exert a strong polarizing action on neighboring halide anions. From Pitzer's[19] calculations on solutions of sodium in sodium chloride, it is clear that the amplitude of vibration of the localized electron is approximately 3.0×10^{-8} cm, which, with the polarizability of the chloride ion 2.82×10^{-24} cm^3, indicates a mean-square dipole sufficient to generate in a dilute solution of sodium in sodium chloride a polarization energy of the order of -30 kJ/mol. This invalidates part of Pitzer's calculation, for he assumes, in effect, that the potential energy of the solution of the metal in its halide, referred to the gaseous ions and electrons, is equal to the coulombic energy calculated on the assumption that the electrons, like the anions, are anchored at their lattice sites. From the Na–NaCl phase diagram he calculated the heat of dilute solution of 1 mol of sodium in sodium chloride to be 76 kJ, which he assumed to be equal to the energy of conversion of sodium to a hypothetical ionic form, with its potential energy equal to the coulombic energy with reference to the gaseous ions and electrons. Actually, however, such a hypothetical ionic sodium would evolve about 30 kJ on dissolving in the sodium chloride, so that the energy of conversion of sodium to the ionic form is $76 + 30 = 106$ kJ, which is equal, within the limits of estimation, to the electrostatic energy of conversion to the ionic form. This proves that there is no important change in the kinetic energy of the metallic electron when it becomes localized.

From a thermodynamic analysis of his measurements of the miscibility gaps in mixtures of alkali metals with their halides, Bredig[20] calculated that the variation of excess free energy with composition in these systems is similar to that in mixtures of nonpolar liquids. This regular-solution behavior indicates a reciprocal relation between solvent and solute, the same physicochemical factors determining the two enthalpies of dilute solution. It has been pointed out that, when due allowance is made for the polarization of the anion by the localized electron, the experimental evidence for solutions of sodium in sodium chloride shows that there is no important change in the kinetic energy of the electrons when they become localized. It remains, therefore, to find a deductive explanation of the regular-solution relationship between the two enthalpies of dilute solution in the Na–NaCl system.

There are two terms in the enthalpy of dilute solution of sodium in sodium chloride. For one of them, the qualitative explanation is simple. The polarization of the surrounding solvent anions by a solute electron when the metal is the solute will be matched by the polarization of a solute anion by the surrounding solvent electrons when the metal is the solvent.

The other term in the enthalpy of solution of sodium in its chloride is the difference between the actual electrostatic energy of sodium metal and that calculated on the assumption that all the electrons are anchored at their lattice sites. To show that this term is closely related to a corresponding term in the electrostatic energy change when sodium chloride is dissolved in sodium requires that both systems should be described within the same frame of reference. The terminology of metal physics is so different from that of molten salt chemistry as to imply that electrons are qualitatively different from anions. As Pitzer points out, when sodium is dissolved in sodium chloride the electron can be regarded as an anion; when sodium chloride is dissolved in sodium, the anion cannot be regarded as being part of an electron gas. The common language must therefore be that of electrochemistry.

The zero point kinetic energy of conduction electrons in an alkali metal is

$$E_k = (3h^2/10m)(3N/2\pi V)^{2/3} \tag{26}$$

where h is Planck's constant, m in the mass of an electron, and V is the volume available for N electrons. When the small fraction of the volume unavailable to the electrons is calculated on the assumption that for this purpose the effective radius of the sodium cation is 0.76×10^{-8} cm, the kinetic energy of the metallic electrons at 0 K is calculated as 489 kJ mol^{-1}, while the enthalpy of decomposition of 1 mol of sodium into gaseous cations and electrons is 605 kJ, so that the electrostatic energy is

$$E_e = -489 - 605 = -1094 \text{ kJ mol}^{-1} \tag{27}$$

If it is assumed that in the ionic form of the metal the cations occupy 0 0 0 and $\frac{1}{2}\frac{1}{2}\frac{1}{2}$ sites on a cubic lattice while the anionic electrons occupy

$0 \frac{1}{2} \frac{1}{2}$ and $\frac{1}{2} 0 \, 0$ sites, the Madelung energy of the ionic form of sodium at 0 K is

$$E_M^i = -2.961 \, N_e^2/l = -974 \text{ kJ mol}^{-1} \tag{28}$$

where l is the lattice parameter, so that

$$E_M^i - E_e = -974 + 1094 = 120 \text{ kJ mol}^{-1} \tag{29}$$

This difference of 120 kJ for solid sodium at 0 K is regarded as being consistent with an estimate of around 106 kJ for liquid sodium at 1100 K.

In the actual metal the time-averaged distribution of electrons is approximately uniform over the available space. If this time-averaged distribution were attained with a random distribution at any instant, the electrostatic energy would be less negative than it would be if the electrons remained fixed at their lattice points. The sign of the energy difference in equation (29) shows that the electrons coordinate their motions so as to avoid coming near to each other; in the extreme case, with perfect coordination, the electrons could be regarded as occupying sites on an independent body-centered cubic lattice that moves randomly with respect to the cation lattice. If the whole volume were available to the electrons, the Madelung factor would be 1.820 for a random distribution of electrons and 3.640 for a body-centered cubic lattice of electrons moving randomly as a unit, but these factors have to be corrected for the excluded volume around the cation nucleus.

The volume available to the localized electron according to Pitzer's calculations indicates that for excluding electrons the ionic radii must be less than the Pauling radii. It is assumed that the electrons are excluded from spheres, of radius $r = 0.76 \times 10^{-8}$ cm, around each cation site. The mean electrostatic energy of each of these excluded electrons would have been estimated as

$$-3e^2/2r + 2.84e^2/l = -5.48e^2/l \tag{30}$$

and the fraction of the electrons wrongly assumed to have been within these spheres would have been

$$8\pi r^3/l^3 = 0.0490 \tag{31}$$

so that these electrons were wrongly assigned an electrostatic energy of

$$-5.48 \times 0.0490 N_e^2/l = -0.269 N_e^2/l \tag{32}$$

and the Madelung factor of 1.820 has to be amended to

$$(1.820 - 0.269)/(1.000 - 0.049) = 1.631 \tag{33}$$

giving the Madelung energy for the randomly moving electrons as

$$E_M^q = -1.631 N_e^2/l = -537 \text{ kJ mol}^{-1} \tag{34}$$

Similarly, for the randomly moving body-centered cubic lattice of electrons, the Madelung energy is

$$E_M^c = -3.545 N_e^2/l = -1166 \text{ kJ mol}^{-1} \tag{35}$$

In equation (35) the superscripted c is used to indicate that the motions of the electrons are coordinated, while in equation (34) the superscript g is used to indicate that the relative motion of the electrons is random, as might be taken to be implied by the term "electron gas."

From equations (27) and (34),

$$E_e - E_M^g = -557 \text{ kJ mol}^{-1} \tag{36}$$

From equations (27) and (35),

$$E_e - E_M^c = 72 \text{ kJ mol}^{-1} \tag{37}$$

The Madelung energy comes much closer to the actual electrostatic energy when it is assumed that the electrons maintain their relative positions. That is to say, the metallic electrons tend to keep in phase with each other as they traverse their quantum-dictated course. The same could be said of the orbital electrons in monatomic ions; the mutual polarization by fluctuating dipoles, which constitutes van der Waals energy as interpreted by London, can be regarded as the tendency of two sets of electrons to keep in phase with each other as they traverse their quantum-dictated course, thereby making the interionic electrostatic energy more negative than it would be if the ions were unpolarizable point changes. The interelectronic and interionic processes are analogous, but, in relation to the intercationic distances, the free metallic electrons generate larger dipoles and are less restricted in the phasing of their quantum-dictated motions than the bound orbital electrons, so that, while the orbital electrons of the ions have a slight tendency to keep in phase with each other, the metallic electrons have a strong tendency to keep in phase with each other. Thus, while from the thermochemical viewpoint electrons can be regarded as anions that generate large fluctuating dipoles, the tendency for neighboring dipoles to be in phase modifies the vectorial additivity when these dipoles are jointly polarizing an anion.

Factors determining the two enthalpies of dilute solution can first be discussed with the simplifying assumption that the electron and the chloride ion are of equal size. When sodium is dissolved in sodium chloride negative electrostatic energy is engendered by the polarizing action of the central electron on the surrounding chloride ions, while when sodium chloride is dissolved in sodium negative electrostatic energy is engendered by the polarizing action of the surrounding electrons on the central chloride ion. If the dipoles generated by the electrons were randomly oriented, simple additivity of polarization energy would apply, as in normal van der

TABLE 3. The Two Dilute Solutions in the
Na–NaX System

NaX	NaCl	NaBr	NaI
V_{NaX}, cm^3	38	45	55
V_{Na}, cm^3	31	31	31
Na in NaX, kJ $\Delta H°$	76	76	68
NaX in Na, kJ $\Delta H°$	64	57	59

Waals interaction, but, as these dipoles tend to be in phase with each other, vectorial summation results in a larger polarization energy for several dipoles cooperatively acting on a single anion than for a single dipole acting on several anions.

With normal van der Waals forces operating the two enthalpies of dilute solution have the same ratio as the molar volumes of the solutes. The correlation between the motions of the metallic electrons enhances the negative energy term when the halide is the solute. This effect is illustrated in Table 3 for the three systems for which Pitzer calculated both enthalpies of dilute solution from the miscibility gaps. Although these three salts have higher molar volumes than sodium, in all three cases the enthalpy of dilute solution is lower when the salt is the solute.

References

1. T. Førland, *Nor. Tek. Vitenskapsakad.* **2** (thesis 4) (1957).
2. J. Lumsden, *Thermodynamics of Molten Salt Mixtures*, Academic, New York (1966), p. 76.
3. K. Furukawa, *Discuss. Faraday Soc.* **32**, 53 (1961).
4. J. E. Mayer, *J. Chem. Phys.* **1**, 270 (1933).
5. J. L. Holm and O. J. Kleppa, *J. Chem. Phys.* **49**, 2425 (1968).
6. L. S. Hersh and O. J. Kleppa, *J. Chem. Phys.* **42**, 1309 (1965).
7. T. Førland, *Discuss. Faraday Soc.* **32**, 122 (1961).
8. M. E. Melnichak and O. J. Kleppa, *J. Chem. Phys.* **52**, 1790 (1970).
9. O. J. Kleppa and M. E. Melnichak, Conference Int. Thermodynam. Chim. (C.R.) 1975, Vol. 3, p. 148.
10. J. Lumsden, *Discuss. Faraday Soc.* **32**, 168 (1961).
11. P. Debye, *Polar Molecules*, Chemical Catalog, New York (1929), p. 70.
12. A. Buchler, J. L. Stauffer, W. Klemperer, and L. Wharton, *J. Chem. Phys.* **39**, 2299 (1963).
13. A. Buchler, J. L. Stauffer, and W. Klemperer, *J. Am. Chem. Soc.* **86**, 4544 (1964).
14. G. N. Papatheodorou and O. J. Kleppa, *J. Chem. Phys.* **47**, 2014 (1967).
15. G. N. Papatheodorou and O. J. Kleppa, *J. Chem. Phys.* **51**, 4624 (1969).
16. K. Grjotheim, J. L. Holm, and J. Malmo, *Acta Chem. Scand.* **24**, 77 (1970).
17. H. Flood and S. Urnes, *Z. Elektrochem.* **59**, 834 (1955).
18. J. Lumsden, *Thermodynamics of Molten Salt Mixtures*, Academic, New York (1966), p. 248.
19. K. Pitzer, *J. Am. Chem. Soc.* **84**, 2025 (1962).
20. M. A. Bredig and H. R. Bronstein, *J. Phys. Chem.* **64**, 64 (1960).

Index